Environmental Science and Engineering

The ultimate goal of this series is to contribute to the protection of our environment, which calls for both profound research and the ongoing development of solutions and measurements by experts in the field. Accordingly, the series promotes not only a deeper understanding of environmental processes and the evaluation of management strategies, but also design and technology aimed at improving environmental quality. Books focusing on the former are published in the subseries Environmental Science, those focusing on the latter in the subseries Environmental Engineering.

Himanshu Gupta · Sughosh Madhav ·
Soniya Dhiman · Ritu Mishra
Editors

Emerging Contaminants in Water and Wastewater

Sources and Substances

Springer

Editors
Himanshu Gupta
Department of Chemistry, School of Sciences
IFTM University
Moradabad, Uttar Pradesh, India

Soniya Dhiman
Helmholtz-Zentrum Dresden-Rossendorf
Helmholtz Institute Freiberg for Resource Technology
Dresden, Germany

Sughosh Madhav
Department of Civil Engineering
Jamia Millia Islamia
New Delhi, India

Ritu Mishra
Department of Botany
University of Allahabad
Prayagraj, Uttar Pradesh, India

ISSN 1863-5520 ISSN 1863-5539 (electronic)
Environmental Science and Engineering
ISBN 978-3-031-82578-1 ISBN 978-3-031-82579-8 (eBook)
https://doi.org/10.1007/978-3-031-82579-8

This Springer imprint is published by the registered company Springer Nature Switzerland AG
The registered company address is: Gewerbestrasse 11, 6330 Cham, Switzerland

Contents

Chapter 1
Emerging Contaminants in Water and Wastewater: An Introduction

Himanshu Gupta, Soniya Dhiman, Sughosh Madhav, and Ritu Mishra

Abstract Emerging contaminants (ECs) are the part of industrial wastewaters and also reaches the water systems through various existing mechanisms. The presence of ECs in the water and wastewaters is highly toxic to human beings and environment both as well as these disturbs the ecological balance. ECs are sub-divided into different categories involving pharmaceuticals, polyaromatics, nanomaterial, personal care product etc. The presence of various ECs and their metabolites in different aquatic organisms has been reported in various studies. The amounts of ECs in the water as well as wastewaters are detected and reported in various recent researches, but proper guidelines or regulations in particular are not yet available. The emerging contaminants in the aqueous media are of particular concern to the environmentalists across the globe. The present chapter focuses on the ECs in water and wastewaters of different industries and their sources in order to remediate the water streams.

Keywords Wastewater · Water · Emerging contaminants · Environment · Toxicity

H. Gupta (✉)
Department of Chemistry, School of Sciences, IFTM University, Lodhipur Rajput, Moradabad, Uttar Pradesh 244102, India
e-mail: hims.research@gmail.com

S. Dhiman
Department of Biochemical Engineering and Biotechnology, Indian Institute of Technology Delhi, New Delhi, India

Helmholtz-Zentrum Dresden-Rossendorf, Helmholtz Institute Freiberg for Resource Technology, Bautzner Landstraße 400, Dresden 01328, Germany

S. Madhav
Department of Civil Engineering, Jamia Millia Islamia, New Delhi, India

R. Mishra
Department of Botany, University of Allahabad, Prayagraj, Uttar Pradesh, India

H. Gupta et al. (eds.), *Emerging Contaminants in Water and Wastewater*, Environmental Science and Engineering, https://doi.org/10.1007/978-3-031-82579-8_1

1.1 Introduction

Emerging contaminants (ECs), are contaminants which tends to be toxic or hazardous towards the human health or environment and continued to be reported by new methods of their pollution, but strict regulations are not available to control them. ECs are categorized in the form of pharmaceuticals, nanomaterials, biocides, polycyclic aromatic hydrocarbon, personal care product, plasticizers, flame retardants etc. Various industries release their wastewaters in the water streams, which is a major source of ECs in the environment. Different researchers are recently been focused on the assessment of ECs in aqueous streams such as water and wastewaters (Gupta & Dhiman, 2023; Khan et al., 2023; Kumar & Gupta, 2020; Madhav et al., 2020; Malule et al., 2020; Rani et al., 2022). The development of different recent industrial segments to ease the humans is releasing large quantities of ECs to the environment and damaging the ecological balance (Gupta, 2018; Singh et al., 2018, 2022). As the sources and toxic effects of emerging contaminants are relatively new, the researchers, scientists, non-governmental as well as governmental organizations are concerned towards the formulation of particular guidelines for each of the ECs in the environment. The absence of specific guidelines and lack of information regarding ECs is more hazardous in the developing countries (Rehman et al., 2015). A large number of microbes and toxic compounds are present in untreated and waste treatment plants of different industries (Agunbiade & Moodley, 2014).

1.2 ECs in Wastewater and Water

ECs are categorized into different types based on various parameters such as on the basis of their applications (Flores et al., 2017). The different categories of ECs include pharmaceuticals, nanomaterials, biocides, polycyclic aromatic hydrocarbon, personal care product, plasticizers, flame retardants etc. (Fig. 1.1).

1.2.1 Pharmaceutical Wastes as ECs in Water Bodies

A large quantity of pharmaceuticals is being utilized by the humans for the treatment of different new as well as old kinds of diseases. The improper disposal of pharmaceutical wastes and untreated wastewaters of pharmaceutical industries has resulted in the deposition of such contaminants in water bodies. The presence of these ECs leads to the enhancement of environmental toxicity and health hazards to the living organisms (Gojkovic et al., 2019; Tiwari et al., 2017). The regulations related to pharmaceutical wastes are still not clear at many places. Various pharmaceuticals such as penicillins, tetracyclines, ranitidine, propranolol, carbamazepine, ibuprofen, paracetamol etc. have been reported in water systems of different countries across the

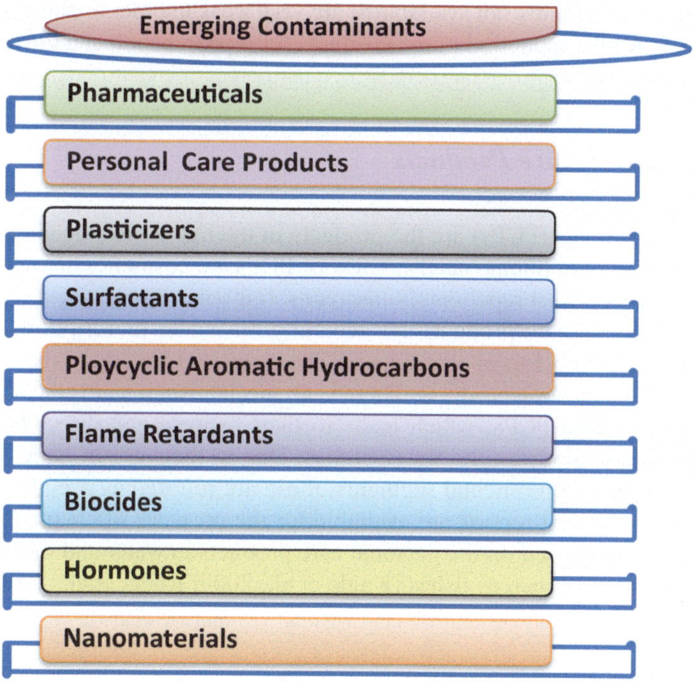

Fig. 1.1 Representation of selected emerging contaminants

globe. Pharmaceuticals are organic chemicals which may be degraded photochemically and generate intermediates, which might be more toxic than the parent chemical in water. More researches should be done in order to assess the pharmaceuticals and their intermediates in wastewater (Rivera-Utrilla et al., 2013).

1.2.2 Plasticizers as ECs

Around 30 thousand chemicals are being used as plasticizers in different products such as medical tools, child toys, packaging and products based on polyvinyl chloride (Cook & Halden, 2020; Godwin, 2017). A major market of plasticizers is covered by the phthalate plasticizers (Qadeer et al., 2022). Plasticizers are found to be toxic in literature and observed to be causing various hazardous effects such as endocrine disruption, embryonic or fetal disruptions, developmental defects, bioaccumulation and carcinogenicity (Huang et al., 2019; Jurewicz & Hanke, 2011; Mankidy et al., 2013). Ferguson et al. (2017) reported 8-isoprostane in the urine of pregnant woman. Therefore, plasticizers are one of emerging contaminants which needs more focus in order to reduce the health hazards caused by them. The regulations for the applications of plasticizers are currently not available or are very weak in order to diminish the use of plasticizers for the reduction of toxic effects for human health and environment.

1.2.3 Personal Care Products

Personal care products (PCPs) are the products in the market for the application to care various aspects of human needs. A few of PCPs are cosmetics, hair, dental and skin care products, insect repellents, sunscreens, fragrances, soaps etc. (Ebele et al., 2017). The personal care products are generally not under prescriptions from the medical practioners, doctors or pharmacists. These are used directly by the consumers as per their needs for these products. There is no mechanism towards the application and disposal of these PCPs, which leads to their overuse and direct flow in the environment, particularly in the water bodies. Due to their use in large quantities and release in the environmental segments, these are referred as pseudo-persistent in environment. Several reports are available for the presence of chemicals as well as their metabolites based on the personal care products in water and the organisms living in water bodies such as fishes (Kagle et al., 2009; Rudel et al., 2006; Subedi et al., 2012).

1.2.4 Polycyclic Aromatic Hydrocarbon (PAHs)

PAHs are a series of compounds composed of fused rings of benzene and these are present in the environment, particularly in water systems (Gupta & Gupta, 2016, 2018). These are priority pollutants as listed by different international agencies such as "United States Environmental Protection Agency (USEPA) and European commission (EU). These reach to water and wastewater systems due to some direct or indirect mechanisms and natural or anthropogenic activities. Eruptions of volcanoes and fires in the forests give rise to natural PAHs emission, whereas, oil spillage, industrial fumes and burning of fuels, smoke and contaminated wastewater give rise to anthropogenic PAHs emission (Gupta & Dhiman, 2023; Gupta & Gupta, 2015). The deterioration of aquatic life and water bodies is reported to be from the PAHs released due to domestic as well as industrial waste. The transport of PAHs from industrial fumes in the atmosphere, also results in deposition of fumes on aquatic and terrestrial surfaces (Abhilash et al., 2013; Gupta & Gupta, 2015). Various researchers have found these pollutants in wastewater or water (Gonzalez et al., 2012; Gupta, 2015; Gupta & Kumar, 2020; Olenycz et al., 2015; WlodarczykMakula, 2005).

1.2.5 Surfactants as ECs

Surfactants are generally molecules with hydrophobic and hydrophilic groups, which are tail of hydrocarbons (non-polar) and head of ions (usually polar), respectively (Mungray & Kumar, 2009). The persistence of surfactants in the water media is toxic to the environment and living organisms. A large quantity of surfactants is reported to be present in domestic, industrial and municipal wastewaters causing severe hazards to the environment (Fig. 1.2). In Zhang et al. (1999) reported 300 mg/L and 1–10 mg/L surfactants in industrial and domestic wastewaters, respectively. The presence of surfactants in municipal wastewaters may become the source for their availability in drinking water and other water bodies, resulting to the toxicity in the aquatic ecosystem to human beings (Bautista-Toledo et al., 2014; Camacho-Muñoz et al., 2014).

1.2.6 Biocides as ECs

Biocides are chemicals which are used to destroy toxic organisms and control them into non-toxic forms. These are applied for the disinfection of industrial, domestic as well as healthcare items (Maillard, 2005). These include disinfectants, quaternary ammonium salts, topical agents, chlorine-releasing agents, biguanides and peroxygens (Wesgate et al., 2016). The application of biocides has been implemented onto a variety of material like surgical scrubs, socks, soap and mouthwash etc. (Gilbert et al., 2002). These are of different types which include disinfectants, algaecides, preservatives, rodenticides and insecticides etc. (Henry et al., 2012). Biocides are reported to be toxic for various non-targeted organisms, therefore, the biocides are considered as emerging contaminants by the environmentalist and their application as well as toxicity is being monitored in different cases (Maund et al., 2012).

1.2.7 Flame Retardants

The combustion process is delayed or prevented by some compounds referred as flame retardants (FR). The different flame retardants are based to halogenated organics, phosphorus, nitrogen and inorganic compounds. FRs based to phosphorus has been reported to be used in furnitures, building substances, floor polishes, electronics and textile industry (Du et al., 2015; Reemtsma et al., 2008). Different studies in literature discussed the impact of FRs on the environment and living beings (Greaves & Letcher, 2014). As compared to high molecular weight FRs, low weight and volatile FRs gets dispersed in the environmental matrices and causes contamination (Cristale et al., 2016; Veen & Boer, 2012). Due to the high availability of FRs in water and wastewaters, the aquatic organisms are primarily harmed by these

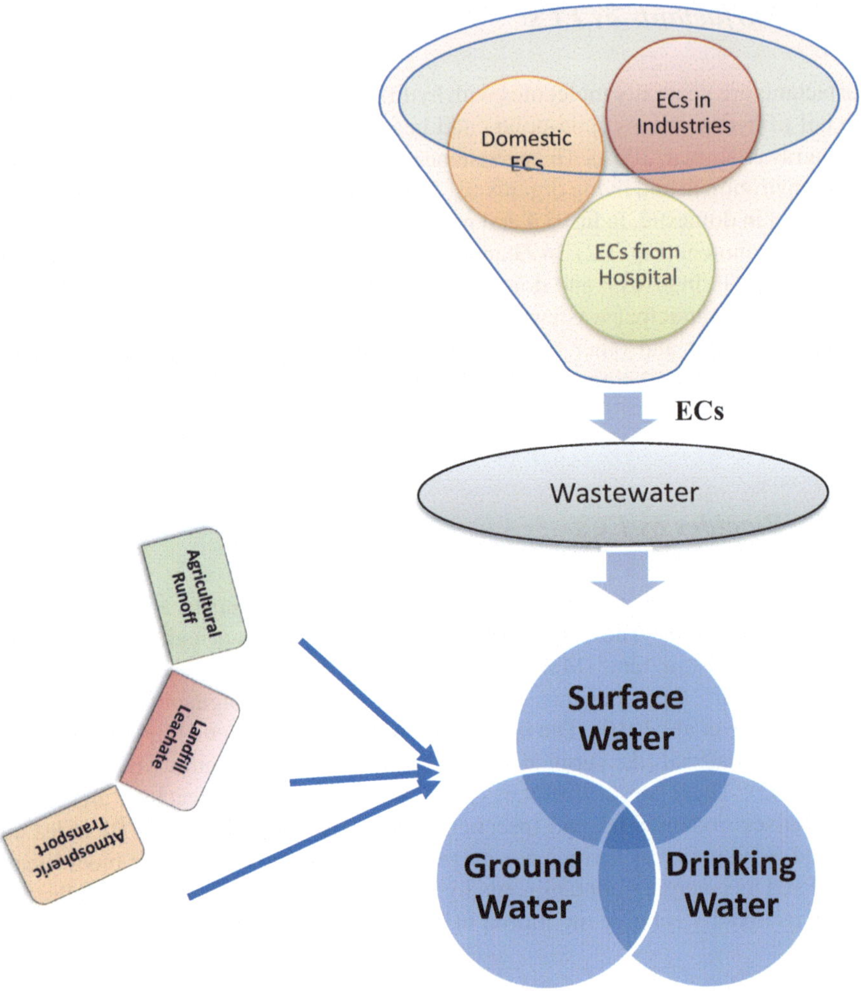

Fig. 1.2 Flow of emerging contaminants into wastewater and water

compounds, whereas, the other organism including humans are harmed due to their presence in surface or drinking water (Wei et al., 2015).

1.2.8 Hormones as ECs

Hormones are some naturally occurring and synthesized chemicals that are significantly influencing the health of the living organisms. The hormonal characteristics of endocrine disruption are found to be responsible for fish hermaphroditism. The application of hormonal therapy as well as their natural occurrence was reported to be increased in the recent years, which leads to their enhancement in the wastewaters and water (Adeel et al., 2017; Purdom et al., 1994).

1.2.9 Nanomaterials as ECs

Nanomaterials are substances with atleast one dimension < 100 nm. Nanomaterials are used in biomedical, pharamaceuticals, cosmetics as well as electronic applications. Due to the smaller sizes and diverse applications, these tend to contaminate the environment (Khan et al., 2019). The water contamination through the nanomaterials can be caused due to the direct release, unintentional spillage, wastes and effluents of treatment plants as well as through the run off into the water bodies (Klaine et al., 2008). Nanomaterials usually withstand in water for longer durations due to smaller particle size and gravity settlement. Their biodegradable nature is still a matter of deep research, as a few nanomaterials found to degrade after long durations. Numerous remediation processes involves nanomaterials, which due to the high mobility reaches to drinking water and cause health and environmental hazards (EPA, 2010).

Though a huge amount of ECs are being detected in water as well as wastewaters, the strategies of risk assessment, guidelines or regulations and laws are not available at all places. Also, if these exist, the implementation is very weak. Thus, a thorough study on each of the emerging contaminants is required alongwith the risk assessment, regulation, laws and their implementation strategies to protect environment and human health.

1.3 Conclusion

The present study focuses on the ECs in water and wastewaters of different industries and their sources in order to remediate the water streams. The development of different recent industrial segments to ease the humans is releasing large quantities of ECs to the environment and damaging the ecological balance. Pharmaceuticals are one of ECs, which may be degraded photochemically and generate intermediates, which might be more toxic than the parent chemical in water. Several reports are available for the presence of chemicals as well as their metabolites based on the personal care products in water and the organisms living in water bodies such as fishes. The deterioration of aquatic life and water bodies is reported to be from the PAHs released due to domestic as well as industrial waste. Biocides are reported to be toxic for various non-targeted organisms, therefore, the biocides are considered as emerging contaminants by the environmentalist. Nanomaterials usually withstand in water for longer durations due to smaller particle size and gravity settlement. Thus, a thorough study on each of the emerging contaminants is required alongwith the risk assessment, regulation, laws and their implementation strategies to protect environment and human health. Thus, a thorough study on each of the emerging contaminants is required alongwith the risk assessment, regulation, laws and their implementation strategies to protect environment and human health.

References

Abhilash, P. C., Dubey, R. K., Tripathi, V., Srivastava, P., Verma, J. P., & Singh, H. B. (2013). Remediation and management of POPs-contaminated soils in a warming climate: Challenges and perspectives. *Environmental Science and Pollution Research, 20*(8), 5879–5885.

Adeel, M., Song, X., Wang, Y., Francis, D., & Yang, Y. (2017). Environmental impact of estrogens on human, animal and plant life: A critical review. *Environment International, 99*, 107–119.

Agunbiade, F. O., & Moodley, B. (2014). Pharmaceuticals as emerging organic contaminants in Umgeni River water system, KwaZulu-Natal, South Africa. *Environmental Monitoring and Assessment, 1*, 7273–7291.

Bautista-Toledo, M. I., Rivera-Utrilla, J., Méndez-Díaz, J. D., Sánchez-Polo, M., & Carrasco-Marín, F. (2014). Removal of the surfactant sodium dodecylbenzenesulfonate from water by processes based on adsorption/bioadsorption and biodegradation. *Journal of Colloid and Interface Science, 418*, 113–119.

Camacho-Muñoz, D., Martín, J., Santos, J. L., Aparicio, I., & Alonso, E. (2014). Occurrence of surfactants in wastewater: Hourly and seasonal variations in urban and industrial wastewaters from Seville (Southern Spain). *Science of the Total Environment, 468–469*, 977–984.

Cook, C. R., & Halden, R. U. (2020). Plastic waste and recycling. In *Ecological and health issues of plastic waste*. Academic Press Cambridge, MA, USA

Cristale, J., Hurtado, A., Gómez-Canela, C., & Lacorte, S. (2016). Occurrence and sources of brominated and organophosphate flame retardants in dust from Di_erent indoor environments in Barcelona, Spain. *Environmental Research, 149*, 66–76.

Du, Z. K., Wang, G. W., Gao, S. X., & Wang, Z. Y. (2015). Aryl organophosphate flame retardants induced cardiotoxicity during zebra fish embryogenesis: By disturbing expression of the transcriptional regulators. *Aquatic Toxicology, 161*, 25–32.

EPA. (2010). *Fact Sheet: Emerging contaminats—nanomaterials*. United States Environmental Protection Agency, EPA 505-F-10-008.

Ebele A. J., Abdallah M. A. E., & Harrad S. (2017). Pharmaceuticals and personal care products (PPCPs) in the freshwater aquatic environment. *Emerging Contaminants, 3*, 1–16.

Ferguson, K. K., Chen, Y. H., Vanderweele, T. J., McElrath, T. F., Meeker, J. D., & Mukherjee, B. (2017). Mediation of the relationship between maternal phthalate exposure and preterm birth by oxidative stress with repeated measurements across pregnancy. *Environmental Health Perspectives, 125*, 488–494.

Flores, C., Francisco, G., Jimenez, D., & Jose, A. (2017). *Emerging pollutants: Origin, structure and properties*. December https://doi.org/10.1002/9783527691203.

Gilbert, P., McBain, A. J., & Bloomfield, S. F. (2002). Biocide abuse and antimicrobial resistance: Being clear about the issues. *Journal of Antimicrobial Chemotherapy, 50*, 137–139.

Godwin A. D. (2017) Plasticizers. In Appl Plast Eng Handbook 2nd Ed: 533−553.

Gojkovic, Z., Lindberg, R. H., Tysklind, M., & Funk, C. (2019). Northern green algae have the capacity to remove active pharmaceutical ingredients. *Ecotoxicology and Environmental Safety, 170*, 644–656.

González, D., Ruiz, L. M., Garralón, G., Plaza, F., Arévalo, J., Parada, J., Pérez, J., Moreno, M., & Gómez, M. A. (2012). Wastewater polycyclic aromatic hydrocarbons removal by membrane bioreactor. *Desalination and Water Treatment, 42*, 94–99.

Greaves, A. K., & Letcher, R. J. (2014). Body compartment distribution in female Great Lakes herring gulls and in ovo transfer of sixteen bioaccumulative organophosphate flame retardants. *Environmental Science and Technology, 48*, 7942–7950.

Gupta, H., & Dhiman, S. (2023). Synthetic nanoparticle-based remediation of soils contaminated with polycyclic aromatic hydrocarbons. In U. Shanker, C. M. Hussain & M. Rani (Eds.), *Handbook of green and sustainable nanotechnology*. Springer, Cham. https://doi.org/10.1007/978-3-030-69023-6_102-1.

Gupta, B., & Gupta, H. (2015). Iron oxide mediated degradation of mutagen pyrene and determination of degradation products. *International Journal of Environmental Science and Development, 6*(12), 908. https://doi.org/10.7763/IJESD.2015.V6.720

Gupta, H. (2015). Removal of phenanthrene from water using activated carbon developed from orange rind. *International Journal of Scientific Research in Environmental Sciences, 3*(7), 248–255.

Gupta, H. (2018). PAH determination in effluent and sludge samples of paper industry. *Environmental Technology & Innovation, 9*, 115–121. https://doi.org/10.1016/j.eti.2017.11.009

Gupta, H., & Gupta, B. (2016). Adsorption of polycyclic aromatic hydrocarbons on banana peel activated carbon. *Desalination and Water Treatment, 57*(20), 9498–9509.

Gupta, H., & Gupta, B. (2018). Vehicular tire as potential adsorbent for the removal of polycyclic aromatic hydrocarbons. *Polycyclic Aromatic Compounds, 38*(4), 354–368.

Gupta, H., & Kumar, R. (2020). Distribution of selected polycyclic aromatic hydrocarbons in urban soils of Delhi, India. *Environmental Technology & Innovation, 17*, 100500.

Henry, M., Requier, F., Rollin, O., Odoux, J. F., Aupinel, P., Aptel, J., Tchamitchian, S., & Decourtye, A. (2012). A common pesticide decreases foraging success and survival in honey bees. *Science, 336*, 348–350.

Huang, S., Li, J., Xu, S., Zhao, H., Li, Y., Zhou, Y., Fang, J., Liao, J., Cai, Z., & Xia, W. (2019). Bisphenol A and bisphenol S exposures during pregnancy and gestational age—A longitudinal study in China. *Chemosphere, 237*, 124426.

Jurewicz, J., & Hanke, W. (2011). Exposure to phthalates: Reproductive outcome and children health. A review of epidemiological studies. *International Journal of Occupational Medicine and Environmental Health, 24*, 115–141.

Kagle, J., Porter, A. W., Murdoch, R. W., Cancel, G. R., & Hay, A. G. (2009). Biodegradation of pharmaceutical and personal care products. *Advances in Applied Microbiology, 67*, 65–108.

Khan, I., Saeed, K., & Khan, I. (2019). Nanoparticles: Properties, applications and toxicities. *Arabian Journal of Chemistry, 12*, 908–931.

Khan, W. U., Ahmed, S., Dhoble, Y., & Madhav, S. (2023). A critical review of hazardous waste generation from textile industries and associated ecological impacts. *Journal of the Indian Chemical Society, 100*(1), 100829. https://doi.org/10.1016/j.jics.2022.100829

Klaine, S. J., Alvarez, P. J. J., Batley, G. E., Fernandes, T. E., Handy, R. D., Lyon, D. Y., Mahendra, S., McLaughlin, M. J., & Lead, J. R. (2008). Nanoparticles in the environment: Behavior, fate, bioavailability and effects. *Environmental Toxicology and Chemistry, 27*(9), 1825–1851.

Kumar, A., & Gupta, H. (2020). Activated carbon from sawdust for naphthalene removal from contaminated water. *Environmental Technology & Innovation, 20*, 101080.

Madhav, S., Kumar, A., Kushawaha, J., Ahamad, A., Singh, P., & Dwivedi, S. B. (2020). Geochemical assessment of groundwater quality in Keonjhar City, Odisha, India. *Sustainable Water Resources Management, 6*, 1–11. https://doi.org/10.1007/s40899-020-00395-7

Maillard, J. Y. (2005). Antimicrobial biocides in the healthcare environment: Efficacy, usage, policies, and perceived problems. *Therapeutics and Clinical Risk Management, 1*, 307–320.

Malule H. R., Murillo, D. H. Q., & Duque, D.M. (2020) Emerging contaminants as global environmental hazards. A bibliometric analysis. *Emerging Contaminants, 6*, 179–193.

Mankidy, R., Wiseman, S., Ma, H., & Giesy, J. P. (2013). Biological impact of phthalates. *Toxicology Letters, 217*, 50–58.

Maund, S. J., Campbell, P. J., Giddilings, J. M., Hamer, M. J., Henry, K., Pilling, E. D., Warinton, J. S., & Wheeler, J. R. (2012). Ecotoxicology of synthetic pyrethroids. In N. Matsuo & T. Mori

(Eds.), *Pyrethroids: From Chrysanthemum to modern industrial insecticide: Topics in current chemistry, 314: Berlin* (pp. 137–165). Springer.

Mungray, A. K., & Kumar, P. (2009). Fate of linear alkylbenzene sulfonates in the environment: A review. *International Biodeterioration and Biodegradation, 63*, 981–987.

Olenycz, M., Sokolowski, A., Niewinska, A., Wolowicz, M., Namiesnik, J., Hummel, H., & Jansen, J. (2015). Comparison of PCBs and PAHs levels in European coastal waters using mussels from the Mytilus edulis complex as biomonitors. *Oceanologia, 57*, 196–211.

Purdom C. E., Hardiman P. A., Bye V. V. J., Eno N. C., Tyler C. R., & Sumpter J. P. (1994) Estrogenic effects of effluents from sewage treatment works: Chemistry and ecology. *8*, 275–285.

Qadeer, A., Kirsten, K. L., Ajmal, Z., Jiang, X., & Zhao, X. (2022). Alternative plasticizers as emerging global environmental and health threat: Another regrettable substitution? *Environmental Science and Technology, 56*, 1482–1488.

Rani, L., Srivastav, A.L., Kaushal, J., Grewal, A.S., & Madhav, S. (2022). Heavy metal contamination in the river ecosystem. In *Ecological significance of river ecosystems* (pp. 37–50). Elsevier

Reemtsma, T., Quintana, J. B., Rodil, R., García-Lopez, M., & Rodríguez, I. (2008). Organophosphate flame retardants and plasticizers in water and air I. Occurrence and fate. *Trends in Analytical Chemistry, 27*, 727–737.

Rehman M. S., Rashid N., Ashfaq M., Saif A., Ahmad N., & Han J. I. (2015) Global risk of pharmaceutical contamination from highly populated developing countries. *Chemosphere, 138*, 1045–1055.

Rivera-Utrilla, J., Sanchez-Polo, M., Ferro-Garcia, M. A., Prados-Joya, G., & Ocampo-Perez, R. (2013) Pharmaceuticals as emerging contaminants and their removal from water. A review. *Chemosphere*. https://doi.org/10.1016/j.chemosphere.2013.07.059

Rüdel, H., Bohmer, W., & Schroter-Kermani, C. (2006). Retrospective monitoring of synthetic musk compounds in aquatic biota from German rivers and coastal areas. *Journal of Environmental Monitoring, 8*(8), 812–823. (19)

Singh, S., Gupta, H., Dhiman, S., & Sahu, N. K. (2022). Decontamination of cationic dye brilliant green from the aqueous media. *Applied Water Science, 12*(4), 61. https://doi.org/10.1007/s13201-022-01596-5

Singh, S., Parveen, N., & Gupta, H. (2018). Adsorptive decontamination of rhodamine-B from water using banana peel powder: A biosorbent. *Environmental Technology and Innovation, 12*, 189–195. https://doi.org/10.1016/j.eti.2018.09.001

Subedi, B., Du, B., Chambliss, C. K., Koschorreck, J., Rudel, H., Quack, M., Brooks, B. W., & Usenko, S. (2012). Occurrence of pharmaceuticals and personal care products in german fish tissue: A national study. *Environmental Science and Technology, 46*, 9047–9054.

Tiwari, B., Sellamuthu, B., Ouarda, Y., Drogui, P., Tyagi, R. D., & Buelna, G. (2017). Review on fate and mechanism of removal of pharmaceutical pollutants from wastewater using biological approach. *Bioresource Technology, 224*, 1–12.

Veen, I. V. D., & Boer, J. D. (2012). Phosphorus flame retardants: Properties, production, environmental occurrence, toxicity and analysis. *Chemosphere, 88*, 1119–1153.

Wei, G. L., Li, D. Q., Zhuo, M. N., Liao, Y. S., Xie, Z. Y., Guo, T. L., Li, J. J., Zhang, S. Y., & Liang, Z. Q. (2015). Organophosphate flame retardants and plasticizers: Sources, occurrence, toxicity and human exposure. *Environmental Pollution, 196*, 29–46.

Wesgate, R., Grasha, P., & Maillard, J. Y. (2016). Use of a predictive protocol to measure the antimicrobial resistance risks associated with biocidal product usage. *American Journal of Infection Control, 44*, 458–464.

Wlodarczyk-Makula, M. (2005). The loads of PAHs in wastewater and sewage sludge of municipal treatment plant. *Polycyclic Aromatic Compounds, 25*, 183–194.

Zhang, C., Valsaraj, K. T., Constant, W. D., & Roy, D. (1999). Aerobic biodegradation kinetics of four anionic and nonionic surfactants at sub- and supra-critical micelle concentrations (CMCs). *Water Research, 33*, 115–124.

Chapter 2
Sensors and Biosensors for Emerging Contaminants in Industrial Wastewater

Anshu Kumar, Kumari Seema, and Ambika Kumar

Abstract Accurate detection of newly discovered pollutants is essential for maintaining the quality and safety of water, due to the substantial hazards they pose of great concern to living kingdom. Compact and affordable sensors that may be deployed in the field provide opportunities for identifying contaminants at their source and enabling comprehensive monitoring and regulation of water quality. Regrettably, the availability of these sensors and the potential applications for them continue to be restricted. This Perspective critically examines emerging pollution detecting technology. The main barriers to adoption of these technologies include high costs, unstable functioning, real-world implementation issues, and insufficient stakeholder input in sensor development. This Perspective analyses current sensing technologies for contamination detection by critically examining them. It also examines the main obstacles to these technologies' mainstream adoption. High costs, unpredictable performance, real-world deployment issues, and stakeholder involvement in sensor development are among these problems. Overall, these obstacles ultimately hinder the successful implementation and adoption of these technologies on a larger scale. In order to understand the outstanding scientific hurdles, which requires to be resolved before implementing these tools in practical scenarios, we showcase unique sensing systems and examine crucial strategies for overcoming these obstacles. More precisely, this involves improving the ability to detect and distinguish substances, increasing the longevity and effectiveness, and optimizing the functionality in real-world water scenarios. This chapter also discusses other important factors, which promotes collaborations between academia and industry, maximizing the value of data through data sharing initiatives, and encouraging workforce development. All of these factors are discussed in detail. The Perspective provides a comprehensive framework that may be utilized for the construction, modification, and deployment of water

A. Kumar
Department of Chemistry, T. N. B. College, Bhagalpur, Bihar 812007, India

K. Seema
Department of Chemistry, J. D. Women's College, Patna, Bihar 800022, India

A. Kumar (✉)
Department of Chemistry, Bhagalpur National College Bhagalpur, Bihar 812007, India
e-mail: kumarambika.1115@gmail.com

H. Gupta et al. (eds.), *Emerging Contaminants in Water and Wastewater*, Environmental Science and Engineering, https://doi.org/10.1007/978-3-031-82579-8_2

quality sensors with the objective of quickly and accurately detecting, examining, monitoring, and managing pollution.

Keywords Industrial wastewater · Emerging pollutants · Sensors · Water quality

2.1 Introduction

Emerging pollutants in water, which include inorganic, organic, and biological components, pose a substantial risk to both human health and the environment (Arman et al., 2021; Geissen et al., 2015; Gupta & Gupta, 2016; Kumar et al., 2023; Madhav et al., 2020; Pereira et al., 2015; Singh et al., 2018, 2022; Vasilachi et al., 2021). There are large differences in the levels of these pollutants that can be attributed to a variety of factors, including geographical location, sources of contamination, environmental circumstances, and remediation methods. There is a high incidence of the detection of organic and inorganic pollutants in industrial areas (Archer et al., 2017; Gupta, 2018; Gupta & Kumar, 2020; Khan et al., 2023; Padhye et al., 2014; Wilson & Ashraf, 2018; Zhao et al., 2018). The quantities of these compounds can vary from nanograms per litre (ng/L) to micrograms per litre (μg/L), indicating a broad spectrum of levels. Moreover, there is an increasing acknowledgment that microplastics and artificial nanomaterials are emerging contaminants that can potentially have Detrimental impacts on ecosystems and the well-being of the general population. The amounts of compounds that are often found in drinking water in the United States, such as atenolol, carbamazepine, estrone, and sulfamethoxazole, were typically found to be below 10 ng/L. However, there were a few instances in which higher levels were seen. According to the results of the study, the concentration of Tris(2-chloroethyl) phosphate (TCEP) in the water was 120 ng/L, whereas the amounts of atrazine in the source water, finished water, and dispersed water ranged from 30 to 50 ng/L (Benotti et al., 2009; Gworek et al., 2020; Schaider et al., 2014). Efficient and timely testing is crucial for the effective supervision of water quality in potentially contaminated locations. Managing water supplies and responding to emergencies in a timely manner are both hampered by the absence of reliable detection tools. For the purpose of analyzing various new pollutants in water, conventional procedures have been demonstrated to be highly effective (Carmona & Picó, 2018; Rodriguez et al., 2007; Tang et al., 2019). In most cases, these conventional methods are characterized by a low quantification limit (LOQ). Though, due to the fact that these operations require the utilization of specialized equipment and staff who have received extensive training, they can be both costly and labor-intensive. This renders them inappropriate for the purpose of field monitoring. Existing techniques necessitate the manual collection of samples and their subsequent transportation to centralised laboratories for analysis, a process that is both laborious and unwieldy. Furthermore, the majority of samples require meticulous processing using perilous solvents that possess the capacity to pollute the environment. In addition, there is a dearth of data from environmental surveys and methodologies that are efficient in identifying the presence of a wide variety of pollutants in the field. By providing solutions that are both cost-effective and efficient, as well as demonstrating remarkable precision in both space and time, the application of sensor technology has the potential to dramatically improve water quality monitoring operations (Adrián et al., 2009; Greb et al., 2018). Sensors employ molecular receptors, which can be either biological or chemical. These receptors are meticulously designed to respond in a certain way to designated targets. Conventional sensors are defined by the integration of molecular receptors, transducers, and user interfaces that are purposefully designed to fulfil various objectives (Fig. 2.1) (Ateia et al., 2024).

These sensors have the ability to measure several pollutants that are important for evaluating water quality, including heavy metals, pesticides, pharmaceuticals, endocrine-disrupting chemicals (EDCs), and biological agents such as bacteria or viruses (Adegoke et al., 2017; Bezbaruah & Kalita, 2010; Gavrilescu et al., 2015; Rathi et al., 2021). There has been a dearth of market, residential, public, and regulatory compliance framework application of sensing technology for water quality monitoring, despite substantial advancements in academic research in this area (Andres et al., 2018; Goldman et al., 2007; Jan et al., 2021; Street et al., 2014). Multiple technical articles and assessments are accessible regarding sensors utilised for water monitoring. These publications specifically highlight the elements of sensors that can identify emerging pollutants (Ahmed et al., 2020; Altenburger et al., 2015; Banna et al., 2014; Hojjati-Najafabadi et al., 2022; Pellerin et al., 2016; Richardson & Ternes, 2018; Zulkifli et al., 2018). These sensors have versatile applications in

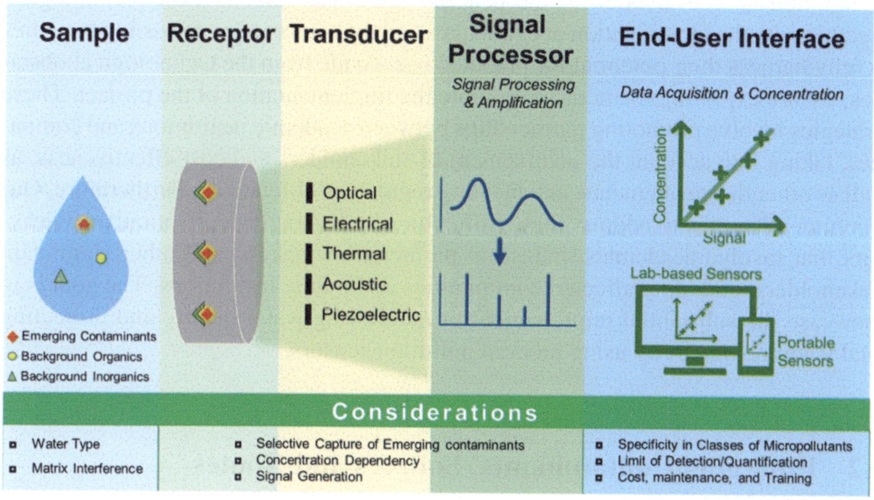

Fig. 2.1 The concept and primary constituents of an integrated sensor device designed to identify developing water contaminants. Adapted with permission from Ateia et al., 2024. Sensors for Emerging Water Contaminants: Overcoming Roadblocks to Innovation. Environmental Science & Technology. Copyright 2024 American Chemical Society

many settings, including field sites, residences, and treatment facilities. This argument centres on the current state of these technologies in relation to advancements in water quality testing. Our priority is to focus on technologies that offer efficient analysis and can easily transition from laboratory to field settings for fast implementation. Despite several studies on water pollutant detection sensors, their practical implementation for regular monitoring has been restricted. The majority of research in the field of creating contamination sensors does not get past the proof-of-concept or laboratory validation phases. Few of these research studies have been validated in real life, and an even smaller fraction have been successfully commercialised and widely adopted. A number of water sensors that are able to detect significant physicochemical characteristics, including temperature, turbidity, dissolved oxygen, and pH, have previously been successfully introduced to the market (Alimorong et al., 2020; Garrido-Momparler & Peris, 2022; Murphy et al., 2015; Sohrabi et al., 2021; Steinegger et al., 2020). In stark contrast to the circumstances that currently exist, this is the case. When it comes to the reasons behind the failure to successfully introduce emerging contaminant sensors to the market or their restricted adoption in the industry, public, and regulatory sectors, there has been a dearth of organized discourse. This is despite the fact that there has been a notable advancement in research and a more profound comprehension of essential principles. This viewpoint provides an overview of the main scientific, technological, regulatory, and economic obstacles that need to be overcome in order to successfully transform research-based sensor advancements into functional devices for field applications. This analysis primarily focuses on the practical aspects of these technologies. This analysis entails a comprehensive evaluation of the benefits and constraints of these technologies, together with the identification of challenges and the exploration of possible remedies to fully harness their potential for practical use. Aside from the technological obstacles, we are exploring methods to promote the implementation of the project. These strategies involve promoting partnerships between academic institutions and companies, taking into account the involvement of stakeholders and cost-effectiveness, as well as other design elements, and meeting regulatory obligations. Furthermore, Our recommendation is to address these difficulties by leveraging collaborative endeavours that involve academics, industrial partners, lawmakers, and other significant stakeholders, including affected communities and public institutions. The goal is to showcase the substantial improvement in monitoring water quality and protecting vital water resources by using modern pollution sensors.

2.2 Emerging Contaminants' Sources and Routes into Water Sources

Emerging contaminants (ECs) can be found in wastewater from Waste Water Treatment Plants (WWTPs), industrial establishments, healthcare facilities, residential areas, and agricultural runoff. Landfill liquid also contains ECs (Fig. 2.2) (Murgolo et al., 2017; Norvill et al., 2016). It is possible for the contaminants to originate from a variety of sources, which may include both particular and diffuse origins. Furthermore, they have the ability to persist in the aqueous medium ranging from parts per trillion (ppt or ng/L) to parts per billion (ppb or μg/L) (Kapelewska et al., 2018). While EPs have existed in the atmosphere for millennia, their significance has only grown in the last twenty years as a result of breakthroughs in analytical methods. In this context, the techniques that are being discussed are solid-phase extraction (SPE), solid-phase microextraction (SPME), high-performance liquid-chromatography-coupled mass spectrometry (HPLC–MS), liquid chromatography-mass spectrometry tandem mass spectrometry (LC–MS/MS) and ultra-high-performance liquid chromatography (UHPLC). These approaches aid in the detection of environmentally persistent chemicals (EPs) found in water (including groundwater and surface water), soil, atmosphere, and sediments at low concentrations (Galindo et al. 2019; Lofrano et al., 2020). Table 2.1 presents recent instances of developments in analytical techniques for detecting emerging contaminants (ECs) in water samples. It is commonly recognised that WWTP effluents are the main source of ECs released into the environment. WWTPs, or wastewater treatment plants, gather wastewater from both industrial and residential regions (Sangion & Gramatica, 2016). The effectiveness of removing emerging contaminants (ECs) ranges from 20 to 50% during the initial phase of wastewater treatment plants (WWTPs), increases to between 30 and 70% during the secondary treatment, and surpasses 90% during the tertiary treatment (Rout et al., 2021). As a result, the tertiary treatment operations in wastewater treatment plants (WWTPs) are considered to be the most important for eliminating ECs. However, achieving complete elimination of EC is still not possible. Once discharged into wastewater treatment facilities (WWTPs), emerging contaminants (ECs) typically undergo minimal transformation or may be metabolised into forms that are considerably more harmful and toxic than their initial state (Noguera-Oviedo & Aga, 2016). These contaminants exhibit persistent transport properties that hinder their degradation over time. In addition, they will be observed in particular domains in which they have not been engaged or employed in the past. These pollutants have the capacity to swiftly disperse through water and air over a variety of locations. Diagram The provided information demonstrates the dissemination of ECs from their origins to receptors throughout their life cycle. Electrochemical cells (ECs) can also be formed by the decomposition of pharmaceutical waste and the degradation of organic substances in neutral environments (Rasheed et al., 2019).

Drinking water can also include detectable amounts of biological micropollutants (Piai et al., 2020). An elevation in the occurrence of leaching processes, caused by the excessive use of animal manure and fertilizers in agricultural activities, results in a higher concentration of environmental pollutants in wastewater (Szogi et al., 2015). As a result of the permanent presence of pesticide metabolites in wastewater, a number of researches are related with examining these compounds (Rasheed et al., 2019). The following are examples of other sources of environmental contaminants (ECs): personal care items, cosmetic components, medications, pain relievers, household compounds, chemical pollutants, and flame retardants. The amount of

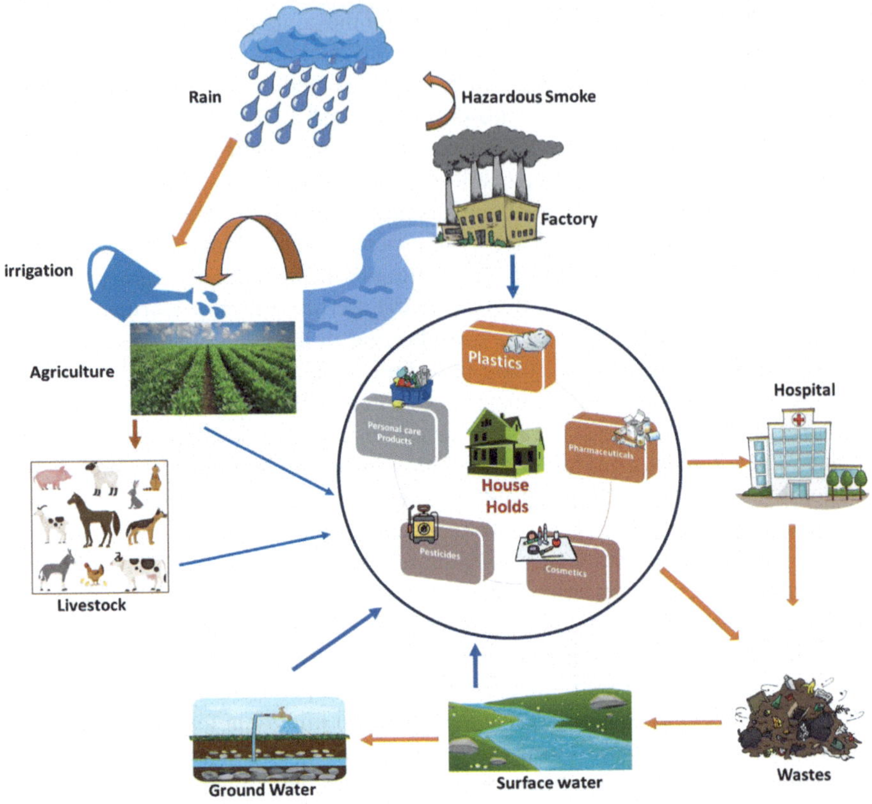

Fig. 2.2 Source and receptor of emerging contaminants and its life cycle

some contaminants that are present is mostly determined by the existence of the principal sources of such contaminants (Hasan et al., 2019). Pharmaceutical pollutants are predominantly present in wastewater generated by pharmaceutical industries, hospitals, and other healthcare institutions (Chander et al., 2016; Patel et al., 2019). Wastewater treatment plants (WWTPs) are a major contributor of excessive amounts of organic and chemical micropollutants.

Toxins once removed from wastewater, these toxins are either disposed of or burned (Zur et al., 2018). This process can be described as the volatilization, dispersion, and assimilation of toxins among multiple locations. Despite the presence of persistent pollutants in wastewater treatment plants, surface water continues to serve as a crucial and reliable supply for a significant number of individuals. On the other hand, freshwater, which encompasses both surface water and groundwater, is also known to contain trace levels of micropollutants that are created by aquatic inhabitants. Chemical substances, both natural and manmade, that fall within the category of ECs are recognised as potential hazards (Table 2.2). However, comprehensive information regarding all the different ECs is still lacking. There is a wide variety

Table 2.1 Comparison of different sensors for the identification of emerging contaminants (ECs) Yuan et al., (2023)

Emerging contaminants	Instruments	Concentration range
Antibiotics	High pressure liquid chromatography—mass spectroscopy	0.1–3.6 ng/L
	Fiber optic microfluidic sensors	A few mg/L
	Electrochemical sensors	0.079 μg/L–17.02 mg/L
	Fiber optic sensors	0.02–0.1 mg/mL
Bisphenol A	Supercritical–fluid chromatography coupled with diode-array detection	0.10–1.59 μg/L
	Fiber optic microfluidic sensors	330 ± 70 aM
	Electrochemical sensors	18.72 μg/L
	Fiber optic sensors	215 pM
Pharmaceuticals	Ultrahigh pressure liquid chromatography–tandem mass spectrometry	0.2 ng/L
	Fiber optic microfluidic sensors	10 nM
	Fiber optic sensors	1.20–1.34 nm
	Electrochemical sensors	2.486 μg/L
Endocrine disrupters	Gas chromatography—gas chromatography setup	< 1 mg/L
	Fiber optic microfluidic sensors	0.35 ppm
	Electrochemical sensors	4.67 μg/L–18.827 mg/L
	Fiber optic sensors	1.5×10^{-3} ng/ml
Perfluorinated compounds	Liquid chromatography tandem mass Spectrometry	0.013–0.248 μg/L
	Fiber optic microfluidic sensors	0.05 μg/L
	Electrochemical sensors	0.04 nM
	Fiber optic sensors	0.9–5 nm/ppm

of molecules that make up ECs. Some examples of these compounds are pharmaceuticals, personal care products (PCPs), endocrine disrupting chemicals (EDCs), steroids, hormones, surfactants and surfactant metabolites, flame retardants, pesticides, industrial additives, nanomaterials, and fuel additives (Stefanakis & Gramatica, 2016).

Furthermore, alongside the route of consumer product chemicals, specific substances in the aquatic environment are derived from production procedures and are introduced into the environment through the dissemination of industrial products. While there are other significant classifications of industrial chemicals, such phthalate ester plasticizers, perfluorinated compounds, and polybrominated biphenyl ether

Table 2.2 List of prominent ECs

Emerging contaminant groups	Examples
Pharmaceuticals	
Analgesics, anti-inflammatory drugs	Diclofenac, Ibuprofene, fenoprofen, paracetamol
Human antibiotics and veterinary	Amoxicillin, erytromycine, lincomycin, sulfamethaxozole,
Antiseptics	Chlorophene, Triclosan
β-blockers	Sotalol, Metoprolol, atenolol,
Lipid regulators	Gemfibrozil, Bezafibrate
Personal care products (PCPs)	
Fragrances	Nitro, polycyclic musks, phthalates
Endocrine Disrupting Chemicals (EDCs)	Octylphenols, nonylphenols, Di(2-ethylhexyl)phthalate
Flame retardants	Polybrominated diphenyl ethers (PBDEs)
Perfluoronated compounds	Perfluorotoctane sulfonates, perfluoroctanoic acid

flame retardants (Loganathan et al., 2020), this chapter does not offer an exhaustive examination of these substances.

2.3 Effects of Emerging Pollutants on Human Health and the Environment

There has been a significant amount of research conducted on the effects that ECs have on animals; however, the impacts that they have on people are still being investigated. The figure illustrates this further. Even minute quantities of electronic cigarettes (ECs) can pose significant risks to the health of a human being (Fig. 2.3) (Ahmed et al., 2021). The primary cause of this exposure is the ingestion of these substances. This phenomenon is known as biomagnification, which involves the progressive buildup of chemicals in living organisms and the following rise in concentration of these compounds as they move up the food chain, especially in animals occupying the highest trophic level. Unfortunately, there is a lack of relevant research about the adverse effects of heavy metals, endocrine-disrupting chemicals (EDCs), and bisphenol-A (BPA). The majority of these compounds are produced as a result of the discharge of surface water, seepage, wastewater treatment facilities (WWTPs), and landfill sites (Murray et al., 2010; Thomas & Hilton, 2004).

An endocrine disturbance disrupts the functioning of the endocrine system in both humans and animals by imitating, inhibiting, or altering the actions of hormones. Endocrine Disrupting Chemicals (EDCs) are external substances that have a limited degree of interaction with hormones produced naturally within the human body.

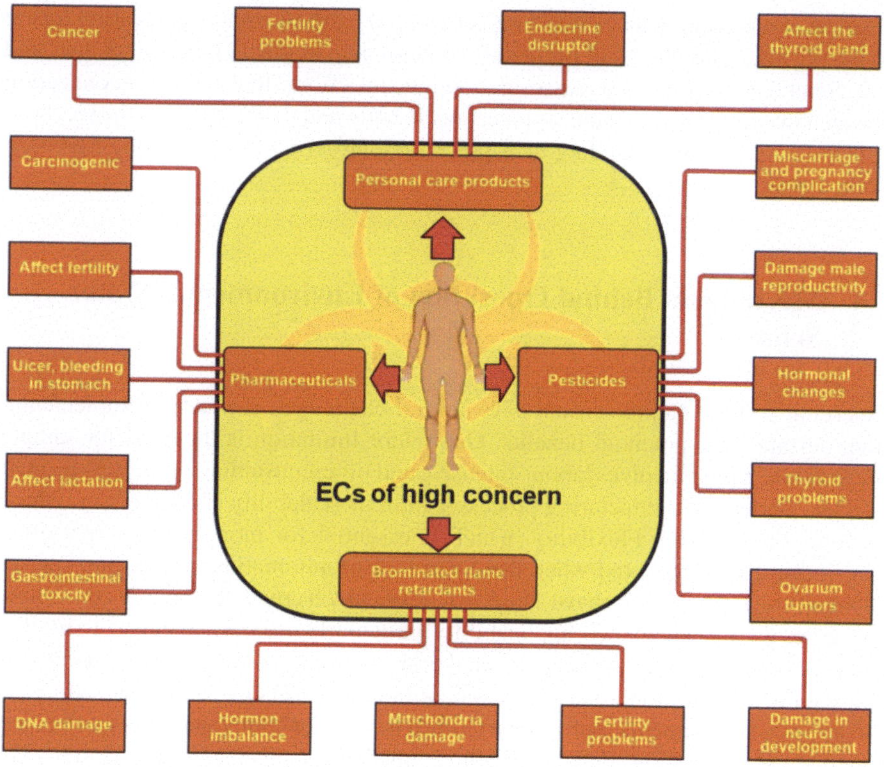

Fig. 2.3 The adverse consequences and detrimental repercussions on human well-being caused by ECs. Reprinted with permission from Ahmed et al., (2021), Copyrights (2021) Elsevier, license no. 5818850984659

Moreover, there is compelling evidence establishing a connection between exposure to endocrine-disrupting chemicals (EDCs) and a decline in male sperm count, as well as an elevated susceptibility to testicular, prostate, ovarian, and breast cancer, along with reproductive dysfunctions (Bolong et al., 2009; Lei et al., 2015; Yan et al., 2010). Routledge et al. (1998) reported that exposure in young individuals has been linked to a retardation in brain development and a reduction in intelligence quotient (IQ). Removing these contaminants is challenging due to their existence at exceedingly low concentrations. Nevertheless, given that these substances have the potential to cause adverse effects on human health, including toxicity over an extended period of time and disruption of the endocrine system, immediate action is of the utmost importance (Rosal et al., 2010). Empirical data has demonstrated the presence of a multitude of ailments (Rosal et al., 2010). Several hormones, including estradiol, estrone, estriol, ethinyl estradiol, and dysfunctional reproductive tissue, have been detected in animals (Boxall, 2012).

Both pharmaceuticals and personal care items can influence the processes of evolution, reproduction, and adaption that organisms experience in their habitats. The creation of a comprehensive database is hindered by the complex and varied nature of the physical and chemical properties of these contaminants in the air, water, and soil. Scientists have conducted several significant efforts to establish a connection between these harmful substances entering the body and their impact on the receptor, with the aim of comprehending their potential long-term effects (Halling-Sørensen et al., 1998).

2.4 Mechanism Behind Operation of Environmental Water Sensors

Emerging contamination sensors have several challenges in practical implementation, despite their potential benefits. One major limitation is the need for sample preparation, which involves labour-intensive and time-consuming processes. Its high cost and intricate manufacturing processes limit its scalability and efficiency, which prevents broader use. Flexibility, which is essential for monitoring in real time, can frequently be hindered when there is a lack of automation and remote-control capabilities. These sensors have a high price tag, which creates budgetary constraints and raises worries about their applicability, particularly in environments with limited resources and when they are implemented on a large scale across extensive geographical regions. In addition, their effectiveness in real-world environmental settings is hindered by impediments such as interference from other compounds and the demand for frequent calibration, both of which diminish their precision and reliability. In order to completely utilize environmental sensors, it is imperative to tackle these intricate difficulties. This entails improving their capacity to recognize and manage emerging environmental contaminants as well as their smooth integration into a variety of monitoring scenarios. Currently, researchers are focused on creating a variety of sensors to find new pollutants in water (Amin et al., 2014;Baruah & Dutta, 2009; Glasgow et al., 2004). The most popular types of these sensors are optical and electrochemical ones, however they all employ various mechanisms and techniques. The sensors utilize recognition components, including antibodies, aptamers, molecularly imprinted polymers (MIPs), and a range of other natural and synthetic receptors, to precisely detect and identify contaminants (Naseri et al., 2020). Biosensors, such as immunosensors and aptasensors, rely on biomolecular identification through affinity or catalytic binding processes. This reliance allows biosensors to demonstrate exceptional selectivity for their specific uses (Gonzalez-Gonzalez et al., 2022). Still, maintaining immobilized bioreceptor's stability and effectiveness over long stretches of time in real-world situations is a challenging undertaking. These sensors use binding events to produce electrical or optical signals that are used to detect the presence of contaminants. Optical sensors employ light instead of physical contact to investigate and gather data from water samples, eliminating the physical exposure of expensive sensing components. Because of this feature, optical sensors are a great option for online, non-destructive, real-time sensing and require less maintenance (Zainurin et al., 2022). Techniques such as voltammetry, amperometry, and impedance are utilized by electrochemical sensors in order to provide monitoring of redox activities that take place at electrode surfaces. Additionally, in comparison to optical sensors, these sensors frequently exhibit a higher level of sensitivity (Sivaranjanee et al., 2022). Furthermore, electrochemical sensors are unaffected by turbidity or the presence of molecules that absorb light, and they can be made smaller for usage in the local vicinity. Portable platforms including microfluidic systems, lateral flow devices, tiny and multiplexed probes, commonly referred to as "lab on a chip" gadgets, have the capability to combine both electrochemical and optical modalities. In order to remotely evaluate big bodies of water, these platforms can also be linked to digital platforms or satellite photography (Farré, 2020).

Below is a detailed summary of these many kinds of sensors and detection techniques. Optical sensors generate a measurable light signal from light and the target analyte. These sensors accurately detect new contaminants' electrical or vibrational characteristics and light extinction or refractive index changes when they cling to desirable molecules. Figure 4a shows how colorimetric sensors detect changes in light wavelengths as they move through a detecting medium when a chemical is detected. Cost-effective colorimetric sensors have demonstrated efficacy in detecting a diverse array of contaminants, including pesticides, heavy metals, bisphenol A, and other deleterious compounds. In spite of this, these sensors frequently fall short of meeting regulatory standards because of their restricted quantitative capabilities and limited sensitivity. Refractive sensors, such as the ones shown in Fig. 4b, can detect changes in the angle of incoming light that result in a decrease in the intensity of reflected light, "surface plasmon resonance (SPR) dip," when the analyte attaches to the surface of sensor.

Refractometric sensors have shown to be extremely accurate and sensitive in detecting various types of water pollutants; yet, they necessitate advanced optical apparatus and are susceptible to interference from complicated water compositions. When the target analyte is present, fluorescent sensors are employed to evaluate changes in the fluorophore's lifetime or intensity. These sensors can't detect more than one analyte at once, and they have other drawbacks including photobleaching. On the other hand, sensors that are based on surface-enhanced Raman spectroscopy (SERS) are able to detect the presence of the target by measuring the Raman scattering that is released by the probe molecules or the agents that are being targeted. For molecular or cellular analysis of water contaminants and pathogens, these sensors provide excellent sensitivity and selectivity. They do, however, need plasmonic substrates and pricey Raman spectrometers. Electrochemical sensors measure current and voltage from redox processes that include the material being investigated. Electrochemical sensors measure conductance, voltage, and current (Fig. 4e–i). More recently, single particle nano sensors have been produced (Hassan et al., 2022; Sivaranjanee et al., 2022). Electrode surfaces can be improved by adding various recognition components, such as molecularly imprinted polymers, enzymes, aptamers, and antibodies, in order to boost the specificity for the analytes that are required (Fig. 4e–h). Various electrochemical sensors have been created to identify pesticides, phenols, heavy metals, and medicines. Biosensors selectively bind target chemicals using biological recognition components such as enzymes, aptamers, and antibodies, resulting in a detectable electrical or visual signal. Biosensors have been employed in the detection of many water pollutants, including heavy metals, pesticides, antibiotics, and substances that disturb the hormone balance. When it comes to the application of biosensing technology in real-world settings, there are a few obstacles that present themselves. These challenges include the fact that the bioreceptor has a long-lasting nature and the possibility that it could be damaged or hindered by other toxins or elements in water. Target analytes can bind or adsorb onto a sensing surface, changing mechanical properties like viscoelasticity or mass loading. Acoustic wave sensors are capable of detecting these changes. Quartz crystal microbalances (QCMs) and surface acoustic wave (SAW) devices are commonly employed as sensors for detecting acoustic waves. Ion concentrations in water samples are measured via an electro-chemical sensor known as an ion-selective electrode (ISE). ISEs work with a selective membrane that is intended to target particular ions, like nutritional or heavy metal ions. The Nernst equation states that the voltage differential across the ion-selective membrane equals the solution ion concentration. Several paper-based potentiometric sensors for heavy metals, including silver, cadmium, sodium, and chloride, have been produced utilising ion-selective electrodes. These sensors are designed to be affordable and can detect heavy metals. Microfluidic sensors offer quick and effective water quality measurement by combining fluid manipulation, sample processing, and sensing components into a tiny device. Microfluidic sensors manage liquids and execute complicated analytical operations in tiny devices using laminar flow, capillary forces, and diffusion in microscopic channels. Lab-on-a-chip (LOC) devices use microfluidics and numerous sensor technologies to detect several environmental analytes in water samples (Narayanamurthy et al., 2020; Sharma & Sharma, 2022).

Lab-on-a-chip systems combine chemical/optical/electrochemical sensors with sample processing components like filtering or pre-concentration mechanisms and microfluidic channels for convenient and effective analysis (Jaywant & Arif, 2019; Nelis, 2020) The evaluation of water quality characteristics on a wide scale, such as turbidity, algal blooms, and temperature changes, can be accomplished through the use of remote sensing techniques, which involve the utilisation of satellite pictures or aerial photography. The practice of analysing the electromagnetic radiation that is reflected or emitted from bodies of water in order to derive information on the physical, chemical, or biological features of the water bodies is referred to as remote sensing (Jaywant & Arif, 2019).

The introduction of cost-effective optical chemical and biological sensors into smartphone cameras, color-reading apps, or image processing software is made possible by the use of smartphone reading devices. These devices come loaded with an intuitive user interface that makes it possible to operate them in a straightforward and uncomplicated manner (Boada et al., 2019).

Nonetheless, obtaining high precision value and quickness in analysis is bit difficult and challenging. When it comes to detecting a variety of substances, lateral flow devices (LFDs) offer a solution that is not only quick and inexpensive, but also simple. The use of lateral flow devices, or LFDs for short, has increased in many different types of healthcare facilities. These tools offer qualitative analysis for a wide range of uses, such as COVID and pregnancy tests (Noviana et al., 2021).

Nevertheless, its application in the environmental field has not received much attention. Lateral flow devices (LFDs) provide qualitative data that can be readily interpreted without the need for any additional tools, within a time frame of 5–30 min, depending on the particular configuration. These gadgets are specifically designed to be easily operated by those in the community who have not undergone official training, thanks to their straightforwardness. Optimizing development timelines and implementing effective techniques to advance market-ready sensors.

To make portable sensors, immobilize chemical or biological receptors, optimize a physical transducer, and communicate with signal amplification and processing devices. Portable sensors need power, hardware, and software to convert data for field usage. Advanced sensors may connect to the IoT and use wireless communication modules, autonomous sampling and data transmission, and advanced data analytics to improve decision-making (Fig. 2.5). Despite remote sensing and big data frameworks for water monitoring, chemical and biological sensing technologies in the Internet of Things (IoT) are limited (Nvs & Saranya, 2020; Vaseashta et al., 2021).

These concerns involve the lack of integration between several components, which obstructs comprehensive analysis and operation by non-technical users. Furthermore, there is a constraint on the precision and regularity of measurements in real-life scenarios, leading to inadequate data at suitable levels of detail. Nonetheless, there have been previous cases where wireless, cost-effective smartphone-based portable sensors have been effectively created to map the existence of nitrite (Xu et al., 2020) and mercury (Wei et al., 2014) in water throughout different periods and locations.

The capacity of various instruments to monitor water quality is demonstrated by a variety of cases. For the purpose of developing systems that are ready for use in the field and putting them into operation on a broad scale, it is essential to conduct a comprehensive analysis that must meet the criteria of portability and cost efficacy. The development of sensing technologies relies heavily on a solid foundation of fundamental scientific knowledge, which is frequently derived from publicly or government-funded research conducted at universities.

The goal of designing a sensor is to provide precise readings, which accurately represent the actual analyte's concentration, when used in real-life scenarios. It is essential to stick to the specific criteria of the desired application situation to successfully execute this assignment. Additionally, it is essential to make certain that the task is completed within a reasonable budget and without exceeding the allotted amount of time.

There are four basic processes involved in the process of creating sensors. These steps are as follows: (1) establishing the requirements of the stakeholders; (2) carrying out performance testing; (3) verifying the analytical approach; and (4) carrying out field deployment while simultaneously carrying out continuing performance reviews.

The primary purpose is to ensure that the sensors being created are acceptable for the working environment in which they will be used, while also meeting market needs and stakeholder expectations. The parameters for sensitivity, such as the limit of detection (LOD) and the limit of quantification (LOQ), must be correctly established in order to fulfil the requirements described above. This will ensure that the expansion of production can be achieved efficiently and economically. It is also crucial for sensor designers to engage in proactive cooperation with diverse stakeholders to incorporate user requirements into the development process at this initial stage.

Following the preceding stage, the technique includes the practical actions of constructing the sensor, selecting and assembling recognition components, transducers, and detecting algorithms, and producing the sensor itself. The use of modern materials and technologies, such as nanoscale materials and micro- and nanofabrication, has demonstrated that the size of sensors used to detect environmental contaminants can be reduced while improving their performance (Nikoleli et al., 2018; Pena-Pereira et al., 2021). Inclusion of nanomaterials have enhanced sensitivity owing to their large surface-to-volume ratio and specific optical properties (Sreejith et al., 2024). Advancements in nano-sensing technology hold great potential for conducting environmental investigation (Bashir et al., 2022). Nevertheless, the early versions of nanotechnology-based sensors need enhancement to tackle issues related to non-specific adsorption, possible aggregation, and stability of nanomaterials. Furthermore, it is critical to improve the scalability of design components and the use of these sensors in complex environmental samples. This is necessary. Sophisticated technologies are essential for reducing the size of sensors, facilitating their deployment, and automating both sensor operations and data interpretation. Additionally, they enable the detection of newly emerging pollutants in real-time. At this point, the sensors are put through a battery of tests that include both standard samples and those that have been intentionally polluted. This collection of samples contains varying quantities of the emerging pollutants that are the primary focus of research. Experimental procedures are used to determine important parameters such as the limit of detection (LOD), limit of quantification (LOQ), measurement range, reaction time, and cost. The criteria for the next phase of development must be consistent with the objectives defined in the previous stage.

Next Stage of the process involves the refinement and validation of the prototype, following the identification of the market and customer expectations and the creation of an initial prototype. In addition, a preliminary minimal viable product (MVP) is constructed and evaluated in the laboratory. As part of this procedure, we will conduct tests on actual samples that have been thoroughly checked and accurately represent the desired application scenario. In addition to cutting-edge sensors, we will use more traditional approaches like LC–MS. The goal is to obtain essential parameters for analysing emerging contaminants, including recovery, accuracy, and precision. At this stage, it is imperative to carry out interlaboratory examinations to demonstrate the resilience and extensive applicability of the research. In order to achieve precise analysis of developing contaminants, it is crucial to detect and minimize the presence of key interfering components. One way to accomplish this is by use sample pretreatment procedures, such as membrane filtration, which enhances selectivity. The quantitative performance has the potential to be greatly enhanced by improving the connections between sensors and actual sample. The results are validated by several independent laboratories over a significant period. Following the successful

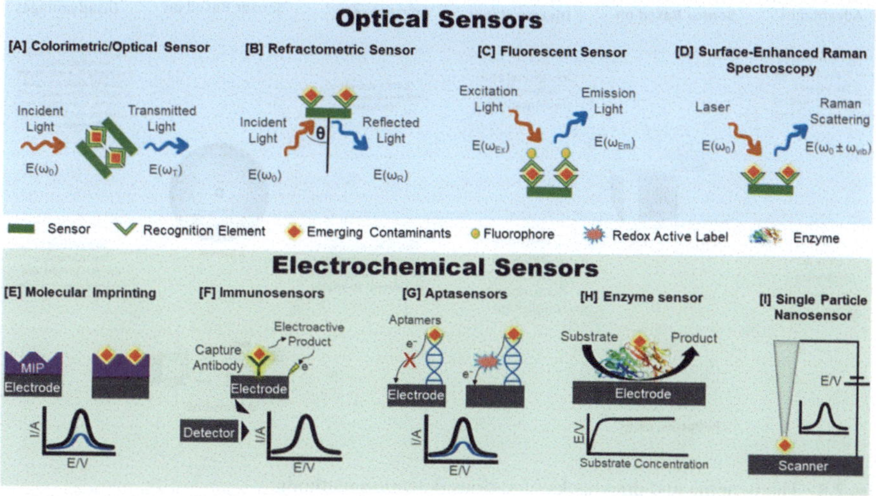

Fig. 2.4 Typical optical and electrochemical sensors for emerging contaminant detection include colorimetric (**a**), refractometric (**b**), fluorescent (**c**), surface-enhanced Raman spectroscopic (**d**) transduction, molecular imprinting (**e**), immunosensors (**f**), aptasensors (**g**), enzyme sensors (**h**), and single particle collisions (**i**). (Adapted with permission from Ateia et al., 2024. Sensors for Emerging Water Contaminants: Overcoming Roadblocks to Innovation. Environmental Science & Technology. Copyright 2024 American Chemical Society)

completion of the field validation procedure, we will proceed to develop standardised operating processes tailored to end clients. To make more improvements, initial user testing may necessitate the development of tailored algorithms for data processing and display. This may be necessary for future enhancements. These changes are intended to improve data interpretation, particularly for those with limited subject matter expertise.

A wide variety of detection strategies have been established as a result of careful analysis and preparation, with the procedures that were discussed before being taken into consideration. As a result of the low interference that they produce, fibre optic sensors are frequently utilised for real-time monitoring. This is accomplished through the utilisation of detecting methods such as fluorescence, surface plasmon resonance (SPR), absorbance, and refractive index (RI). This section provides an examination of the advantages and disadvantages of a variety of optical detection methods that are frequently employed to detect contamination in water (Fig. 2.4).

2.5 Absorption Detection

The absorption spectra of various materials are distinct from one another due to the fact that their chemical structures absorb electromagnetic radiation to varied degrees. This is because the degree to which various chemical structures absorb light varies.

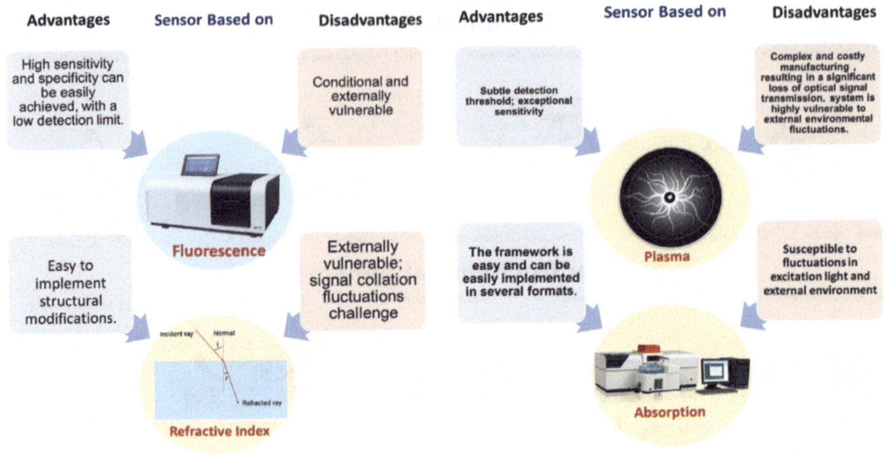

Fig. 2.5 The benefits and drawbacks of various detection methods

Analysing substances involves using a method called absorbance testing to measure the amount and quality of the substance's light-absorbing capacity. Absorbance detection is the standard technique for measuring in sensors that utilise microfluidic fibre optics. These sensors can simultaneously detect the concentration and type of the substance being tested by examining the spectrum for the position and strength of the absorption peak. Generally, absorbance-based devices are intended with a comparatively longer optical path to increase detection sensitivity. But, due to the device's minuscule size, the use of fibre optic microfluidic systems is limited in terms of light absorption distance. As a result, the detection sensitivity is rather reduced. To overcome this issue, researchers created a sensor that combines optical fibres and microfluidics components. To reduce the quantity of light lost and interference from the outside, the construction was designed to increase the number of reflections that occur within the microchannel structure. This approach also significantly reduces the sample quantity (Zhu et al., 2017). In contrast to fluorescence detection, absorbance detection stands out due to its higher sensitivity, repeatability, and user-friendliness. Researchers can modify different microchannel topologies to accommodate different absorber cell lengths, which in turn increases detection accuracy and addresses the problem of inadequate absorber cell lengths. The overall testing capacity is reduced since the microchannel has an exceedingly limited detecting region and path.

2.6 Surface Plasmon Resonance Detection

While different substances may have different resonance angles for activating surface plasmons, the response intensity is proportional to the amount of the activated material clinging to the surface. The development of surface plasmon resonance

(SPR) sensors were based on the underlying sensing principle of response strength fluctuation (Zhao et al., 2019).

By combining the fibre surface's receptive area with a microfluidic chip, distant and real-time measurements are made possible. Surface plasmon resonance (SPR), a method that takes advantage of the unique characteristics of optical fibre, allows for this enhanced responsiveness. Researchers have found that etched sensing structures, FBG, TFBG, or D-type are better at producing the SPR effect when the fibre resonance angle is considered (Sun et al., 2016; Tomyshev et al., 2019).

The sensitivity of the reflecting SPR sensor was enhanced by applying a fifty nanometre thick layer of gold to the surface of the fibre, surpassing the sensitivity of most multimode SPR sensors. However, SPR sensors have faced a significant obstacle in accurately identifying specific adsorption effects.

2.7 Fluorescence Detection

In order to produce fluorescence in a probe that is attached to the surface of the core, microfluidic fibre optic fluorescence sensors often make use of a rapid wave. Following that, the fluorescence, which is responsible for carrying detection information, is conveyed through the fibre, and it is finally recorded by the detector (the detector). The analysis of data pertaining to the target analyte is accomplished through the process of fluorescence sensing, which involves the detection of changes in fluorescence intensity, fluorescence lifespan, and fluorescence wavelength. The laser-induced fluorescence (LIF) detection method is now the most widely used method for detecting fluorescence in microfluidics. (Duca et al., 2022; Liénard et al., 2020). The inclusion of optical fibre within the device, which acts as both a light source and a means for receiving signals during the detecting phase, considerably reduces the difficulties that are often faced. (Gupta et al., 2014, 2015; Sadeghi et al., 2013). The use of a tapered nanofiber structure allows the identification of extremely small concentrations of biological molecules. This is enabled by the ability to detect volumes as small as femtoseconds. Fibre optic fluorescent microfluidic sensors have stronger sensing capabilities than traditional fluorescence detection technologies.

2.8 Refractive Index Detection

The interaction between the light and the external medium influences the properties of transmitted light. Among these features are the light's intensity, size, and location. The detection of the refractive index (RI), which allows the identification of the sample, is based on the measurement of oscillations in the light signal. The versatility of RI detection makes it useful in many different types of detection scenarios, as it can assess substances without labels. In this context, the most common detection principles used are refraction, interference, optical path reflection, and sudden field stimulation. The detection methods that use RI are more flexible than those that use other methodologies. These methods are sometimes called "non-invasive" or "markerless" treatments because they don't include markers. The integration of fibre optic sensing elements onto microfluidic chips, including interferometers, Fabry–Perot cavities, and fibre gratings, has led to an increase in the variety of viable applications. However, there are situations where the impact of sample absorption cannot be ignored, and accurately measuring complex sample solutions becomes difficult (Yang et al., 2019).

2.9 Obstacles for Sensors from the Laboratory to the Market

There has been a lot of pushbacks from both technically and non-technically inclined people when it comes to getting proof-of-concept sensors that can identify water toxins into actual use. In order to offer accurate and precise analysis that is also dependable over time and easy to operate, it is necessary to overcome technical challenges. To improve assessment precision, sensor validation must be performed under environmentally appropriate circumstances and using well-established analytical procedures. Although a lot of prototypes may operate well equipped laboratory, their results in real-world scenarios may be less impressive (Taverniers et al., 2004a, 2004b).

Sensors that use artificial or organic receptors are particularly susceptible to stability and selectivity issues. This is due to the high detection ability and susceptibility to decay that these sensors exhibit. The inability to function in complex water compositions, as well as the adsorption of substances without specificity, are examples of issues that make navigating these problems even more difficult. Cost, scalability, and manufacturability are some of the other variables that must be carefully considered throughout the design phase in order to overcome all of the barriers to the widespread use of sensors. Unfortunately, there are not enough resources available to increase sensor manufacture and validate performance under environmental conditions. This is due to the fact that basic scientific research is typically the emphasis of academic laboratories rather than applied development. There are difficulties in verifying and guaranteeing the calibre of sensors as well. A large number of published research use standard solutions or, at most, purposefully tainted samples. The proposed sensors are rarely tested with genuine environmental samples, and even fewer are validated against established analytical procedures. The accuracy and precision of the sensors are reported in compliance with stringent quality assurance and control criteria. Researchers working on developing a colorimetric-based sensor confront a variety of technical challenges, including a lack of specificity, inadequate sensitivity at the parts per million (ppm) level, the influence of the water matrix, and the low specificity of electrochemical-based sensors. In addition, electrochemical sensors have problems due to electroactive interference. When compared to other types of sensors, SERS-based sensors have issues such as low repeatability, the water matrix effect, and difficulty scaling up operations. Furthermore, it suggests potential areas for development that could help to successfully transition prototype sensors into commercially viable devices are able of correctly identifying developing pollutants in water systems.

The main non-technical problems include interdisciplinary teamwork, market forces, university entrepreneurship, and alliances with companies, sectors, and regulatory or governmental bodies. The sensors built may not suit the demands of many stakeholders. Without a universal sensing technique or toolset, substantial debate is needed to identify locations where these sensors may be efficiently employed for varied purposes. However, other barriers to progress in this field include the absence of well-established procedures, lack of market validation, inconsistent reproducibility between laboratories, and vaguely defined accuracy criteria. These problems are essential to sensor validation. The commercialization of sensors intended to detect newly emerging contaminants has slowed down as a result of investors' and users' diminished trust due to the absence of stringent validation procedures. Because of the limited scale of the fast-increasing pollution monitoring business and a scarcity of personnel competent in entrepreneurship and commercialization, the frequency of knowledge transfer and firms based on sensor-based is hindered. The increasing incidence of unchecked emerging contaminants, which have unknown consequences for the environment and public health, is another significant barrier. To establish safety criteria, comprehensive and complex environmental investigations are also required. In addition, numerous types of water, including as drinkable water, surface water, and wastewater effluents, are subject to special legislation that are designed to reduce the number of contaminants that are present in the water. For the most part, however, these guidelines remain inadequately specified for recently discovered pollutants. As a result, a considerable number of these compounds have yet to be allocated a sensitivity objective or a specific circumstance for use in sensor development. This lack of clarity also stifles progress and prevents a thorough evaluation of emerging sensor technology. Despite the creation and validation of various sensors, conventional laboratory procedures continue to be used for the vast majority of investigations. Currently, the focus of sensing is mostly on improving ability through the implementation of novel materials such as carbon based nanotubes (CNT), rather than prioritising customer needs, industry requirements, manufacturing processes, or market expectations. There is an acknowledged necessity for sensors in some situations, such as with the recently identified pollutants known as Per- and Polyfluoroalkyl Substances (PFAS). Nevertheless, the disclosed sensors do not offer the required sensitivity and selectivity. When compared to the low parts per trillion recommended levels given by the Environmental Protection Act (EPA), there are instances where the discrepancy is substantial, ranging up to several orders of magnitude (e.g., 10 parts per million using a colorimetric paper-based approach). While certain MIP-based electrochemical sensors are sensitive enough to identify compounds at as low as 20 parts per trillion (ppt), they are not highly selective, and other pollutants like as humic acid and chloride can interfere with their performance. Delays in determining the requirements and needs for testing these sensors may result from this. Rather than depending just on scholarly publications, it is critical to prioritize intellectual property (IP) protection through strategic investment in order to expedite the process of translating research discoveries into practical applications. According to recent study, there has been a concurrent decline in patent filings for nano sensors and an increase in publications about nano sensors. This drop can be linked to a lack of interest in investigating sensor validation, prototyping, and field testing, as well as a scarcity of research funding. Industry usually finds these activities more interesting than academia. The preponderance of fundamental research, which is primarily motivated by scientific curiosity rather than stakeholder or business objectives, may act as a barrier to quick technology transfer. To address this difficulty, concerned personals must collaborate to create a feedback system that works in a loop. This is especially important for those who live in underprivileged, resource-constrained areas where chemical exposures have a significant impact. To ensure that new detection technologies effectively meet the needs of communities that are disproportionately affected by water contaminants but lack the necessary analytical instruments for detection, it is critical to include key stakeholders and disadvantaged groups in the technology development process. Participating in this collaborative process gives stakeholders the chance to offer their input throughout the early stages of design, ensuring that the device's parameters meet market demands. In addition to technological issues, overcoming difficulties requires addressing company strategy, stakeholder effect, law, legal considerations, and human and financial resource availability. These components working together will speed up the transition from proof-of-concept devices to field deployment. Studies carried out in the field have confirmed this (Loureiro et al., 2020; Markovic & Bagherzadeh, 2018).

2.10 Important Strategic Goals and Best Techniques for Getting Beyond Challenges

In spite of the tremendous research and development efforts that have been made to identify novel toxins in surface water, the sensors that have been developed by researchers continue to face a significant gap in terms of their marketability. In order to drive innovation from the laboratory to the market, major improvements may be accomplished by applying effective methods and creating objectives that are attainable. This will allow for the overcoming of both technical and non-technical obstacles.

1. **Enhancing the accuracy and specificity of existing sensor systems.**
    a. The use of unique synthetic receptors and biorecognition components like aptamers and peptides has the potential to increase the system's selectivity.
    b. It is feasible to improve the transducer's transduction and sensing capabilities by using hybrid materials that contain both organic and inorganic components on the transducer side.

c. The detection capability could be increased by combining multiple transduction mechanisms on a single platform. Optical and electrochemical technologies are two examples of such approaches.
d. Using in-line sample pretreatment modules has the potential to drastically reduce fouling on sensor surfaces. Membrane filtering and hydrogen peroxide oxidation are two examples of such modules.

2. **Integrating real-time, on-location monitoring capabilities.**

   a. More modern designs that combine wireless connectivity and Internet of Things technology may be able to bypass this limitation.
   b. We will be able to expand the capabilities of networked sensing devices that employ energy-efficient wireless communication protocols such as LoRaWAN and Sigfox, if we build alliances with Internet of Things (IoT) specialists. This allows for the instantaneous transmission and processing of data, which is very useful in software applications involving distant sensing.
   c. Data visualisation, analysis, and administration require the creation of user-friendly software systems capable of real-time water monitoring.
   d. Calibration is the most critical step in addressing this problem. Real-time monitoring tasks necessitate the creation and implementation of calibration systems that can respond to changes in environmental parameters such as temperature, pH, salinity, and humidity. These systems must be in place.

3. **Reduce the costs and complexity associated with current analytical method**
   When it comes to large-scale water quality monitoring, sensing research aims to design platforms that are not only cost-effective, but also portable and compact. These systems will facilitate and speed analysis.

   a. Our capacity to produce sensor components on a large scale while remaining cost-effective is enabled by the use of cutting-edge manufacturing techniques such as screen printing and 3D printing.
   b. Chip-based sensing devices should be easy to integrate into current water quality monitoring infrastructure. These systems may additionally include sample devices to speed up the analysis process.
   c. It is important to align prototype designs and performance aims with market studies to guarantee that the created sensors can meet market expectations.
   d. Stakeholders, such as water utilities and regulatory agencies, must actively participate in the technology evaluation and confirmation process. This input is critical for establishing which contaminants are most important, prioritising them, and allocating resources for platform development based on those priorities.

4. **Maintain robustness, efficacy, and suitability in complicated environments**
   The fundamental goal of sensor research is to assure the development of long-lasting sensors that can work at their best in a variety of environmental situations and water compositions.

a. To acquire accurate analysis, calibration approaches that account for the impact of matrix effects as well as the different environmental conditions encountered during field applications must be developed.
b. The introduction of cutting-edge chemicals, such as antifouling coatings, can considerably reduce the development of biofouling and ensure the sensors' capacity to perform successfully for an extended period of time.
c. It is critical to assess the performance and stability of sensors in a number of environmental matrices, such as wastewater, surface water, and groundwater, in order to determine their general usefulness in a variety of settings.

5. **Initial prototype design covers component integration, portability, and scalability**

   When designing portable water sensors for field applications, affordability, mobility, and manufacturability must all be considered. These are essential challenges that must be addressed in order to attain widespread acceptance and commercial success. When it comes to developing a link between academic sensing research and corporate practice, it is critical to address these qualities from the start of the design process. Increasing sensor performance is a popular academic topic, especially when it comes to lowering the limit of detection (LOD) and limit of quantification (LOQ).

6. **Develop standard testing methods & collaborate to evaluate sensor performance efficiency**

   a. The creation of globally acknowledged benchmarks for the performance of ambient water sensors, as well as the testing procedures that they employ, is critical to ensuring their practicability and marketability.
   b. To allow for meaningful comparisons between different sensor systems, widely understood performance standards must be established. These criteria should include sensitivity, selectivity, and reaction time.

7. **Raising awareness of water pollutants and their risks**

   Citizen science initiatives that include state-of-the-art sensor technologies may greatly increase public awareness of environmental issues and participation in monitoring these areas. For sensing equipment designed to be user-friendly and appealing to individuals with low prior knowledge, citizen science activities are a good fit. To raise awareness about the significance of recently identified water contaminants and inspire participation in monitoring programmes, it is possible to form partnerships with educational institutions and local organisations and to host educational sessions, workshops, and online seminars. By supporting citizen science initiatives, urging public participation, and making use of their input, we can improve our understanding of newly discovered water contaminants and effectively deal with them.

8. **Enhance employees' education and training**

   a. To promote sensor innovation, cooperation across scientific fields is necessary. These professions include researchers, product development engineers, data scientists, businesspeople, and policymakers.

b. To enable academic researchers, it is critical to promote interdisciplinary training and provide opportunities for them to participate in internships and collaborate with practitioners who specialise in various aspects of emerging contaminants. As a result, kids will develop the necessary skills (Wright et al., 2007).

There is a limited number of commercial biosensors that are appropriate for environmental monitoring. These biosensors are available. The bulk of biosensors now on the market require significant improvements in terms of sensitivity, selectivity, functionality, and other qualities before they can be used in wastewater treatment plants. The addition of efficient bio-recognition elements, such as whole microbes or biomolecules, has the potential to increase biosensor performance in detecting trace amounts of chemicals, especially in non-transparent fluids like wastewater. BOD biosensors are widely acknowledged as the primary sensing devices for evaluating the quality of water and wastewater. In 1983, Nisshin Denki (Electric) Co. Ltd. introduced the first commercially viable BOD biosensor, as reported by Iranpour et al. (1997), (Iranpour et al., 1997). Korbi Co., Ltd. unveiled a commercial biosensor that utilizes electrochemically-powered microbial fuel cell technology to measure biochemical oxygen demand (BOD) (Jouanneau et al., 2012).

A novel biosensor employing a mediator-less microbial fuel cell has been created to measure the concentrations of biodegradable organic materials in wastewater. The biosensor shows great promise for future commercial use in monitoring biochemical oxygen demand (BOD) (Anam et al., 2017). Surface plasmon resonance (SPR) biosensors, such as the β-SPR biosensor manufactured by SENSIA in Spain, are frequently employed for the surveillance of environmental samples. They have a high level of efficacy in detecting Carbaryl. Texas Instruments Inc. has developed the Spreeta 2000, a portable and affordable SPR-sensor that employs nanotechnology to identify various biomolecule pollutants in soil or water (Chinowsky et al., 2003). Label-free cantilever sensors, such as VeriScan 3000 and Biocom, are specifically developed to detect and measure more than 100 distinct substances in environmental samples.

In order to detect and remove copper nanoparticles on an industrial scale, Correa and his colleagues developed two innovative biosensing systems. These systems utilize the deceased biomass of Rhodotorula mucilagin, Hypocrea lixii, and Trichoderma koningiopsis. The researchers obtained patents for these systems in 2014 (Corrèa et al., 2014). A patent has been obtained for the monitoring of organic elements in water and wastewater fluids by sensing methane released from the biodegradation of organic components. A patent has been granted for the detection and identification of macromolecules and pathogens at exceptionally low concentrations, demonstrating high sensitivity. The experiment entailed the application of certain bioreceptors onto a substrate composed of single-walled carbon nanotubes (SWNT) and two electrodes. The addition was accomplished through non-covalent bonding. The study was undertaken by April and her colleagues in 2018 (April et al., 2018) (Table 2.3).

Table 2.3 Commercial tools for detecting contaminants in wastewater and environmental fluids

Instrument	Analyte	Transducing system	Manufacturer
BOD-2000	BOD	Electrochemica (Amperometric)	Nisshin Denki Kosaku KK, Japan
BOD biosensor 7842-Microbe Sensor Series	BOD	Electrochemical (Potentiometric)	DKK-TOA Corporation, Japan
ARAS BOD sensor	BOD	Electrochemical (Amperometric)	Dr. Lange GmbH, Germany
BOD-Module	BOD	Electrochemical (Amperometric)	Prufgeratewerk Medingen GmbH, Germany
HABS-2000	BOD	Electrochemical (Amperometric)	Korbi Co., Ltd., Korea
Microtox	Organic materials, HMs	Bioluminescence	Azur Environmental, US
ToxAlert	Different toxicants	Bioluminescence	Merck, Germany
LUMIStox	Different toxicicants	Bioluminescence	Dr Lange GmbH, Germany
Spreeta 2000	Different toxicicants	SPR	Texas Instruments Inc., US
β-SPR	Carbaryl	SPR	SENSIA, Spain
CellSense TM	HMs	Amperometric	Outotec oyj, Finland
VeriScan 3000 system	Adaptable for different molecules	Microcantilever	Protiveris Inc., US
BioCom chip	Adaptable for different molecules	Microcantilever	Kalinex Inc., US
BIAcore series	Sulfonamides Microorganisms	SPR	Biacore AB, Sweden
SPR-CELLIA	Microorganisms	SPR	Nippon Laser & Electronics Lab., Japan

Reprinted with permission from Ejeian et al., (2018), Copyrights (2018) Elsevier, license no. 5818870666235

2.11 Consequences and Future Prospects

The detection of newly developing contaminants in a timely manner and with precision is necessary in order to guarantee the maintenance of water quality. Nevertheless, there are a number of obstacles that stand in the way of the innovation of novel sensors. It is feasible to create novel nano-based materials to improve sensor performance and obtain much better performance. For the foreseeable future, the primary focus of research should be on the creation of new recognition components that exhibit a high degree of selectivity when binding to specific targets in complicated water models. This should be the primary emphasis of the research. If you want to enhance stability and reduce the amount of time it takes to analyse something, you should investigate various receptors and recognition elements that can provide a binding affinity that is both faster and more accurate. Reducing the amount of sample preparation and improving accuracy may be possible through the development of sensors, the addition of separation units, and the streamlining of measuring operations using electrodes or supporting materials that are inexpensive. When it comes to the construction of electrochemical and optical sensors for field deployment, having accessible instruments such as hand-held potentiostats and spectrometers that are connected to smartphones is quite beneficial. This approach lowers the expenses of the equipment while also making operation more convenient. When combined with modern data analytics, vibrational sensors such as SERS-based sensors are able to obtain precise quantitative information from a wide range of newly discovered pollutants that are present in complex water matrices. The combination of Internet of Things technology and advanced data analytics makes it feasible to perform data monitoring that is conducted in real time, continuously, and independently. It is possible to develop scalability in order to attain commercial viability. This can be accomplished by enhancing manufacturing techniques and apparatus and ensuring that sensor performance is maintained in difficult environmental conditions.

References

Adegoke, O., Dabrowski, J.M., Montaseri, H., Nsibande, S. A., Petersen, F., & Forbes, P. B. C. (2017). Development of novel fluorescence sensors for the screening of emerging chemical pollutants in water. *Water Research Commission Technical Report.*

Adrián, J., Fernández, F., Muriano, A., Obregon, R., Ramón-Azcon, J., Tort, N., & Marco, M. P. (2009). Biosensors for pharmaceuticals and emerging contaminants based on novel micro and nanotechnology approaches. In *Biosensors for environmental monitoring of aquatic systems: Bioanalytical and chemical methods for endocrine disruptors* (pp47–68).

Ahmed, S. F., Mofijur, M., Nuzhat, S., Chowdhury, A. T., Rafa, N., Uddin, M. A., Inayat, A., Mahlia, T. M. I., Ong, H. C., Chia, W. Y., & Show, P. L. (2021). Recent developments in physical, biological, chemical, and hybrid treatment techniques for removing emerging contaminants from wastewater. *Journal of Hazardous Materials, 416*, 125912.

Ahmed, U., Mumtaz, R., Anwar, H., Mumtaz, S., & Qamar, A. M. (2020). Water quality monitoring: From conventional to emerging technologies. *Water Supply, 20*(1), 28–45.

Alimorong, F. M. L. S., Apacionado, H. A. D., & Villaverde, J. F. (2020, December). Arduino-based multiple aquatic parameter sensor device for evaluating pH, turbidity, conductivity and temperature. In *2020 IEEE 12th International Conference on Humanoid, Nanotechnology, Information Technology, Communication and Control, Environment, and Management (HNICEM)* (pp. 1–5). IEEE.

Altenburger, R., Ait-Aissa, S., Antczak, P., Backhaus, T., Barceló, D., Seiler, T. B., Brion, F., Busch, W., Chipman, K., de Alda, M. L., & de Aragão Umbuzeiro, G. (2015). Future water quality monitoring—Adapting tools to deal with mixtures of pollutants in water resource management. *Science of the Total Environment, 512*, 540–551.

Anam, M., Yousaf, S., Sharafat, I., Zafar, Z., Ayaz, K., & Ali, N. (2017). Comparing natural and artificially designed bacterial consortia as biosensing elements for rapid non-specific detection of

organic pollutant through microbial fuel cell. *International Journal of Electrochemical Science, 12*(4), 2836–2851.
Andres, L., Boateng, K., Borja-Vega, C., & Thomas, E. (2018). A review of in-situ and remote sensing technologies to monitor water and sanitation interventions. *Water, 10*(6), 756.
April, G. U., Yildirim, N., Lee, J., Cho, H., & Busnaina, A., & Northeastern University Boston. (2018). *Nanotube-based biosensor for pathogen detection.* U.S. Patent Application 15/537,080.
Archer, E., Wolfaardt, G. M., & Van Wyk, J. H. (2017). Pharmaceutical and personal care products (PPCPs) as endocrine disrupting contaminants (EDCs) in South African surface waters. *Water Sa, 43*(4), 684–706.
Arman, N. Z., Salmiati, S., Aris, A., Salim, M. R., Nazifa, T. H., Muhamad, M. S., & Marpongahtun, M. (2021). A review on emerging pollutants in the water environment: Existences, health effects and treatment processes. *Water, 13*(22), 3258.
Ateia, M., Wei, H., & Andreescu, S. (2024). Sensors for emerging water contaminants: Overcoming roadblocks to innovation. *Environmental Science & Technology.*
Banna, M. H., Imran, S., Francisque, A., Najjaran, H., Sadiq, R., Rodriguez, M., & Hoorfar, M. (2014). Online drinking water quality monitoring: Review on available and emerging technologies. *Critical Reviews in Environmental Science and Technology, 44*(12), 1370–1421.
Baruah, S., & Dutta, J. (2009). Nanotechnology applications in pollution sensing and degradation in agriculture: A review. *Environmental Chemistry Letters, 7*, 191–204.
Bashir, O., Bhat, S. A., Basharat, A., Qamar, M., Qamar, S. A., Bilal, M., & Iqbal, H. M. (2022). Nano-engineered materials for sensing food pollutants: Technological advancements and safety issues. *Chemosphere, 292*, 133320.
Benotti, M. J., Trenholm, R. A., Vanderford, B. J., Holady, J. C., Stanford, B. D., & Snyder, S. A. (2009). Pharmaceuticals and endocrine disrupting compounds in US drinking water. *Environmental Science & Technology, 43*(3), 597–603.
Bezbaruah, A. N., & Kalita, H. (2010). Sensors and biosensors for endocrine disrupting chemicals: State-of-the-art and future trends. *Treatment of Micropollutants in Water and Wastewater*, 93–127.
Boada, M., Lazaro, A., Villarino, R., & Girbau, D. (2019). Battery-less NFC sensor for pH monitoring. *IEEE Access, 7*, 33226–33239.
Bolong, N., Ismail, A. F., Salim, M. R., & Matsuura, T. (2009). A review of the effects of emerging contaminants in wastewater and options for their removal. *Desalination, 239*(1–3), 229–246.
Boxall, A. B. A. (2012). *New and emerging water pollutants arising from agriculture.*
Carmona, E., & Picó, Y. (2018). The use of chromatographic methods coupled to mass spectrometry for the study of emerging pollutants in the environment. *Critical Reviews in Analytical Chemistry, 48*(4), 305–316.
Chander, V., Sharma, B., Negi, V., Aswal, R. S., Singh, P., Singh, R., & Dobhal, R. (2016). Pharmaceutical compounds in drinking water. *Journal of Xenobiotics, 6*(1), 5774.
Chinowsky, T. M., Quinn, J. G., Bartholomew, D. U., Kaiser, R., & Elkind, J. L. (2003). Performance of the Spreeta 2000 integrated surface plasmon resonance affinity sensor. *Sensors and Actuators b: Chemical, 91*(1–3), 266–274.
Corrêa, B., Nascimento, C. A. O., Salvadori, M.R., Universidade de Sao Paulo, U. S. P., Vale S. A., (2014). *Process for obtaining copper nanoparticles from a fungus selected between hypocrea lixii and trichoderma koningiopsis and use of fungi selected between hypocrea lixii and trichoderma koningiopsis in bioremediation of wastewater and production of copper nanoparticles.* U.S. Patent Application 14/297,417.
Corrèa, B., Nascimento, C. A. O., & Salvadori, M. R., Universidade de Sao Paulo USP, & Vale, S. A. (2014). *Process for obtaining copper nanoparticles from rhodotorula mucilaginosa and use of rhodotorula mucilaginosa in bioremediation of wastewater and production of copper nanoparticles.* U.S. Patent Application 14/297,379.
Duca, Z. A., Speller, N. C., Cato, M. E., Morbioli, G. G., & Stockton, A. M. (2022). A miniaturized, low-cost lens tube based laser-induced fluorescence detection system for automated microfluidic analysis of primary amines. *Talanta, 241*, 123227.

Ejeian, F., Etedali, P., Mansouri-Tehrani, H. A., Soozanipour, A., Low, Z. X., Asadnia, M., Taheri-Kafrani, A., & Razmjou, A. (2018). Biosensors for wastewater monitoring: A review. *Biosensors and Bioelectronics, 118*, 66–79.

Farré, M. (2020). Remote and in situ devices for the assessment of marine contaminants of emerging concern and plastic debris detection. *Current Opinion in Environmental Science & Health, 18*, 79–94.

Galindo-Miranda, J. M., Guízar-González, C., Becerril-Bravo, E. J., Moeller-Chávez, G., León-Becerril, E., & Vallejo-Rodríguez, R. (2019). Occurrence of emerging contaminants in environmental surface waters and their analytical methodology—A review. *Water Supply, 19*(7), 1871–1884.

Garrido-Momparler, V., & Peris, M. (2022). Smart sensors in environmental/water quality monitoring using IoT and cloud services. *Trends in Environmental Analytical Chemistry, 35*, e00173.

Gavrilescu, M., Demnerová, K., Aamand, J., Agathos, S., & Fava, F. (2015). Emerging pollutants in the environment: Present and future challenges in biomonitoring, ecological risks and bioremediation. *New Biotechnology, 32*(1), 147–156.

Geissen, V., Mol, H., Klumpp, E., Umlauf, G., Nadal, M., van der Ploeg, M., van de Zee, S. E., & Ritsema, C. J. (2015). Emerging pollutants in the environment: A challenge for water resource management. *International Soil and Water Conservation Research, 3*(1), 57–65.

Glasgow, H. B., Burkholder, J. M., Reed, R. E., Lewitus, A. J., & Kleinman, J. E. (2004). Real-time remote monitoring of water quality: A review of current applications, and advancements in sensor, telemetry, and computing technologies. *Journal of Experimental Marine Biology and Ecology, 300*(1–2), 409–448.

Goldman, J., Ramanathan, N., Ambrose, R. F., Caron, D., Estrin, D., Fisher, J., Gilbert, R., Hansen, M., Harmon, T. C., Jay, J. A., & Kaiser, W. J., (2007). *Distributed sensing systems for water quality assessment and management.*

Gonzalez-Gonzalez, R. B., Flores-Contreras, E. A., Gonzalez-Gonzalez, E., Torres Castillo, N. E., Parra-Saldivar, R., & Iqbal, H. M. (2022). Biosensor constructs for the monitoring of persistent emerging pollutants in environmental matrices. *Industrial & Engineering Chemistry Research, 62*(11), 4503–4520.

Greb, S., Dekker, A., & Binding, C. (2018). *Earth observations in support of global water quality.*

Gupta, H. (2018). PAH determination in effluent and sludge samples of paper industry. *Environmental Technology & Innovation, 9*, 115–121. https://doi.org/10.1016/j.eti.2017.11.009

Gupta, H., & Gupta, B. (2016). Adsorption of polycyclic aromatic hydrocarbons on banana peel activated carbon. *Desalination and Water Treatment, 57*(20), 9498–9509.

Gupta, H., & Kumar, R. (2020). Distribution of selected polycyclic aromatic hydrocarbons in urban soils of Delhi, India. *Environmental Technology & Innovation, 17*, 100500.

Gupta, V. K., Mergu, N., Kumawat, L. K., & Singh, A. K. (2015). Selective naked-eye detection of magnesium (II) ions using a coumarin-derived fluorescent probe. *Sensors and Actuators b: Chemical, 207*, 216–223.

Gupta, V. K., Singh, A. K., & Kumawat, L. K. (2014). Thiazole Schiff base turn-on fluorescent chemosensor for Al^{3+} ion. *Sensors and Actuators b: Chemical, 195*, 98–108.

Gworek, B., Kijeńska, M., Zaborowska, M., Wrzosek, J., Tokarz, L., & Chmielewski, J. (2020). Occurrence of pharmaceuticals in aquatic environment—A review. *Desalination and Water Treatment, 184*, 375–387.

Halling-Sørensen, B. N. N. S., Nielsen, S. N., Lanzky, P. F., Ingerslev, F., Lützhøft, H. H., & Jørgensen, S. E. (1998). Occurrence, fate and effects of pharmaceutical substances in the environment—A review. *Chemosphere, 36*(2), 357–393.

Hasan, M. K., Shahriar, A., & Jim, K. U. (2019). Water pollution in Bangladesh and its impact on public health. *Heliyon, 5*(8).

Hassan, M. H., Khan, R., & Andreescu, S. (2022). Advances in electrochemical detection methods for measuring contaminants of emerging concerns. *Electrochemical Science Advances, 2*(6), e2100184.

Hojjati-Najafabadi, A., Mansoorianfar, M., Liang, T., Shahin, K., & Karimi-Maleh, H. (2022). A review on magnetic sensors for monitoring of hazardous pollutants in water resources. *Science of the Total Environment, 824*, 153844.

Iranpour, R., Straub, B., & Jugo, T. (1997). Real time BOD monitoring for wastewater process control. *Journal of Environmental Engineering, 123*(2), 154–159.

Jan, F., Min-Allah, N., & Düştegör, D. (2021). Iot based smart water quality monitoring: Recent techniques, trends and challenges for domestic applications. *Water, 13*(13), 1729.

Jaywant, S. A., & Arif, K. M. (2019). A comprehensive review of microfluidic water quality monitoring sensors. *Sensors, 19*(21), 4781.

Jouanneau, S., Durand, M. J., & Thouand, G. (2012). Online detection of metals in environmental samples: Comparing two concepts of bioluminescent bacterial biosensors. *Environmental Science & Technology, 46*(21), 11979–11987.

Kapelewska, J., Kotowska, U., Karpińska, J., Kowalczuk, D., Arciszewska, A., & Świrydo, A. (2018). Occurrence, removal, mass loading and environmental risk assessment of emerging organic contaminants in leachates, groundwaters and wastewaters. *Microchemical Journal, 137*, 292–301.

Khan, W. U., Ahmed, S., Dhoble, Y., & Madhav, S. (2023). A critical review of hazardous waste generation from textile industries and associated ecological impacts. *Journal of the Indian Chemical Society, 100*(1), 100829. https://doi.org/10.1016/j.jics.2022.100829

Kumar, N., Kumar, B., Gupta, H., & Gupta, A. (2023). Development and evaluation of cellulose/graphene-oxide based composite for removing phenol from aqueous solutions. *Polymers, 15*(3), 572. https://doi.org/10.3390/polym15030572

Lei, M., Zhang, L., Lei, J., Zong, L., Li, J., Wu, Z., & Wang, Z. (2015). Overview of emerging contaminants and associated human health effects. *BioMed Research International, 2015*(1), 404796.

Liénard, T., Furter, J. S., Taverna, M., Pham, H. V., Hauser, P. C., & Mai, T. D. (2020). Modular instrumentation for capillary electrophoresis with laser induced fluorescence detection using plug-and-play microfluidic, electrophoretic and optic modules. *Analytica Chimica Acta, 1135*, 47–54.

Lofrano, G., Libralato, G., Meric, S., Vaiano, V., Sacco, O., Venditto, V., Guida, M., & Carotenuto, M. (2020). Occurrence and potential risks of emerging contaminants in water. In *Visible light active structured Photocatalysts for the removal of emerging contaminants* (pp. 1–25). Elsevier.

Loganathan, B. G., Ahuja, S., & Subedi, B. (2020). Synthetic organic chemical pollutants in water: origin, distribution, and implications for human exposure and health. In *Contaminants in our water: Identification and remediation methods* (pp. 13–39). American Chemical Society.

Loureiro, S. M. C., Romero, J., & Bilro, R. G. (2020). Stakeholder engagement in co-creation processes for innovation: A systematic literature review and case study. *Journal of Business Research, 119*, 388–409.

Madhav, S., Ahamad, A., Singh, A. K., Kushawaha, J., Chauhan, J. S., Sharma, S., & Singh, P. (2020). Water pollutants: sources and impact on the environment and human health. *Sensors in Water Pollutants Monitoring: Role of Material*, 43–62.

Markovic, S., & Bagherzadeh, M. (2018). How does breadth of external stakeholder co-creation influence innovation performance? Analyzing the mediating roles of knowledge sharing and product innovation. *Journal of Business Research, 88*, 173–186.

Murgolo, S., Yargeau, V., Gerbasi, R., Visentin, F., El Habra, N., Ricco, G., Lacchetti, I., Carere, M., Curri, M. L., & Mascolo, G. (2017). A new supported TiO_2 film deposited on stainless steel for the photocatalytic degradation of contaminants of emerging concern. *Chemical Engineering Journal, 318*, 103–111.

Murphy, K., Heery, B., Sullivan, T., Zhang, D., Paludetti, L., Lau, K. T., Diamond, D., Costa, E., & Regan, F. (2015). A low-cost autonomous optical sensor for water quality monitoring. *Talanta, 132*, 520–527.

Murray, K. E., Thomas, S. M., & Bodour, A. A. (2010). Prioritizing research for trace pollutants and emerging contaminants in the freshwater environment. *Environmental Pollution, 158*(12), 3462–3471.

Narayanamurthy, V., Jeroish, Z. E., Bhuvaneshwari, K. S., Bayat, P., Premkumar, R., Samsuri, F., & Yusoff, M. M. (2020). Advances in passively driven microfluidics and lab-on-chip devices: A comprehensive literature review and patent analysis. *RSC Advances, 10*(20), 11652–11680.

Naseri, M., Mohammadniaei, M., Sun, Y., & Ashley, J. (2020). The use of aptamers and molecularly imprinted polymers in biosensors for environmental monitoring: A tale of two receptors. *Chemosensors, 8*(2), 32.

Nelis, J. (2020). *Development of smartphone hyphenated colorimetric, plasmonic and electrochemical biosensors for food contaminant detection* (Doctoral dissertation, Queen's University Belfast).

Nikoleli, G. P., Siontorou, C. G., Nikolelis, D. P., Karapetis, S., & Bratakou, S. (2018). Prototype biosensing devices: Design and microfabrication based on nanotechnological tools for the rapid in the field detection of food toxicants and environmental pollutants. *Nanotechnology and Biosensors*, 1–28.

Noguera-Oviedo, K., & Aga, D. S. (2016). Lessons learned from more than two decades of research on emerging contaminants in the environment. *Journal of Hazardous Materials, 316*, 242–251.

Norvill, Z. N., Shilton, A., & Guieysse, B. (2016). Emerging contaminant degradation and removal in algal wastewater treatment ponds: Identifying the research gaps. *Journal of Hazardous Materials, 313*, 291–309.

Noviana, E., Ozer, T., Carrell, C. S., Link, J. S., McMahon, C., Jang, I., & Henry, C. S. (2021). Microfluidic paper-based analytical devices: From design to applications. *Chemical Reviews, 121*(19), 11835–11885.

Nvs, B., & Saranya, P. L. (2020). Water pollutants monitoring based on Internet of Things. In *Inorganic pollutants in water* (pp. 371–397). Elsevier.

Padhye, L. P., Yao, H., Kung'u, F. T., & Huang, C. H. (2014). Year-long evaluation on the occurrence and fate of pharmaceuticals, personal care products, and endocrine disrupting chemicals in an urban drinking water treatment plant. *Water Research, 51*, 266–276.

Patel, M., Kumar, R., Kishor, K., Mlsna, T., Pittman, C. U., Jr., & Mohan, D. (2019). Pharmaceuticals of emerging concern in aquatic systems: Chemistry, occurrence, effects, and removal methods. *Chemical Reviews, 119*(6), 3510–3673.

Pellerin, B. A., Stauffer, B. A., Young, D. A., Sullivan, D. J., Bricker, S. B., Walbridge, M. R., Clyde, G. A., Jr., & Shaw, D. M. (2016). Emerging tools for continuous nutrient monitoring networks: Sensors advancing science and water resources protection. *JAWRA Journal of the American Water Resources Association, 52*(4), 993–1008.

Pena-Pereira, F., Bendicho, C., Pavlović, D. M., Martín-Esteban, A., Díaz-Álvarez, M., Pan, Y., Cooper, J., Yang, Z., Safarik, I., Pospiskova, K., & Segundo, M. A. (2021). Miniaturized analytical methods for determination of environmental contaminants of emerging concern—A review. *Analytica Chimica Acta, 1158*, 238108.

Pereira, L. C., de Souza, A. O., Bernardes, M. F. F., Pazin, M., Tasso, M. J., Pereira, P. H., & Dorta, D. J. (2015). A perspective on the potential risks of emerging contaminants to human and environmental health. *Environmental Science and Pollution Research, 22*, 13800–13823.

Piai, L., Blokland, M., van der Wal, A., & Langenhoff, A. (2020). Biodegradation and adsorption of micropollutants by biological activated carbon from a drinking water production plant. *Journal of Hazardous Materials, 388*, 122028.

Rasheed, T., Bilal, M., Nabeel, F., Adeel, M., & Iqbal, H. M. (2019). Environmentally-related contaminants of high concern: Potential sources and analytical modalities for detection, quantification, and treatment. *Environment International, 122*, 52–66.

Rathi, B. S., Kumar, P. S., & Vo, D. V. N. (2021). Critical review on hazardous pollutants in water environment: Occurrence, monitoring, fate, removal technologies and risk assessment. *Science of the Total Environment, 797*, 149134.

Richardson, S. D., & Ternes, T. A. (2018). Water analysis: Emerging contaminants and current issues. *Analytical Chemistry, 90*(1), 398–428.

Rodriguez-Mozaz, S., de Alda, M. J. L., & Barceló, D. (2007). Advantages and limitations of on-line solid phase extraction coupled to liquid chromatography–mass spectrometry technologies versus biosensors for monitoring of emerging contaminants in water. *Journal of Chromatography A, 1152*(1–2), 97–115.

Rosal, R., Rodríguez, A., Perdigón-Melón, J. A., Petre, A., García-Calvo, E., Gómez, M. J., Agüera, A., & Fernández-Alba, A. R. (2010). Occurrence of emerging pollutants in urban wastewater and their removal through biological treatment followed by ozonation. *Water Research, 44*(2), 578–588.

Rout, P. R., Zhang, T. C., Bhunia, P., & Surampalli, R. Y. (2021). Treatment technologies for emerging contaminants in wastewater treatment plants: A review. *Science of the Total Environment, 753*, 141990.

Routledge, E. J., Sheahan, D., Desbrow, C., Brighty, G. C., Waldock, M., & Sumpter, J. P. (1998). Identification of estrogenic chemicals in STW effluent. 2. In vivo responses in trout and roach. *Environmental Science & Technology, 32*(11), 1559–1565.

Sadeghi, R., Karimi-Maleh, H., Khalilzadeh, M. A., Beitollahi, H., Ranjbarha, Z., & Zanousi, M. B. P. (2013). A new strategy for determination of hydroxylamine and phenol in water and waste water samples using modified nanosensor. *Environmental Science and Pollution Research, 20*, 6584–6593.

Sangion, A., & Gramatica, P. (2016). PBT assessment and prioritization of contaminants of emerging concern: Pharmaceuticals. *Environmental Research, 147*, 297–306.

Schaider, L. A., Rudel, R. A., Ackerman, J. M., Dunagan, S. C., & Brody, J. G. (2014). Pharmaceuticals, perfluorosurfactants, and other organic wastewater compounds in public drinking water wells in a shallow sand and gravel aquifer. *Science of the Total Environment, 468*, 384–393.

Sharma, B., & Sharma, A. (2022). Microfluidics: Recent advances toward lab on chip applications in bioanalysis. *Advanced Engineering Materials, 24*(2), 2100738.

Singh, S., Gupta, H., Dhiman, S., & Sahu, N. K. (2022). Decontamination of cationic dye brilliant green from the aqueous media. *Applied Water Science, 12*(4), 61. https://doi.org/10.1007/s13201-022-01596-5

Singh, S., Parveen, N., & Gupta, H. (2018). Adsorptive decontamination of rhodamine-B from water using banana peel powder: A biosorbent. *Environmental Technology & Innovation, 12*, 189–195. https://doi.org/10.1016/j.eti.2018.09.001

Sivaranjanee, R., Kumar, P. S., Saravanan, R., & Govarthanan, M. (2022). Electrochemical sensing system for the analysis of emerging contaminants in aquatic environment: A review. *Chemosphere, 294*, 133779.

Sohrabi, H., Hemmati, A., Majidi, M. R., Eyvazi, S., Jahanban-Esfahlan, A., Baradaran, B., Adlpour-Azar, R., Mokhtarzadeh, A., & de la Guardia, M. (2021). Recent advances on portable sensing and biosensing assays applied for detection of main chemical and biological pollutant agents in water samples: A critical review. *TrAC Trends in Analytical Chemistry, 143*, 116344.

Sreejith, S., Ajayan, J., Radhika, J. M., Reddy, N. U., & Manikandan, M. (2024). Recent advances in nano biosensors: An overview. *Measurement*, 115073.

Stefanakis, A. I., & Becker, J. A. (2016). A review of emerging contaminants in water: classification, sources, and potential risks. *Impact of Water Pollution on Human Health and Environmental Sustainability*, 55–80.

Steinegger, A., Wolfbeis, O. S., & Borisov, S. M. (2020). Optical sensing and imaging of pH values: Spectroscopies, materials, and applications. *Chemical Reviews, 120*(22), 12357–12489.

Street, A., Sustich, R., Duncan, J., & Savage, N. (Eds.) (2014). *Nanotechnology applications for clean water: Solutions for improving water quality*. William Andrew.

Sun, Y. S., Li, C. J., & Hsu, J. C. (2016). Integration of curved D-type optical fiber sensor with microfluidic chip. *Sensors, 17*(1), 63.

Szogi, A. A., Vanotti, M. B., & Ro, K. S. (2015). Methods for treatment of animal manures to reduce nutrient pollution prior to soil application. *Current Pollution Reports, 1*, 47–56.

Tang, Y., Yin, M., Yang, W., Li, H., Zhong, Y., Mo, L., Liang, Y., Ma, X., & Sun, X. (2019). Emerging pollutants in water environment: Occurrence, monitoring, fate, and risk assessment. *Water Environment Research, 91*(10), 984–991.

Taverniers, I., De Loose, M., & Van Bockstaele, E. (2004). Trends in quality in the analytical laboratory. I. Traceability and measurement uncertainty of analytical results. *TrAC Trends in Analytical Chemistry, 23*(7), 480–490.

Taverniers, I., De Loose, M., & Van Bockstaele, E. (2004). Trends in quality in the analytical laboratory. II. Analytical method validation and quality assurance. *TrAC Trends in Analytical Chemistry, 23*(8), 535–552.

Thomas, K. V., & Hilton, M. J. (2004). The occurrence of selected human pharmaceutical compounds in UK estuaries. *Marine Pollution Bulletin, 49*, 436–444.

Tomyshev, K. A., Tazhetdinova, D. K., Manuilovich, E. S., & Butov, O. V. (2019). Ultrastable combined planar-fiber plasmon sensor. *Physica Status Solidi (A), 216*(3), 1800541.

Vaseashta, A., Gevorgyan, G., Kavaz, D., Ivanov, O., Jawaid, M., & Vasović, D. (2021). Exposome, biomonitoring, assessment and data analytics to quantify universal water quality. *Water Safety, Security and Sustainability: Threat Detection and Mitigation*, 67–114.

Vasilachi, I. C., Asiminicesei, D. M., Fertu, D. I., & Gavrilescu, M. (2021). Occurrence and fate of emerging pollutants in water environment and options for their removal. *Water, 13*(2), 181.

Wei, Q., Nagi, R., Sadeghi, K., Feng, S., Yan, E., Ki, S. J., Caire, R., Tseng, D., & Ozcan, A. (2014). Detection and spatial mapping of mercury contamination in water samples using a smart-phone. *ACS Nano, 8*(2), 1121–1129.

Wilson, M., & Ashraf, M. A. (2018). Study of fate and: Transport of emergent contaminants at waste water treatment plant. *Environmental Contaminants Reviews, 1*(10.26480).

Wright, M., Hmieleski, K. M., Siegel, D. S., & Ensley, M. D. (2007). The role of human capital in technological entrepreneurship. *Entrepreneurship Theory and Practice, 31*(6), 791–806.

Xu, K., Chen, Q., Zhao, Y., Ge, C., Lin, S., & Liao, J. (2020). Cost-effective, wireless, and portable smartphone-based electrochemical system for on-site monitoring and spatial mapping of the nitrite contamination in water. *Sensors and Actuators b: Chemical, 319*, 128221.

Yan, S., Subramanian, S. B., Tyagi, R. D., Surampalli, R. Y., & Zhang, T. C. (2010). Emerging contaminants of environmental concern: Source, transport, fate, and treatment. *Practice Periodical of Hazardous, Toxic, and Radioactive Waste Management, 14*(1), 2–20.

Yang, Z., Xia, L., Li, S., Qi, R., Chen, X., & Li, W. (2019). Highly sensitive refractive index detection based on compact HSC-SPR structure in a microfluidic chip. *Sensors and Actuators a: Physical, 297*, 111558.

Yuan, Y., Jia, H., Xu, D., & Wang, J. (2023). Novel method in emerging environmental contaminants detection: Fiber optic sensors based on microfluidic chips. *Science of the Total Environment, 857*, 159563.

Zainurin, S. N., Wan Ismail, W. Z., Mahamud, S. N. I., Ismail, I., Jamaludin, J., Ariffin, K. N. Z., & Wan Ahmad Kamil, W. M. (2022). Advancements in monitoring water quality based on various sensing methods: a systematic review. *International Journal of Environmental Research and Public Health, 19*(21), 14080.

Zhao, Y., Tong, R. J., Xia, F., & Peng, Y. (2019). Current status of optical fiber biosensor based on surface plasmon resonance. *Biosensors and Bioelectronics, 142*, 111505.

Zhao, Y., Ye, L., & Zhang, X. X. (2018). Emerging pollutants–part I: Occurrence, fate and transport. *Water Environment Research, 90*(10), 1301–1322.

Zhu, J. M., Shi, Y., Zhu, X. Q., Yang, Y., Jiang, F. H., Sun, C. J., Zhao, W. H., & Han, X. T. (2017). Optofluidic marine phosphate detection with enhanced absorption using a Fabry-Pérot resonator. *Lab on a Chip, 17*(23), 4025–4030.

Zulkifli, S. N., Rahim, H. A., & Lau, W. J. (2018). Detection of contaminants in water supply: A review on state-of-the-art monitoring technologies and their applications. *Sensors and Actuators b: Chemical, 255*, 2657–2689.

Żur, J., Piński, A., Marchlewicz, A., Hupert-Kocurek, K., Wojcieszyńska, D., & Guzik, U. (2018). Organic micropollutants paracetamol and ibuprofen—toxicity, biodegradation, and genetic background of their utilization by bacteria. *Environmental Science and Pollution Research, 25*, 21498–21524.

Chapter 3
Pathological Contaminants in Wastewater from Hospital Waste: Implications and Mitigation Strategies

Deepak Chandra Joshi, Sonima Prasad, Anuraag Sharma, and Nirmal Joshi

Abstract Hospital wastewater presents a complex mixture of contaminants derived from various healthcare activities, encompassing patient care, laboratory testing, and surgical procedures. This chapter focuses on elucidating the molecular composition of pharmaceutical contaminants within hospital wastewater, recognizing their implications and outlining mitigation strategies. Significant quantities of pharmaceutical compounds, such as diagnostic chemicals and antibiotics, are present in hospital effluent. These chemicals are frequently discharged into wastewater streams through patient excretion, leftover drugs, or inadequate metabolization. In addition, medical treatments introduce biological contaminants such as blood and tissues, which further complicate the composition of waste water. Understanding the molecular intricacies of pharmaceutical contaminants is pivotal for assessing environmental risks and designing effective treatment solutions. Advanced molecular technologies are crucial in the analysis of these pollutants, the identification of their sources, and the monitoring of treatment activities. Researchers can use mass spectrometry and chromatography techniques to identify certain pharmaceutical chemicals and measure their quantities in samples of wastewater. The ramifications of pharmaceutical pollutants go beyond environmental concerns, affecting both human and ecological health. Unchecked release of these compounds into aquatic environments can lead to antibiotic resistance development and disruption of aquatic ecosystems. Mitigation strategies must encompass advanced treatment processes tailored to the removal of pharmaceuticals, such as activated carbon adsorption and advanced oxidation processes.

D. C. Joshi (✉)
Department of Pharmacy, School of Chemical Sciences and Pharmacy, Central University of Rajasthan, Bandar Sindri, Ajmer, Rajasthan, India
e-mail: deepakjoshi024@gmail.com

S. Prasad
Chandigarh University, Ludhiana-Chandigarh State Highway, Gharuan, Mohali, Punjab, India

A. Sharma
Invertis Institute of Pharmacy, Invertis University, Bareilly, Uttar Pradesh, India

N. Joshi
Faculty of Pharmaceutical Sciences, Amrapali University, Haldwani, Nainital, India

H. Gupta et al. (eds.), *Emerging Contaminants in Water and Wastewater*, Environmental Science and Engineering, https://doi.org/10.1007/978-3-031-82579-8_3

This chapter aligns with the One Health approach, emphasizing the interconnectedness of human, animal, and environmental health. By integrating molecular analysis with mitigation strategies, this research contributes to the development of efficient treatment systems that safeguard human health and promote environmental stewardship in hospital wastewater management.

Keywords Hospital wastewater · Molecular composition · Pharmaceuticals · Biological pollutants · Environmental impact etc.

3.1 Background Introduction

3.1.1 Introduction to Hospital Wastewater Contaminants

Hospital wastewater is a complicated combination, including a wide range of pollutants, each with its own molecular identity and possible environmental effects. This complex matrix results from a variety of operations within healthcare institutions, such as patient care, laboratory testing, and sanitation (Kaur et al., 2020). Understanding the molecular composition of hospital wastewater is critical for determining its environmental risk and creating efficient treatment solutions. Pharmaceutical chemicals are one of the most abundant molecular components of hospital effluent (Majumder et al., 2024). These comprise a wide range of medications given to patients for therapeutic purposes. Once consumed or introduced into the body, these medications undergo metabolic changes, culminating in the excretion of metabolites into the wastewater stream (Yilmaz et al., 2017). The molecular diversity of pharmaceuticals in hospital wastewater includes antibiotics, analgesics, antipyretics, and a variety of other medicinal compounds. Furthermore, diagnostic chemicals used in medical tests have an impact on the molecular landscape of hospital wastewater (Alazaiza et al., 2023). The complex makeup of wastewater includes contrast agents, radioisotopes, and other diagnostic compounds. This is because of the many imaging and lab procedures that are done in healthcare facilities (Pirsaheb et al., 2015). The molecular nature of these compounds raises concerns about their persistence and potential effects on aquatic ecosystems. Medical treatments, such as surgeries and interventions, add biological pollutants to hospital wastewater. Blood, tissues, and other body fluids include a diverse array of proteins, enzymes, and biological components (Pirsaheb et al., 2015). These biological ingredients not only complicate the wastewater but also provide obstacles to effective treatment since they might interfere with standard filtration processes. In addition to medications and biological components, disinfectants and cleaning chemicals can be found in hospital wastewater at the molecular level (Akin, 2016). These compounds, which are intended to maintain a sterile environment in healthcare institutions, add to the total chemical load in wastewater. Understanding the molecular behavior of these compounds is critical for assessing their possible environmental implications and developing wastewater treatment procedures that can effectively remove them (Yadav et al., 2020). As we look deeper into the molecular complexities of hospital wastewater, it becomes clear that this environmental hazard extends beyond the visible to the microscopic level. The interaction of medications, diagnostic agents, biological materials, and cleaning chemicals results in a molecular mosaic that necessitates a sophisticated approach to wastewater treatment (Hocaoglu et al., 2021). Future research and technical advances must focus on understanding these molecular intricacies to build long-term and efficient hospital wastewater treatment systems that protect both human health and the environment. Biological material, cleaning chemicals, scanning agents, and medications all combine to form a difficult problem that requires a high-level strategy to resolve the aims and objectives of effective hospital wastewater management are threefold: to protect the environment, safeguard public health, and ensure regulatory compliance. This involves preventing water, soil, and air contamination through proper wastewater treatment, minimizing disease spread and antibiotic resistance, and adhering to environmental standards to avoid legal consequences. Additionally, efforts focus on resource recovery for sustainability, promoting innovation for efficient management, and raising public awareness about the importance of wastewater treatment for environmental and public health protection (Hocaoglu et al., 2021).

3.1.1.1 Significance of Investigating Pathological Contaminants in Hospital Wastewater

In the intricate web of molecular interactions, the significance of delving into pathological contaminants within hospital wastewater becomes a paramount focus in contemporary scientific endeavors. As we peer into the microscopic realm, the molecular tapestry of hospital wastewater reveals a dynamic landscape that intertwines health, environmental, and technological dimensions. At its core, the molecular-level exploration of hospital wastewater is pivotal due to the unique cocktail of biological agents it harbors (Tsakona et al., 2007). Pathogens, pharmaceutical residues, and other bioactive compounds, each bearing a molecular signature, cascade into the water stream as a consequence of medical procedures, patient care, and the disposal of pharmaceuticals. Unraveling the intricacies of these molecular constituents is fundamental for several reasons (Gautam et al., 2007). One critical aspect lies in the potential public health ramifications. Pathogenic microorganisms, such as bacteria and viruses, manifest at the molecular level with distinct genetic markers (Hocaoglu et al., 2021). Investigating these markers not only aids in understanding the prevalence and diversity of pathogens but also offers insights into the development of antibiotic resistance. By discerning the molecular underpinnings, researchers can identify emerging threats and devise targeted strategies to mitigate the risks of waterborne infections (Azar et al., 2010). Moreover, the molecular analysis of pharmaceutical residues in hospital wastewater is a crucial facet of this investigation. Pharmaceuticals, designed to interact with specific molecular targets within the human body, can persist in the environment and potentially lead to unintended consequences.

Studying these molecular remnants provides a nuanced understanding of the environmental impact, ecological disruptions, and potential long-term effects on aquatic life (Azar et al., 2010). In the realm of molecular biology, the genetic material of pathogens serves as a treasure trove of information. By employing advanced molecular techniques, scientists can trace the origin, transmission routes, and evolution of pathogens within hospital wastewater. This not only aids in developing effective treatment strategies but also in implementing proactive measures to curtail the spread of waterborne diseases (Khan et al., 2021c). Furthermore, the molecular-level investigation of hospital wastewater aligns with the broader goals of sustainable practices and environmental stewardship. Understanding the molecular signatures of contaminants allows for the development of targeted and efficient treatment technologies. These technologies, rooted in molecular principles, can enhance the removal of specific contaminants, thereby minimizing the environmental footprint of hospital wastewater discharge (Kemprai et al., 2021). The significance of investigating pathological contaminants in hospital wastewater extends to the intricate interplay between microbial communities. Molecular analyses shed light on the microbial ecology within these wastewater systems, unraveling the complex relationships between bacteria, viruses, and other microorganisms (Gupta et al., 2009). Understanding these molecular interactions is pivotal for developing strategies to maintain a balanced microbial ecosystem, preventing the proliferation of harmful pathogens and preserving beneficial microbial functions (Siri et al., 2024). Furthermore, the molecular scrutiny of hospital wastewater encompasses the study of virulence factors and antibiotic-resistance genes carried by pathogenic microorganisms. At the molecular level, researchers can dissect the genetic mechanisms that confer resistance, providing crucial information for the development of targeted antimicrobial therapies and strategies to mitigate the spread of antibiotic resistance. Molecular techniques also play a pivotal role in identifying the sources of contamination within hospital wastewater (Ramdani et al., 2018). By analyzing the specific molecular markers associated with different pathogens and pharmaceuticals, researchers can trace the origins of pollutants. This source tracking enables targeted interventions to reduce contamination at its roots, contributing to more effective pollution control measures (Alimba et al., 2017). The advent of advanced molecular technologies, such as next-generation sequencing and quantitative polymerase chain reaction (qPCR), empowers scientists to conduct comprehensive analyses with unprecedented precision. These techniques allow for the simultaneous detection and quantification of multiple molecular targets, providing a holistic understanding of the diverse contaminants present in hospital wastewater (Khan et al., 2020a). Moreover, the molecular-level investigation delves into the fate and transport of contaminants within the wastewater treatment processes. Molecular markers serve as tracers, enabling researchers to monitor the efficiency of treatment methods in removing specific contaminants. This information is instrumental in refining and optimizing wastewater treatment protocols, ensuring the safe discharge of treated water into the environment (Liu et al., 2010). On a broader scale, the molecular insights gained from studying pathological contaminants contribute to the ongoing dialogue surrounding One Health—a holistic approach that recognizes the interconnectedness of human health, animal health, and environmental health

(Parida et al., 2022). By comprehending the molecular intricacies of contaminants in hospital wastewater, researchers contribute to a more integrated understanding of health dynamics, emphasizing the need for collaborative efforts across disciplines (Dargahi et al., 2022).

3.1.1.2 Scope and Objectives

The scope and aims of this research explore into the complex world of molecular dynamics, investigating the dynamic behaviors and interactions of molecules at a fundamental level. At the molecular level, the chapter seeks to understand the intricate dance of atoms inside molecules, as well as the dynamic forces that regulate their motions (Dargahi et al., 2022). One key goal is to examine molecules' dynamic behaviors in diverse settings, considering aspects such as temperature, pressure, and chemical composition. By clarifying molecular dynamics, we hope to get a thorough knowledge of how molecules react to external stimuli and surroundings. Furthermore, the chapter aims to understand the intermolecular forces that govern molecular interactions (Khan et al., 2021b). To do this, forces like as hydrogen bonding, van der Waals forces, and electrostatic interactions must be thoroughly investigated. These forces are critical for sculpting molecule structures and controlling how they move (Verlicchi, 2018). A molecular-level inquiry also requires a thorough examination of chemical processes. The chapter tries to shed light on the molecular processes that enable reactions, including bond formation, cleavage, and the dynamic rearranging of atoms during chemical changes (Verlicchi, 2018). Understanding these processes at the molecular level is critical for improving sciences like organic chemistry, biochemistry, and materials science (Emmanuel et al., 2005). In addition, the scope includes molecular simulations and computational approaches for studying molecular dynamics. The chapter will describe cutting-edge approaches that allow researchers to view and understand molecular processes in unprecedented detail (Emmanuel et al., 2009). Molecular simulations create a virtual laboratory for researching complicated molecular systems, allowing researchers to investigate circumstances that would be difficult or impossible to recreate in a physical context. Finally, the scope and aims of this molecular chapter aim to contribute to a deeper knowledge of molecular dynamics, opening the way for advances in a variety of scientific fields (Wiest et al., 2018). This chapter attempts to build a better appreciation for the tiny world, where atoms and molecules participate in a never-ending dance, forming the foundation of our physical and chemical reality (Wiest et al., 2018).

3.2 Pathogens and Pharmaceutical Residues in Hospital Wastewater

The presence and consequences of microorganisms and pharmaceutical residues in hospital wastewater need a molecular level study of the complex matrix (Mackuľak et al., 2021). Microbial analyses focusing on pathogen genetic material are at the forefront of these molecular studies. Polymerase chain reaction (PCR) and quantitative PCR (qPCR) procedures may identify and quantify certain nucleic acid sequences, revealing the genetic identity of pathogenic organisms. Furthermore, pharmaceutical residues in hospital wastewater require molecular analysis (Mackuľak et al., 2021). Analytical techniques such as liquid chromatography-mass spectrometry (LC–MS) and gas chromatography-mass spectrometry (GC–MS) are critical for identifying and quantifying pharmaceutical substances. By studying the chemical structures of these residues, researchers get insight into their persistence and possible environmental impact (Zhang et al., 2020b). The molecular knowledge of microorganisms and pharmaceutical residues in hospital wastewater is critical for developing successful treatment techniques. Advanced molecular biology tools allow researchers to understand the complex dynamics of microbial ecosystems and detect new diseases (Ahmad et al., 2021). Simultaneously, understanding the molecular properties of pharmaceutical residues assists in the development of tailored removal procedures, therefore mitigating the environmental and public health problems associated with hospital wastewater discharge. This molecular viewpoint emphasizes the value of multidisciplinary methods in handling the challenges of hospital wastewater treatment and environmental protection (Ngigi et al., 2020).

3.2.1 Identification and Quantification of Diverse Contaminants

In the field of environmental and analytical chemistry, identifying and quantifying various pollutants are critical difficulties that require precise molecular study. Modern analytical methods and methodologies are used in this approach (Khan et al., 2023). Each one is designed to simplify the complex molecular fingerprints of pollutants found in different matrices. At its heart, identification is the discerning of individual molecules within a given sample, whereas quantification is the determination of their concentrations. Achieving this at the molecular level necessitates advanced apparatus and procedures with remarkable sensitivity and specificity (Prado et al., 2011). Chromatography, a technique for separating complicated mixtures based on molecular interactions, is a key strategy in this effort. High-performance liquid chromatography (HPLC) and gas chromatography (GC) are two common techniques for identifying and separating pollutants (Prado et al., 2011; Rodríguez-Rodríguez et al., 2023). These approaches use substances' distinctive molecular features, such as size, polarity, or volatility, to enable exact separation and identification. Mass spectrometry (MS) appears as a strong ally when used with chromatography, aiding in the elucidation of molecular structures (Santos et al., 2013). Mass spectrometry, which measures the mass-to-charge ratio of ionized molecules, gives crucial information on the identity of pollutants. Tandem mass spectrometry (MS/MS) improves specificity by fragmenting ions and examining the resultant spectra, allowing for more thorough identification (Santos et al., 2013). Nuclear Magnetic Resonance (NMR) spectroscopy is another valuable technique in the molecular arsenal for contamination analysis. This approach takes advantage of the intrinsic magnetic characteristics of specific nuclei to provide a comprehensive fingerprint of molecular structures (Mullot et al., 2009). NMR spectroscopy is particularly effective in elucidating the structure of organic substances, allowing for the precise detection of impurities. In recent years, advances in molecular biology methods have broadened the toolset for contamination detection. Polymerase Chain Reaction (PCR) and Next-Generation Sequencing (NGS) allow researchers to detect and quantify pollutants at the genomic level. These approaches identify microbial contamination, allowing for better knowledge of the biological components of a sample (Pérez-Alvarez et al., 2018). Quantifying various pollutants entails not only identifying them but also calculating their concentrations. Calibration curves created using established standards are frequently used in quantitative techniques. Accuracy and precision are required for molecular-level measurement, which necessitates stringent calibration and validation methods. To go deeper into the complexities of detecting and measuring various pollutants at the molecular level, it is critical to emphasize the importance of high-resolution mass spectrometry (HRMS) (Ajala et al., 2022). HRMS has transformed contamination analysis by providing unprecedented resolution and mass accuracy. This accuracy enables the differentiation of closely similar molecules while lowering the likelihood of false positives, resulting in a more robust identification procedure. Furthermore, the burgeoning area of metabolomics plays an important role in molecular-level investigation (Le et al., 2016). Metabolomics is the complete study of tiny molecules inside biological systems. Metabolomics uses methods such as liquid chromatography-mass spectrometry (LC–MS) and gas chromatography-mass spectrometry (GC–MS) to simultaneously identify and quantify a wide range of endogenous and exogenous chemicals (Oliveira et al., 2015). The molecular-level examination includes the characterization of pollutants in complex matrices such as water, soil, and air. When used with analytical methods like GC–MS, solid-phase microextraction (SPME) makes it possible to directly extract and study volatile and semi-volatile pollutants from these mixtures. This method provides more direct molecular insight into the makeup of environmental materials. Furthermore, advances in data analysis and informatics have made major contributions to our knowledge of pollutants at the molecular level (Mendoza et al., 2015; Oliveira et al., 2015). Computational technologies, such as cheminformatics and bioinformatics, help to handle vast datasets produced by analytical procedures. Machine learning techniques improve pattern recognition and predictive modeling, enabling more efficient and precise detection and quantification of pollutants based on molecular properties. The molecular investigation of contaminants is not restricted to classical pollutants but also includes new contaminants, including medicines, personal care items, and hormone-disrupting substances. Advanced methods such as liquid chromatography combined with high-resolution mass spectrometry (LC-HRMS) are essential for unraveling the molecular complexity of these new pollutants and providing a thorough assessment of their existence and possible effects (Mendoza et al., 2015).

3.2.2 Analysis of Bacteria

The molecular world of microbial entities, such as bacteria, viruses, fungi, and parasitic species, reveals a panorama of complexity and adaptive tactics. Molecular analysis of *Escherichia coli* bacteria shows the genomic complexity of plasmids, virulence factors, and antibiotic resistance mechanisms (Le et al., 2016). This helps scientists figure out what kinds of bacteria might be good targets for future treatments. *Pseudomonas aeruginosa*, recognized for its durability and toxicity, reveals its molecular secrets through research into biofilm formation, quorum sensing systems, and virulence gene expression (Asfaw et al., 2017). Moving on to viruses, research into the capsid structure and interactions between host cell receptors reveals the molecular mimicry that norovirus, a common cause of gastroenteritis, uses. Adenovirus, on the other hand, demonstrates advanced molecular techniques for genome transport and integration, stressing the complicated interaction between viral proteins and host cellular components (Moges et al., 2014). Moving on to fungi, Candida albicans' molecular arsenal, which includes virulence factor expression and antifungal resistance mechanisms, plays a crucial role in understanding its pathogenicity (Le et al., 2016). Last but not least, parasites like Cryptosporidium and Giardia use complicated molecular processes, focusing on genome analysis, finding virulence factors, and studying how hosts and parasites interact. The molecular characterization of these microbial agents not only improves our understanding of their fundamental biology, but it also lays the groundwork for tailored treatment approaches to infectious disorders (Moges et al., 2014). Molecular analysis of microbial agents, such as bacteria, viruses, fungi, and parasitic organisms, shows a tapestry of complexity at the microscopic level, giving us important information about how they do harm. Starting with bacteria, the molecular complexities of *Escherichia coli* provide an engrossing investigation of its genetic landscape (Kehl et al., 2022). Researchers investigate the structure of its DNA, specifically plasmids, which are tiny circular DNA units that frequently carry critical genes such as those responsible for antibiotic resistance (Kehl et al., 2022). The research includes the discovery of virulence factors, which allows for a more detailed understanding of how E. coli evades the human immune system and causes infections (Zhang et al., 2020b). This molecular information becomes critical in the creation of tailored treatments. Similarly, *Pseudomonas aeruginosa*, which is known for its strength and clinical relevance, reveals its secrets through molecular studies. Biofilm development, a protective matrix that shields bacteria from immune reactions and drugs, becomes a focal point (Mirghani et al., 2022; Zhang et al., 2020a, 2020b). Understanding the quorum sensing systems that P. aeruginosa uses to coordinate its virulence factors enables researchers to better understand the bacteria's intricate communication processes (Fuentefria et al., 2011). The study of virulence gene expression gives information on the molecular switches that contribute to P. aeruginosa pathogenesis, opening the door for new treatment techniques (Magalhães et al., 2016; Wang et al., 2022). Moving on to viruses, the molecular investigation of norovirus, a significant cause of gastroenteritis outbreaks, entails decoding the complexities of its capsid structure (Magalhães et al., 2016; Yan, 2021). This outer protein coat contributes significantly to the virus's capacity to avoid the host's immunological systems. Furthermore, studying the molecular interactions between the virus and host cell receptors provides insight into the virus's infective processes (White et al., 2016). Another viral protagonist, adenovirus, participates in a complicated chemical dance with host cellular components to ensure optimal genome delivery and integration. These molecular mechanisms are critical for the virus's capacity to form chronic infections and cause a wide range of illnesses (Songca, 2022). Candida albicans, a common fungal pathogen, has a diverse molecular landscape. The study includes the expression of virulence factors such as adhesins and release enzymes that aid in tissue penetration. Understanding the molecular basis of antifungal resistance mechanisms in C. albicans is critical for improving treatment methods for invasive candidiasis, a potentially fatal illness. In terms of parasitic agents, Cryptosporidium and Giardia provide insight into their molecular complexity (Mir, 2022). The investigation focuses on parasite genomes, virulence factors, and host-parasite interactions. Understanding these molecular complexities is critical for the development of tailored medicines and prevention strategies against parasitic diseases (Murugaiyan et al., 2022).

3.2.2.1 Bacteria: *Escherichia coli* and *Pseudomonas aeruginosa*

At the molecular level, *Escherichia coli*'s (*E. coli*) intricacies unfold through a detailed examination of its genomic composition. Researchers scrutinize the plasmids within E. coli, small circular DNA structures that often carry genes conferring antibiotic resistance (Baghal Asghari et al., 2021a). This molecular insight is critical to understanding and combating the rising challenges of antibiotic-resistant strains (Karungamye et al., 2023). Additionally, the analysis extends to virulence factors, enabling a deeper comprehension of how E. coli establishes infections and evades the host's immune defenses (Baghal Asghari et al., 2021b).

Pseudomonas aeruginosa, known for its adaptability and pathogenicity, reveals molecular secrets linked to its clinical significance. The formation of biofilms, intricate matrices protecting bacteria from antibiotics and the immune system, is a focal point (Magalhães et al., 2016). The molecular exploration includes quorum sensing systems, which are ways that P. aeruginosa coordinates the expression of virulence factors. These systems show how these bacteria communicate in very complicated ways. Understanding virulence gene expression adds another layer to deciphering the molecular switches contributing to P. aeruginosa's pathogenicity (Shehabi et al., 2011).

3.2.2.2 Viruses: Norovirus and Adenovirus

Norovirus, a prominent cause of gastroenteritis outbreaks, demonstrates its molecular capabilities by analyzing its capsid structure. This outer protein coat not only aids viral entrance but also plays an important function in avoiding host immunological responses. Molecular studies reveal the delicate dance between the virus and host cell receptors, offering information on the mechanisms behind norovirus's infectivity (Prado et al., 2011). Adenovirus, a complex viral family that causes respiratory, ophthalmic, and gastrointestinal diseases, works at the molecular level to guarantee proper genome transport and integration. To create long-term infections, the virus utilizes intricate molecular methods that include interactions with host biological components. Molecular insights into adenovirus genome modification help to better understand its many clinical manifestations and aid in the development of tailored antiviral medicines (Uhrbrand et al., 2017).

3.2.2.3 Fungi: Candida Albicans

Candida albicans, a common fungal disease, has a diverse molecular landscape in its interactions with the host. At the molecular level, virulence factors like as adhesins and secreted enzymes are prominently expressed. These molecular components promote tissue invasion, which is an important stage in the formation of infection (Mir-Tutusaus et al., 2017). Another reason to study the molecular basis of antifungal resistance mechanisms in C. albicans is to learn how to make better treatments for invasive candidiasis (Eze et al., 2016; Mataraci-Kara et al., 2020).

3.2.2.4 Parasitic Agents: Cryptosporidium and Giardia

Comprehensive genomic investigations shed light on the molecular complexities of parasitic agents such as Cryptosporidium and Giardia. The discovery of virulence factors in these parasites enables researchers to better comprehend their pathogenic potential. Molecular investigations also look at the complicated connections between parasites and their hosts, revealing the methods used by these organisms to create and maintain infections (Yamashiro et al., 2019). These findings are critical for designing tailored treatment strategies and prevention measures for parasitic illnesses. In conclusion, the molecular investigations of *Escherichia coli*, *Pseudomonas aeruginosa*, norovirus, adenovirus, Candida albicans, Cryptosporidium, and Giardia provide a thorough grasp of the complex molecular pathways used by these pathogenic pathogens. These findings not only advance basic biological understanding but also pave the path for tailored therapy techniques and preventative measures against a wide range of infectious illnesses (Robertson et al., 2006).

3.2.3 Impact of Pharmaceutical Residues on Wastewater Composition and Ecological Systems

Apprehensions have been raised regarding the molecular ramifications of pharmaceutical residues in effluent and ecological systems, as well as their intersection with environmental impact. While these residues are maintained structurally intact at the molecular level, they infiltrate effluent via incomplete metabolism and excretion (Fatta-Kassinos et al., 2011). Conventional methodologies employed in wastewater treatment frequently encounter difficulties in completely degrading such molecular compounds, which results in their continued presence in aquatic ecosystems. The complicated metabolic and transformational processes that microorganisms in wastewater help with make the molecular fate of pharmaceutical residues even more difficult to predict (Frédéric & Yves, 2014). Once these compounds enter aquatic ecosystems, they could potentially disrupt cellular and biochemical processes by engaging in molecular interactions with non-target species. Persistent molecular structures accumulate in organisms and proliferate via the food chain, resulting in bioaccumulation and biomagnification as secondary effects (Le Corre et al., 2012). In order to effectively mitigate the complex molecular challenges presented by pharmaceutical residues in ecosystems, it is imperative to establish a connection between the disciplines of pharmacology, environmental science, and molecular biology. Concerns have increased regarding the molecular effects of pharmaceutical residues on ecological systems due to their widespread presence in effluent (Daouk et al., 2016). Discontinued metabolic processes and excretion are the molecular origins of these residues. Potentially impeding normal decomposition upon entering wastewater treatment facilities (Daouk et al., 2016; Verlicchi, 2018). The molecular persistence of active pharmaceutical ingredients (APIs) in these treatment facilities poses significant difficulties. The various species of microorganisms that inhabit wastewater treatment facilities primarily regulate the numerous complex metabolic and chemical processes that pharmaceutical residues undergo. Biotransformation processes in wastewater treatment plants change the molecular structures of drugs to make metabolites that may keep their biological activity or lose it (Khan et al., 2021a). This intricate molecular interaction determines the trajectory of pharmaceutical residues and their potential environmental effects. To find good ways to lessen the damage that pharmaceutical residues in effluent do to the environment, we need to fully understand how these molecules move and interact with each other (Escher et al., 2011). Pharmaceutical residues induce molecular-level impacts on dwelling organisms in aquatic ecosystems after their release. Analogous targets might exist in non-target species of aquatic ecosystems, although numerous pharmaceuticals are engineered to interact with particular molecular targets within the human body (Emmanuel et al., 2002). These molecular and cellular interactions may harm physical processes such as growth, reproduction, and the general health of aquatic organisms. Moreover, pharmaceutical residues have been observed to tend to accumulate and strengthen in aquatic food chains. Over time, the tissues of organisms accumulate molecular structures that are hardened against degradation (Lin & Tsai, 2009). A concentration of pharmaceutical residues increases as organisms at the top of the food chain consume organisms at the bottom. This compromises the ecological equilibrium and weakens the position of apex predators. A multidisciplinary approach is necessary to tackle the intricate molecular challenges that are relevant to pharmaceutical residues. For a comprehensive comprehension of the molecular fate of these compounds and their ensuing ecological repercussions, it is vital to bridge the gaps between pharmacology, environmental science, and molecular biology (Emmanuel et al., 2002). Once scientists fully understand the complicated molecular processes at play, they might be able to come up with workable plans and eco-friendly solutions that effectively lower the harmful effects of pharmaceutical residues on the make-up of wastewater and natural systems (Lin & Tsai, 2009).

3.3 Sources and Pathways of Contaminant Introduction

The infiltration of contaminants into our ecosystems presents an ever-present danger, and it is critical for responsible environmental stewardship to have a thorough understanding of the molecular mechanisms and origins of these contaminants (Verlicchi et al., 2010). Contaminants, which originate at the molecular level, can be traced back to both natural and anthropogenic sources. Their generation frequently involves complex biochemical reactions (Emmanuel et al., 2009). Molecular compounds discharged by natural processes, including biogenic activities and volcanic eruptions, possess the potential to induce extensive environmental repercussions. On the other hand, anthropogenic activities like agriculture and industrial processes that involve the degradation or synthesis of particular compounds introduce contaminants into the environment (Khan et al., 2021b). For instance, the incineration of fossil fuels produces nitrogen oxides and polycyclic aromatic hydrocarbons (PAHs), which are molecular byproducts that contribute to the escalation of air and water pollution. An examination of the molecular mechanisms by which contaminants enter the environment reveals an intricate web of interconnections (Escher et al., 2011). When volatile organic compounds (VOCs) from industrial processes meet the air, they change shape at the molecular level. This creates secondary pollutants with their own unique chemical profiles. Contaminants in aquatic ecosystems enter via atmospheric deposition or direct discharge via molecular pathways (Kümmerer, 2001). Once contaminants enter water bodies, molecular interactions between dissolved substances and contaminants determine their fate and mode of transportation. Adsorption at the molecular level of sediment particles is of critical importance as it significantly influences the persistence and bioavailability of contaminants in aquatic environments (Kümmerer, 2001). The soil, which serves as a reservoir for contaminants, is involved in complex molecular interactions that have an impact on microbial transformations, sorption, and desorption (Dinh et al., 2017). At the point where soil, water, and plants meet, contaminants go through complex changes. This creates a changing molecular environment that affects both their ability to move and their potential to build up in the food chain (Khan et al., 2020c).

3.3.1 Understanding the Origins of Pathogens and Pharmaceuticals in Hospital Wastewater

Investigating the molecular complexity of hospital effluent with a focus on the sources of pathogens and pharmaceuticals reveals a crucial aspect of environmental health. The sources of pathogens present in hospital effluent are multifaceted, originating from a range of locations within healthcare establishments (Varela et al., 2014). Infectious agents, such as bacteria and viruses, can get into effluent more easily through patient waste (like feces and urine), medical procedures, and the disposal of contaminated medical materials. Understanding where these pathogens come from at the molecular level is important for creating and using effective wastewater treatment methods. Standard treatment methods might not be able to handle the complicated molecular structures of some pathogens (Frascaroli et al., 2021). Pharmaceutical compounds, when combined with pathogens, become a substantial concern in hospital effluent. The pharmaceuticals found in these effluents are the result of patient excretion, improper disposal of unused medications, and insufficient drug metabolism within the human body (Emmanuel et al., 2005). Pharmaceutical compounds demonstrate a wide array of molecular structures, which presents difficulties for conventional wastewater treatment facilities that might lack the necessary capabilities to efficiently degrade or eliminate these intricate molecules (Khan et al., 2020c). The molecular interactions that take place between these substances and environmental matrices, such as water and detritus, have an impact on the potential ecological impact and persistence of pharmaceuticals as well (Sim et al., 2011). In addition, the interdependence of pharmaceutical and pathogen origins in hospital wastewater emphasizes the necessity for a comprehensive strategy in wastewater management. The comprehension of these origins at the molecular level enables the creation of sophisticated treatment technologies that are customized to tackle the unique difficulties presented by various pathogens and pharmaceutical compounds (Escher et al., 2011). Moreover, it shows how important it is for healthcare facilities to have strict rules for getting rid of pharmaceuticals so that these substances don't get into wastewater systems. By exploring the molecular aspects of hospital wastewater, valuable knowledge is acquired that plays a crucial role in protecting public health and environmental integrity (Escher et al., 2011). This underscores the significance of implementing a sophisticated and molecularly informed strategy for wastewater management in healthcare environments. The dynamic interaction of numerous molecular components in hospital effluent contributes to the contamination landscape's complexity (Escher et al., 2011). Pathogens, which often originate from medical procedures and patient excretions, pose a complex and multifaceted molecular challenge. The combination of their resistance mechanisms and genomic makeup determines the persistence and survival of bacteria and viruses in effluent. A comprehensive comprehension of molecules is critical to developing precise methods of disinfection and treatment that can efficiently inactivate these organisms. Such knowledge guarantees the secure release of treated effluent back into the environment (Klai & Sellamuthu, 2020). Pharmaceuticals present in hospital effluent concurrently introduce an unprecedented array of molecular complexities. Incomplete metabolic processes and the varied structures of pharmaceutical compounds, which originate from patient medications, necessitate the use of sophisticated treatment approaches. This is because regular wastewater treatment plants might have trouble breaking down or getting rid of these complicated molecular entities (Wang et al., 2018). This could lead to the release of active pharmaceutical components into the ecosystem (Casas et al., 2015). The effects that these compounds have on aquatic ecosystems and the way they interact with water molecules show how important it is to create advanced treatment methods that can handle the complicated chemistry that is found in pharmaceutical residues (Khan et al., 2019). A thorough comprehension of the sources of pharmaceuticals and pathogens in hospital effluent transcends the boundaries of healthcare establishments (Khan et al., 2019). The process includes looking at the changes that happen at the molecular level during the treatment phase as well as the creation of residues that may have ecological effects that haven't been fully understood yet. The study of molecules can help us come up with better ways to treat diseases. New technologies that use complex oxidation processes or membrane technologies to target specific molecular structures can be used in these treatments (Lamba et al., 2017).

3.3.2 Examination of Patient Excretions, Disposal Practices, and Treatment Protocols Contributing to Contaminant Influx

The excrement of patients, the manner in which they are treated, and the disposal of items are all essential factors to consider when attempting to comprehend the manner in which pollutants are introduced into various healthcare environments (Yan et al., 2020). In order to do this, it is necessary to conduct research on molecular levels to learn about the biochemical frameworks, metabolism routes, and removal of medicinal chemicals from the body (Bhandari et al., 2023). Some of the waste products that patients produce, such as faeces and urine, include molecular data about the medications and byproducts that are present in their bodies. Scientists are able to examine these samples in order to identify certain molecules, determine the quantities of those compounds, and see how they change chemically (Verlicchi et al., 2013). It is possible to correctly identify and characterize these compounds with the use of sophisticated technologies such as high-performance liquid chromatography and mass spectrometry. The method by which we dispose of things has a significant influence on the molecules that reach the environment (Khan et al., 2020c). The removal of some pollutants is a strength of conventional wastewater treatment facilities; nevertheless, these plants may have difficulty dealing with pharmaceutical contaminants. Recent molecular investigations have shown the presence of these residues and the products of their breakdown in treated wastewater, indicating that we need more effective treatment procedures (Yan et al., 2020). The use of molecular monitoring allows us to trace the movement of these substances via water and get an understanding of the potential damage they might do to the environment (Emmanuel et al., 2005). It is important to note that environmental and pharmaceutical therapies both contribute to the increased molecular complexity of contaminant flow (Le Corre et al., 2012). By studying how medications function in cells and how they degrade inside the body, we are able to have a better understanding of how byproducts may wind up in the environment (Akin, 2016). By studying molecular pharmacokinetics, we are able to get a better understanding of the processes that occur inside the body with regard to medications, such as how they are absorbed, distributed, metabolized, and eliminated (Pariente et al., 2022). We need to improve the remediation procedures in order to clean up molecular pollutants in an adequate manner. Molecularly imprinted polymers, membrane filters, and improved oxidation processes are examples of technologies that are able to target particular molecules (Queen et al., 2022). The investigation of molecules, which demonstrates that they are capable of decomposing and cleaning up medicines and their byproducts, is beneficial to the development of these technologies. It is not simply the particular healthcare locations that have an impact on the environment; the process of patient waste, disposal techniques, and treatment procedures all have an impact on the environment. A molecular-level investigation of these aspects reveals the intricate relationship that exists between medications, environmental health, and human health. We are able to discover a wide variety of medications and byproducts in wastewater by doing research on what patients take out of their bodies. This increases our understanding of people's health, the ways in which they use drugs, and the possible risks associated with exposure. We are able to identify intermediate molecules and essential enzymes involved in the drug metabolism pathways if we have a thorough knowledge of these processes (Khan et al., 2020e). There are a lot of molecular problems that come with disposal strategies, such as discharging treated wastewater into bodies of water. Not only does treated wastewater include direct pharmaceutical inputs, but it also contains transformation products that are the result of environmental activities. Through the use of molecular studies, we are able to observe the behaviour of these chemicals, which provides us with information on their durability, accessibility, and chemical changes in water. One may see the effects of treatment regimens, whether they are pharmacological or environmental, at the level of the molecules (Behnam et al., 2020). By examining the manner in which medications interact with molecular targets, pharmacological therapies attempt to relieve symptoms or cure illnesses by influencing the pathways that are involved in biological processes. By gaining an understanding of these interactions, it is possible to maximize the efficacy of the treatment while simultaneously minimizing its negative effects, such as the production of hazardous byproducts (Queen et al., 2022). The environmental remediation procedures, on the other hand, make use of methods that include the manipulation of molecular structures in order to lessen the negative effects that pharmaceutical pollutants generate. For instance, atomically imprinting polymer compounds allows for the selective absorption of certain medicinal compounds, which offers an exclusive technique for the removal of impurities (Mackuľak et al., 2021). By providing information on exposure and response, molecular biomarkers are able to assist in the evaluation of the ecological impacts of pharmaceutical pollutants. We have a comprehensive grasp of the process by which pollutants enter healthcare systems as a result of our molecular research on patient waste products, dumping techniques, and treatment regimens (Khan et al., 2020d). With this information, medical experts are better able to comprehend medication metabolism and the well-being of patients, and environmental scientists are better able to design environmentally friendly methods to limit the damage that pharmaceutical toxins do to the environment (Verlicchi et al., 2015). It is possible to achieve complete waste management in the healthcare industry by using a molecular approach, which guarantees that the growth of medical technology is in accordance with the preservation of the environment (Ulvi et al., 2022) (Fig. 3.1; Table 3.1).

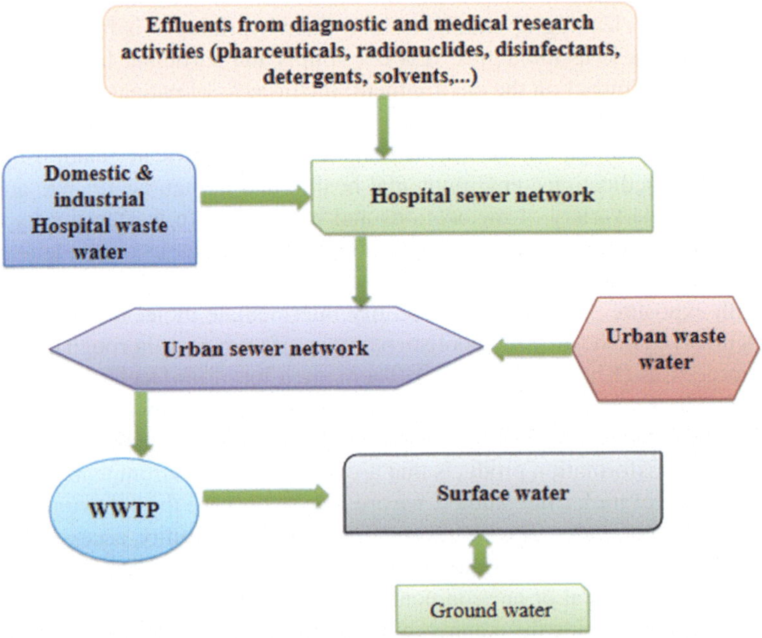

Fig. 3.1 Protocol for the disposal of the hospital wastewater

Table 3.1 Various aspects and features of hospital wastewater

Aspect	Features
Presence and consequences in hospital wastewater	Pathogens and pharmaceutical residues in hospital wastewater require molecular-level study due to their complex matrix. Microbial analyses focus on pathogen genetic material, while analytical techniques like liquid chromatography-mass spectrometry (LC–MS) and gas chromatography-mass spectrometry (GC–MS) are crucial for identifying and quantifying pharmaceutical substances. This molecular knowledge is critical for developing successful treatment techniques and mitigating environmental and public health problems (Cusack et al. 2024; Dandachi et al., 2019)
Identification and quantification of diverse contaminants	Accurately identifying and measuring pollutants at the molecular level is challenging but crucial. Techniques like chromatography, mass spectrometry (MS), and nuclear magnetic resonance (NMR) spectroscopy are key. Polymerase chain reaction (PCR) and Next-Generation Sequencing (NGS) help detect and quantify pollutants at the genomic level (Dargahi et al., 2022; De Almeida Cabral, 2019; Dinh et al., 2017; Domínguez & Meza-Rodriguez, 2019; Emmanuel et al., 2002)
Analysis of bacteria	Molecular analysis of bacteria like *Escherichia coli* and *Pseudomonas aeruginosa* reveals complex genetic landscapes, virulence factors, and antibiotic resistance mechanisms. Understanding these molecular properties aids in developing targeted treatments (Emmanuel et al., 2009; Frascaroli et al., 2021; Frédéric & Yves, 2014; Fuentefria et al., 2011; Gadipelly et al., 2014)
Analysis of viruses	Molecular studies of viruses like norovirus and adenovirus elucidate capsid structures, interactions with host cells, and genome integration mechanisms. This knowledge is crucial for understanding viral infectivity and developing antiviral treatments (Garbacz et al., 2023; Guide, 2010; Gupta et al., 2009; Gurjar et al., 2022)
Analysis of fungi	Candida albicans' molecular arsenal, including virulence factor expression and antifungal resistance mechanisms, aids in understanding its pathogenicity. This knowledge is vital for improving treatment methods for invasive candidiasis (Emmanuel et al., 2009; Kemprai et al., 2021; Kerneis et al., 2019)
Analysis of parasitic agents	Molecular characterization of parasites like Cryptosporidium and Giardia helps in understanding their genomes, virulence factors, and host interactions. This information is critical for developing tailored treatments and prevention strategies (Fatta-Kassinos et al., 2011; Hawken & Snitkin, 2019; Henrici De Angelis et al., 2017)

(continued)

Table 3.1 (continued)

Aspect	Features
Impact of pharmaceutical residues on wastewater and ecological systems	Pharmaceutical residues in effluent and ecosystems have complex molecular ramifications, including bioaccumulation and biomagnification. Understanding these molecular impacts is crucial for effective mitigation strategies (Khan et al., 2019, 2020b, 2021b)
Sources and pathways of contaminant introduction	Contaminants in ecosystems originate from both natural processes and human activities, involving complex biochemical reactions. Understanding these molecular mechanisms helps in responsible environmental stewardship (Kümmerer, 2001; Le et al., 2016)
Understanding the origins of pathogens and pharmaceuticals in hospital wastewater	Pathogens and pharmaceuticals in hospital wastewater come from patient waste, medical procedures, and disposal practices. Molecular analysis is crucial for understanding their impact and developing advanced treatment technologies (Li et al., 2022; Marschall et al., 2023)
Examination of patient excretions, disposal practices, and treatment protocols contributing to influx	Patient excretions contain molecular information about medications and metabolites. Disposal practices and treatment protocols significantly influence the molecular composition of pollutants in the environment. Molecular analysis helps understand these processes and develop efficient treatment methods (Marschall et al., 2023; Mathur et al., 2011)

3.4 Challenges Posed by Multidrug-Resistant Organisms (MDROs)

The fact that multidrug-resistant organisms (MDROs) continue to develop new methods to avoid the therapies that we typically utilize makes them a significant challenge in the field of medicine (Chia et al., 2020). In order to create novel methods of combating MDROs, it is essential to have a solid grasp of the genetic mechanisms underlying the resistance they demonstrate (Daoud et al., 2018).

Genetic Basis of Resistance

The fact that multidrug-resistant organisms (MDROs) continue to develop new methods to avoid the therapies that we typically utilize makes them a significant challenge in the field of medicine (Hawken & Snitkin, 2019). In order to create novel methods of combating MDROs, it is essential to have a solid grasp of the genetic mechanisms underlying the resistance they exhibit (Dandachi et al., 2019).

Mechanisms of Antibiotic Inactivation

In a number of different molecular processes, MDROs have a remarkable ability to inhibit the activity of antibiotics (Cabral & Ag, 2019). Enzymatic breakdown is a frequent approach that these organisms use, since they create enzyme that break or alter the antibiotic molecules, so leaving them inactive. This causes the molecules of antibiotics to become inactive. This kind of molecular warfare, which is carried out at the enzyme level, is an example of the complexity of MDROs in terms of their ability to undermine the effectiveness of therapeutic drugs (Gurjar et al., 2022).

Biofilm Formation

There is a considerable contribution to the robustness of MDROs made by the molecular processes that drive the production of biofilms. The formation of defensive biofilm structures by these microbes often serves as a barrier that prevents antibiotics from penetrating the cell. It is very difficult to understand the cellular pathways of signaling that are responsible for controlling both the development and survival of biofilms. It is possible to utilize them to destroy the protective conditions that bacteria use to become resistance to antibiotics (Daoud et al., 2018; Schmiege et al., 2021).

Adaptive Evolution

Evolutionary adaptation is a unique capability of MDROs, which is attributed to genetic changes and the selection pressure caused by antimicrobial treatment. In order to shed light on how MDROs quickly evolve and acquire resistance mechanisms, which thwarts efforts to remove them, it is necessary to investigate the molecular foundation of evolutionary adaptation. In order to maintain a competitive advantage in the continuing battle against resistant microbes it is essential to fully understand these cellular modifications (Kerneis et al., 2019).

Target Modification and Reduced Drug Permeability

It is common practice for MDROs to use molecular methods that change antibiotic receptors to make them more resistant to the inhibitory properties of the medications. This may include changes to the binding locations on important cellular elements like ribosomes or enzymes, which hinder the ability of antibiotics to communicate with their designated targets. Also, these creatures could make their cell walls even more impenetrable, so antimicrobials can't get in as easily. The complex relationship between MDROs and medicinal substances may be better understood by delving into the cellular components of such modifications (Bekele & Alamnie, 2022).

Interplay of Quorum Sensing in Resistance

To regulate group behaviors like pathogenicity and resistance, bacteria employ the detection of quorums, a molecular communications mechanism. In order to coordinate the regulation of genes that play a role in adaptive responses like as biofilm development and resistance to antibiotics, MDROs use quorum sensing. Examining the molecular mechanisms of MDRO quorum sensing sheds light on the complex receptor and signaling system that controls their sensitivity to outside stimuli (Kelambakkam, 2020).

Epigenetic Factors Influencing Resistance

In MDROs, epigenetic modifications, such as DNA methylation and histone acetylation, aid in gene regulation. To alter the bacteria's resistance to antibiotics, these alterations may activate or deactivate genes. In order to comprehend how MDROs evolve to evade antibiotics, it is helpful to have a better grasp of these regulatory variables (Garbacz et al., 2023).

Antibiotic Tolerance and Persister Cells

One feature of MDROs is antibiotic tolerance, and which occurs when a subset of cells shows less sensitivity to antibiotics. Persister cells, which may enter a latent condition to avoid antibiotic effects, are often associated with this resistance. At the molecular level, there are interesting pathways to investigate about the processes that control persister cell generation and reactivation. These pathways may provide prospective targets for preventing MDRO persistence after therapy (Bowler et al., 2020).

Environmental Influences on Molecular Adaptations

The molecular challenges posed by MDROs are not only shaped by therapeutic interventions but also influenced by environmental factors. MDROs change because of things outside of their cells, like being exposed to antibiotics at levels that are below the threshold of killing or interacting with other microorganisms. Looking into how the molecular responses of MDROs interact with their surroundings gives us a full picture of the things that cause antibiotic resistance to form and spread (Bowler et al., 2020).

3.4.1 *Analysis of MDRO Resilience Against Conventional Treatment Methods*

Because of their outstanding durability, which is seen across a variety of molecular sizes, multidrug-resistant organisms (MDROs) provide a considerable challenge for conventional treatment methods on account of their presence. It is partly due to the genetic plasticity that MDROs possess that they are able to maintain their resistance (Marschall et al., 2023). Through the use of horizontal gene transfer pathways, these living organisms are able to rapidly create and distribute resistance features. These processes make it simpler for resistance genes to enter cells and begin functioning. Since MDROs are able to swiftly overcome the inhibitory effects of traditional antibiotics, their genetic flexibility produces a dynamic environment that allows them to do so. In addition, MDROs are able to make antimicrobial drugs ineffective by adopting a wide variety of molecular methods, which allows them to neutralize the agents (Bartley et al., 2010). This kind of organism often engages in the process of enzymatic degradation. The process includes the production of enzymes that have the ability to alter or break down antibiotics, rendering them ineffective for therapy. By demonstrating their smart and flexible abilities to escape conventional treatments, MDROs have shown their ability to engage in molecular-level enzymatic warfare (De Almeida Cabral, 2019). Furthermore, the deployment of efflux pump systems as a demonstration of the molecular robustness of MDROs takes place. It is because of these pump systems that antibiotics are actively expelled from bacterial cells, which prevents the medications from reaching the targets they are designed to treat. In order to improve the efficiency of this mechanism for expulsion, molecular components of the efflux pumps, such as regulatory elements and transport proteins, are crucial. The addition of this barrier makes antimicrobial therapies more difficult to implement (Cusack et al., 2024). In order to circumvent the effects of antibiotics, MDROs adopt biofilm development as an extra powerful molecular approach. These organisms are able to withstand standard therapies because of the intricate signalling pathways that govern the production and maintenance of biofilms (Guide, 2010). This demonstrates how beneficial MDROs are, as the molecular intricacy of biofilm development, in which MDROs generate barrier environments that prevent medicinal substances from entering, demonstrates. By changing antibiotic targets and reducing drug permeability, MDROs also exhibit resistance to conventional therapy, which is a significant advantage. At the molecular level, these organisms have the ability to alter the binding sites that are present on essential cellular components such as enzymes and ribosomes (Domínguez & Meza-Rodriguez, 2019). Antibiotics have an inhibitory impact on the targets, but this alteration makes them more resistant to that action. While this is happening, MDROs may simultaneously enhance the resistance to infection of their cells' membranes, which would restrict the number of antimicrobial drugs that may enter the cell. By using a combined approach that combines shifting targets and making medicines less permeable, they are able to make their molecular defenses more complicated (Li et al., 2022). Despite the fact that they are receiving normal treatments, this makes it simpler for them to resist and remain alive. Quorum sensing interactions provide another layer of intricacy to MDRO resistance, which is already rather complicated. Through the use of this molecular communication system, MDROs are able to coordinate collective actions, such as the expression of genes that are related to antibiotic resistance. In response to shifts in population density, MDROs are able to make real-time adjustments to the molecular reactions they exhibit (Zhang et al., 2022). response to therapeutic interventions. An additional component that plays a significant role in predicting MDRO resistance is epigenetic variables. Methylation of DNA and acetylation of histones are two examples of changes that have the potential to influence the regulation of genes that are related to antibiotic resistance (Bekele et al., 2023).

We may acquire a better understanding of the additional degrees of resistance that MDROs exhibit against typical therapies if we investigate the molecular pathways that are responsible for the production and activation of persister cells. There are additional environmental elements that have an effect on MDRO resilience at the molecular level. A variety of factors, including interactions with other bacteria and exposure to antibiotics at doses below the lethal threshold, have the potential to produce adaptive responses (Gurjar et al., 2022). They are more difficult to treat using conventional approaches because of the complicated interplay that occurs between the molecular responses of MDROs and their environment, which results in crosstalk. The fact that MDRO is resistant to the most conventional treatment procedures reveals a complicated network of molecular tactics that these organisms use, to put it more simply (Henrici De Angelis et al., 2017). The fact that MDRO is resistant to the most conventional treatment procedures reveals a complicated network of molecular tactics that these organisms use, to put it more simply. In terms of molecular mechanisms, MDROs are capable of performing a wide range of functions, including the modification of antibiotic targets and the reduction of drug permeability, as well as complicated quorum sensing, epigenetic factors, and the maintenance of persister cell viability. In order to design specific therapies that are capable of successfully dealing with the molecular resilience of MDROs and preventing the problems that antibiotic resistance is generating from getting exacerbated in the healthcare business, it is essential to have a solid understanding of these processes (Chia et al., 2020).

3.4.2 Implications of MDRO Presence in Wastewater and Potential Health Risks

Sewage containing multidrug-resistant organisms (MDROs) poses serious risks to human and environmental health. The presence of disinfectants and antibiotics in wastewater creates an environment that is conducive to the survival and growth of MDROs (Chia et al., 2020). Maintaining and growing MDRO populations is ideal here. MDROs can become resistant by molecular changes such biofilms and efflux pump systems. MDROs can spread to larger ecosystems if untreated wastewater is dumped into natural waterways (Boutin et al., 2024). Transmitting through dirty drinking water, agricultural irrigation, and recreational waters is a concern. Through horizontal gene transfer, MDROs can pass their resistance genes to other bacteria, including human pathogens, making diseases harder to treat (Boutin et al., 2024).

Antibiotic residues in water systems may also lead to MDROs and resistance. To prevent pathogen spread, wastewater treatment must include advanced filtration and disinfection (Khan et al., 2020b). MDROs affect wastewater, therefore this is required. Implement strict restrictions and monitor pharmaceutical emissions into wastewater to reduce antibiotic resistance's selective pressure. Understanding the genetic basis of wastewater MDROs helps design focused monitoring, risk assessment, and intervention approaches. MDROs can survive in effluents and wastewater treatment facilities by developing resistant biofilms and transmitting resistance genes (Cohen et al., 2020). These MDROs, when introduced into receiving water bodies, have the potential to operate as a reservoir for antibiotic resistance genes, hence enabling the transmission of these genes across various bacterial species (Cohen et al., 2020). These genetic exchanges, enabled by mobile genetic components like plasmids, hinder antibiotic resistance reduction efforts. The molecular kinetics of gene transfer events must be studied to fully comprehend their environmental effects.

When resistance to several antibiotic classes is co-selected, continual exposure to sub-lethal antibiotic dosages in effluent may promote strain creation and dissemination. This is especially true with high antibiotic doses. MDROs may be more resistant to antimicrobials due to efflux pump systems and other molecular changes. MDROs may affect microbial populations, which can affect nutrient cycling and ecosystem health, in addition to antibiotic resistance. To forecast and mitigate the wider repercussions of MDRO pollution in effluent, one must understand the molecular basis of these ecological changes. To combat MDRO contamination in wastewater, molecular insights must be combined with successful wastewater management approaches (Rankin et al., 2024). These include modern treatment procedures, safe antibiotic administration, and excellent monitoring. Despite multidrug resistance, an all-encompassing strategy for tackling these molecular challenges may help protect water systems and human health (Bartley et al., 2010).

3.5 Mitigation Strategies for Pathological Contaminants:

When it comes to the management of pathological pollutants in wastewater from hospital waste, bioremediation provides a viable solution. To decompose waste without creating dangerous by-products, it entails the application of microbes as the degrading agents. Here are some most important points: Hydrolytic Bacteria: Hydrolytic bacteria have the ability to minimize the elements of organic waste that contribute to water pollution. Those that are not pathogenic play a significant part in the acceleration of the breakdown of biomedical wastes. They do this by reducing the amount of nutrients that are accessible, which in turn inhibits the development of pathogenic bacteria Ethica et al. Biofilms are aggregation of microbial cells that attach to a surface in an aqueous environment. Biofilms are the agent that is responsible for biofilm-mediated remediation. According to 124, bacteria that form biofilms have a high survival rate even when exposed to harsh environmental conditions. These bacteria are also suitable for the sorption and metabolism of organic pollutants. Algae Bioremediation: The process of algae biological remediation includes bio-absorption, bioadsorption, and then biological transformation, which may be accomplished both intra- and extra-cellularly. These technologies have the potential to result in a replacement method of treatment of hospital wastewater in India that is both more cost-effective and less harmful to the natural world Ethica et al. (2018). The application of these approaches, on the other hand, calls for meticulous planning and control in order to guarantee both safety and efficacy (Fig. 3.2).

Material and Methods

In order to purify wastewater, a procedure known as activated sludge is used; this process involves the utilization of microorganisms that are capable of breaking down toxic compounds. The clustering of these organisms results in the formation of flock, which eventually sinks to the bottom of a tank. This leaves behind water that is clean and free of any particles of trash or filth that may be floating in it. Initially, the contaminated water is combined with a liquid that is loaded with these organisms in order to initiate the process. After that, air bubbles are added to this mixture in order to provide the microorganisms with oxygen and to keep them moving. Following

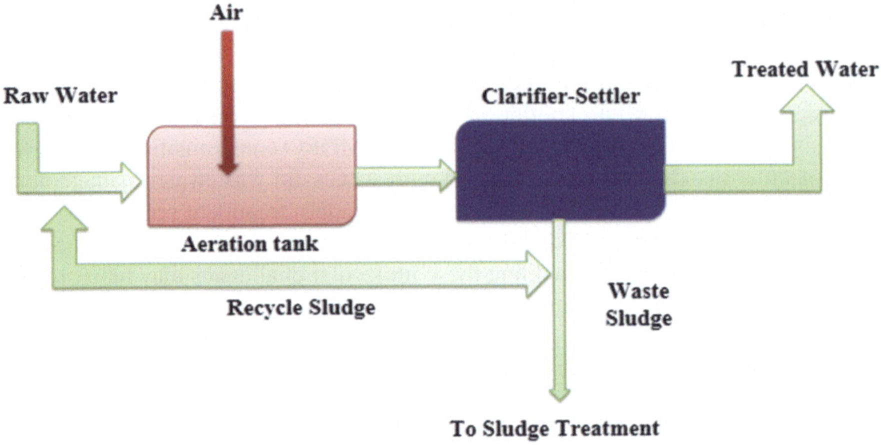

Fig. 3.2 Traditional method of the treatment of hospital wastewater by the activated sludge method

a period, the mixture is transferred to a different tank, where the flock is allowed to settle, and some of the organisms are extracted. The water that has been partly cleaned is subjected to further cleaning, while the particles and organisms that are still present are sent back to the initial tank so that the process may continue. In wastewater, pollutants may take on a variety of forms, including the form of dissolved particles or tiny particles. It is possible that various contaminants, in a variety of forms, will wind up in the tank with the organisms, according to the manner in which the water was treated in the past. In the beginning, the intended purpose of this technique was to clean the water by removing specific kinds of contaminants and filth. However, when other pollutants became a concern, such as an excessive amount of nutrients that caused a disproportionate number of plants to grow in the water, the procedure was updated to eliminate these pollutants as well. It is necessary to remove a wide variety of contaminants from the water, in addition to the nutrients that are present in it. The elimination of substances such as residual pharmaceuticals, chemicals that disturb hormones, insecticides, and heavy metals is of utmost significance Saini et al. (Table 3.2).

3.5.1 Advanced Hospital Wastewater Treatment Technologies Review

Membrane Bioreactors

The Membrane Bioreactor, often known as the MBR, is a methodology that utilizes a membrane in conjunction with the activated sludge process in order to separate sludge from water. Through the process of recirculation, this membrane may be located either within or outside of the bioreactor. It is not necessary to use external settlers or

Table 3.2 Mitigation measure of pathological contaminants in wastewater from hospital waste

Contaminant	Mitigation measure	Description	Challenges	Benefits	References
Pathogens	1. Proper waste segregation practices	Separate infectious waste from non-infectious waste at the point of generation. This includes sharp segregation, infectious material containment, and color-coded bins	Compliance with segregation protocols may require continuous staff training and monitoring	Reduced risk of pathogen transmission to the environment and public health	(Copple, 1989; Chandrappa & Das, 2024)
	2. Effective wastewater treatment systems	Employ advanced treatment methods such as UV disinfection, chlorination, and biological treatment to eliminate pathogens. Regular testing and monitoring are essential to ensure effectiveness	Initial capital investment and operational costs for installing and maintaining treatment systems	Removal of harmful pathogens from wastewater, preventing waterborne diseases	(Bailey et al., 2021)
	3. Regular maintenance and monitoring	Conduct routine inspections of treatment systems, including checking equipment functionality, chemical levels, and treatment efficiency. Prompt repairs and adjustments are necessary to maintain optimal performance	Resource-intensive in terms of time and manpower	Ensures consistent performance of treatment systems and compliance with regulatory standards	(Sehulster et al., 2003)
	4. Disinfection procedures	Implement rigorous disinfection protocols using appropriate agents such as chlorine or ozone. Ensure proper contact time and concentration levels to achieve effective pathogen inactivation	Potential health hazards associated with handling and storing disinfectants	Reduction in microbial load, minimizing the risk of waterborne infections	(Moccia et al., 2020)

(continued)

Table 3.2 (continued)

Contaminant	Mitigation measure	Description	Challenges	Benefits	References
Pharmaceuticals	1. Development of waste management protocols	Establish comprehensive guidelines for the proper segregation, storage, and disposal of pharmaceutical waste, including expired or unused medications. This should involve collaboration with pharmacy departments and waste management experts	Requires coordination between different departments and stakeholders	Prevents pharmaceutical pollution in water bodies and reduces environmental impact	(Nyaga et al., 2020)
	2. Advanced treatment technologies	Utilize specialized treatment processes like membrane filtration or adsorption to remove pharmaceutical compounds from wastewater. Regular maintenance and monitoring are necessary to ensure treatment efficiency	High upfront costs associated with purchasing and operating advanced treatment systems	Decreased presence of pharmaceutical residues in water, safeguarding aquatic ecosystems	(Gadipelly et al., 2014; Loganathan et al., 2023)
	3. Staff education and training	Provide ongoing education and training programs for hospital staff regarding proper pharmaceutical waste handling practices and the environmental impact of improper disposal	Time and resource-intensive to organize and conduct training sessions	Increased awareness leads to improved compliance with waste management protocols	(Manyele & Tanzania, 2004; Mathur et al., 2011)
Chemicals	1. Labeling and safe storage	Clearly label chemical containers with hazard information and store them in designated areas equipped with spill containment measures. Regular inventory checks should be conducted to track chemical usage and disposal	Ensuring consistent adherence to labeling and storage protocols may require continuous monitoring	Minimizes the risk of accidental chemical spills and exposures, protecting human health	(Council et al., 2011)

(continued)

Table 3.2 (continued)

Contaminant	Mitigation measure	Description	Challenges	Benefits	References
	2. Adoption of safer alternatives	Identify and substitute hazardous chemicals with less toxic or non-toxic alternatives wherever feasible. This may require collaboration with procurement departments and chemical suppliers	Limited availability and higher cost of alternative chemicals may pose challenges	Reduces the risk of chemical hazards to human health and the environment	(Tickner & Jacobs, 2016)
	3. Chemical spill prevention	Implement preventive measures such as spill response kits, secondary containment systems, and staff training in spill response procedures to minimize the risk of chemical releases	Regular training and drills are necessary to ensure staff readiness for spill incidents	Mitigates environmental contamination and potential harm to ecosystems	(Abdeljalil et al., 2021)
	4. Integration of treatment systems	Install specialized treatment units like chemical neutralization tanks or oxidation reactors to treat chemical-laden wastewater before discharge. Regular maintenance and monitoring are essential for optimal performance	Complexities in the design and operation of treatment systems may arise	Improves water quality by removing hazardous chemicals, safeguarding public health	(Ewing et al., 1978; Samiksha et al., 2024)

post-treatment processes when using a membrane for both solids and effluent separation. This reduces the requirement for the installation to be compact and allows for larger sludge concentrations. The removal of all suspended and colloidal particulates, which includes microcontaminants, bacteria, and viruses, results in an improvement in the effluent's quality. The biological processes that occur in MBRs are often on par with or even superior to those that occurs in traditional systems. MBRs improve the efficiency of nitrogen removal because to the extended ages of the sludge they process. The process of denitrification may take place either by adding tanks that contain low quantities of oxygen or by aerating the tanks on an intermittent basis. Membrane filtration is a procedure that involves the use of a membrane that is very porous and functions as a barrier. This membrane allows water and solutes to flow through while simultaneously capturing suspended particles and other contaminants. Selectivity and productivity are two factors that determine how successful membrane filtering becomes. Selectivity refers to the degree to which the membrane is able to segregate constituents, while productivity refers to the amount of water that is able to flow through.

The use of Membrane Bioreactors (MBR) for the purpose of purifying wastewater from hospitals has gained popularity. Through the use of membrane filtration and activated sludge treatment, MBR technology eliminates the need for clarifiers or additional filters. In addition to preventing the spread of pathogens and guaranteeing that reuse water is of good quality, the membrane functions as a physical barrier. Numerous studies have shown that MBR is an effective method for cleaning wastewater from hospitals and eliminating pharmaceutical chemicals. A number of different methods have been investigated, and nanofiltration has been shown to be successful in eliminating virtually all of the medications that have been tested. There should be consideration given to the collecting of rainwater in a separate location, the maintenance of a sludge age of more than one hundred days, and the consideration of particular criteria for emission levels in order to accommodate hospital patients and inhabitants in the surrounding area. The full destruction of medicines that have been adsorbed may be achieved by the thermal treatment of waste activated sludge and screens. MBRs have gained popularity in Sewage Treatment Plants (STPs) due to their ability to provide effluents of a better grade that are appropriate for direct discharge, additional treatment, or reuse. There is a lack of clarity on whether or not these technologies are capable of successfully improving the elimination of organic micro pollutants. Additionally, connected growth MBRs provide an alternate method for producing high-quality effluent, compact facilities for treatment, and cost-effective management.

3.5.2 Implementation of Effective Hospital Protocols for Waste Segregation, Disinfection, and Pharmaceutical Disposal

This information is derived from the Bio-medical Waste Rules that were implemented in India in 1998. These rules provide precise instructions on how to properly manage and dispose medical waste. These regulations involve the classification of various forms of waste and the collection of medical waste in a secure manner. An explanation of the management of trash in hospitals is as follows: In the first category, trash from human bodies is either burnt or buried far below, and the bag carrying the waste has an orange color. Waste that comes from animals is either incinerated or buried deep underground in Category 2, therefore there is no particular bag color for this kind of waste. Category 3 waste is either burnt or buried deep underground, and the bag color is yellow or red. This waste comes from laboratories and biotech companies. Sharp things, such as needles, are destroyed, burnt, disinfected, or treated with chemicals in Category 4, and the bag color is a transparent blue-white color. In the fifth category, pharmaceutical medications and hazardous pharmaceuticals are either destroyed or burnt before being deposited in a specialized landfill. The bags from this category are dark in color. Solid waste, such as blood and fluids, is processed with heat, chemicals, or buried, and the bag color is yellow. This category is referred to as classification 6. Category 7: Items that are disposable, such as gloves and masks, are treated with chemicals, heat, or burial, and the color of the bag transforms from blue to white and is transparent. Under Category 8, liquid waste, such as blood, is subjected to chemical treatment and may be discharged into drains. The bag that contains this trash is black in color. The ash that is produced by burning garbage is disposed of in the same manner as ordinary garbage, and there is no particular bag color for this category. Garbage that falls under Category 10 is either processed with chemical or disposed of in a specific dump; the container that contains this garbage is yellow in color. To summarize, the management of waste in hospitals is essential for ensuring the safety of both individuals and the environment. There is a significant shortage of adequate knowledge and appropriate procedures for handling healthcare waste in many underdeveloped nations. The education of individuals and those working in health on the correct way to manage waste is of utmost importance. It is imperative that we shift our mentality about garbage and make certain that everyone is aware of how to appropriately handle it.

3.6 Multidisciplinary Approaches for Sustainable Management

3.6.1 *The Significance of Collaboration in Wastewater Management*

It is essential to have treatment technologies for hospital wastewater since existing approaches often are unable to entirely eradicate emerging contaminants (ECs) and micropollutants (Parida et al., 2022). Consequently, tertiary treatment is required in order to complete the removal of these chemicals from hospital wastewater (HWW). This research investigates a variety of primary, secondary, and tertiary therapy methods that are presently being used for the treatment of HWW, as well as the benefits and drawbacks associated with each of these treatment methods. Utilizing chemical flocculation as the primary method of treatment for hospital effluents allows for the removal of colloids and suspended particles that are difficult to settle. As an example, a large-scale hospital wastewater treatment facility in Korea uses chemical flocculation with the aim of doing this. In a different method, which was used in Barcelona, Spain, the first treatments consisted of coagulation and flocculation, which were then followed by a fluidized bed reactor (FBR) to remove organic debris, sediments, and some elemental contaminants. In this particular investigation, the researchers used flocculants and coagulants, with the average dosages of each being 95 mg/L and ten mg/L, respectively (Mir-Tutusaus et al., 2017). According to Suárez et al., the fundamental process responsible for the removal of micropollutants during primary treatment is sorption. When it comes to adequately removing pollutants, only those with more excellent sorption characteristics are considered beneficial. Compounds that are hydrophobic and have a log Kow value that is larger than one are able to swiftly bond to particles and be removed along with the sludge. On the other hand, hydrophilic chemicals only partly adsorb on these particles, and as a result, they are only partially removed (Parida et al., 2022; Suárez et al., 2008). Primary treatment is successful in eliminating particles and oily matter from wastewater, and it also partly favours the removal of biological oxygen demand (BOD) and chemical oxygen demand (COD). However, it is not typically effective in removing effective concentrations (ECs) from wastewater (Parida et al., 2022).

The management of wastewater from hospitals may be considered an essential component of healthcare facilities in India. This is a list of crucial practices: Management of Water: In order to carry out their daily activities, hospitals need a considerable volume of water at all times. In order to lower the cost of the user facility while simultaneously contributing to the sustainable growth of the country, it is essential that hospitals devote particular thought to the management of water. The discharge of wastewater from hospitals may contribute considerable loads of pollutants and pathogens, such as bacteria that are resistant to antibiotics, to municipal wastewater. Wastewater management is an essential aspect of this process. The management of this effluent is thus critical. Plan for Water Management: A reduction in the amount of water used on campus may be achieved by the implementation of the following water management practices: Addressing water conservation via the three environmental R's is an essential step in water conservation. Decrease: Reduce the amount of water that you use, which is necessary for making more effective use of water. Reuse: Water that has been used once, such as water from a shower, may be recovered and utilized for various uses. Utilize treated or recycled effluent from sewage treatment facilities (STPs) for a variety of purposes, including gardening, cleaning roads, and other similar activities. In order to achieve greater efficiency with water, it is necessary to use metering and measuring in order to assess water consumption and to regulate mechanical equipment. Enhance the facilities that are clean: Install urinals that do not produce moisture, toilets with modest flow rates, and control the flow of outlets. Implementing systems that create wastewater is an example of the use of wastewater-generating systems.

Regulatory Framework The Bio-Medical Rules, which were drafted in 1998 and subsequently updated in 2000 and 2003, serve as the legal framework for the management of the environment in the health sector in India. These rules are applicable to all individuals and organizations that generate and/or handle healthcare waste in any form (Bhalla et al., 2019). Regulations for the Management of Biomedical Waste, 2016: These regulations provide a legislative framework for the ecologically responsible treatment of biological waste that is produced by healthcare institutions (Dehal et al., 2021). Guidelines for Actual Implementation: In accordance with the Bio-Medical Waste Management Rules, 2016 (Dehal et al., 2021), the National Health System Resource Centre offers instructions for the implementation of the management of biomedical waste.

It is common practice in India to discharge wastewater from hospitals straight into municipal sewers without first undergoing any treatment. The fact that conventional sewage treatment facilities are not able to cope with the dangerous compounds that are present in hospital wastewater is the reason why this is an issue. As an outcome of this practice, significant repercussions have arisen, such as the near-extinction of some species of sharks and the transformation of male fish into females as a result of the high quantities of painkillers and birth control hormones that are present in bodies of water. In 1998, India enacted the Biomedical Waste (Management and Handling) Rules, which were later revised in 2016. These rules were designed to solve the problem that was being discussed. It is the responsibility of hospitals to collect, transport, and rid of their waste and effluent in accordance with these standards, which establish guidance. However, in spite of these requirements, a significant number of medical institutions in India continue to get rid of their waste in an incorrect manner. This has resulted in situations in which hospitals have been prohibited from operating or penalized for failing to comply with the standards.

In India, the manner in which hospitals address the management of their wastewater differs from state to state. In some states, such as Punjab and Karnataka, for instance, wastewater from hospitals is cleaned on-site before being released into sewer drains. This is done in accordance with the criteria that were established by the Biomedical Waste (Management and Handling) Rules, 1998. On the other hand, in states such as Maharashtra, the wastewater from hospitals is subjected to a distinct treatment process before being discharged into sewers. According to studies, there is a high likelihood of heavy metal contamination in the sediment of rivers that receive untreated hospital effluent. There is now a rising focus on cleaning wastewater from hospitals on-site in order to limit the spread of hazardous bacteria in municipal sewers (Svebrant et al., 2021). After the water has been subjected to this on-site treatment, which may include a variety of procedures such as sedimentation, coagulation, filtering, and disinfection, the solids will be broken down, and then the water will be released to regular sewage treatment facilities. As a result of the presence of several bacteria that are detrimental to human health, the sludge that is generated in sewage treatment facilities must be treated with extreme caution. Additionally, a number of research have investigated the prevalence of helminth eggs and other hazardous diseases in wastewater, as well as the consequences of these pathogens. With a turnover of $45 billion annually, India's pharmaceutical sector is one of the largest in the world, making it one of the leading makers of pharmaceuticals worldwide. One out of every three medications that are taken all over the globe is manufactured in India, according to estimates. It has been reported that between ten and ninety per cent of the medication that we take gets eliminated from our systems without being completely broken down. This waste may wind up in rivers, lakes, and oceans, where it can cause damage to aquatic life and potentially contribute to the formation of bacteria that are resistant to antibiotics (Svebrant et al., 2021). Despite the fact that India has a sizable pharmaceutical sector, the nation as a whole does not have enough sewage treatment facilities that are up to par. Most of the studies that are being done in this field are focused on the medications themselves rather than the waste materials that are produced as a result of their usage. In order to safeguard aquatic life and the well-being of humans, it is essential to discover methods that may cut down on the quantity of these waste items that end up in the natural world.

3.6.2 Holistic Solutions for Wastewater Contaminants

It is necessary to have a complete plan in order to address the issues that are presented by pollutants in hospital wastewater in India. This strategy should involve the use of sophisticated cleaning technologies, the observance of laws, and the exploration of new concepts. The following are some recommendations: In order to efficiently remove organic debris and medicines from wastewater, it is necessary to make use of modern treatment techniques such as Fenton oxidation, electrocoagulation, and membrane bioreactors (MBRs, among others). These treatments have the potential to improve the water's cleanliness.

Utilize created wetlands, which are natural systems that are capable of successfully treating wastewater. Created wetlands are a viable option. It is possible that a sustainable solution may consist of a multi-stage CW system or a CW system integrated with other therapeutic methods. It is necessary to make use of natural regions in order to purify the water.

Innovative solutions include the use of non-contact technologies such as UV irradiation and hydrogen peroxide vapour, in addition to preventative measures such as antimicrobial surface coatings, in order to lessen the likelihood of disease transmission. The effectiveness of these procedures is much higher than that of other tactics (Khan et al., 2019). In order to ensure compliance with regulations, it is important to adhere to the standards that are given in the Water (Prevention and Control of pollutant) Act, 1974. This act addresses wastewater discharge as a pollutant. In this context, "following the rules" refers to adhering to the regulations on what may be put into the water (Khan et al., 2019). Monitoring of Wastewater: It is important to implement wastewater monitoring, particularly during pandemics such as COVID-19, in order to guarantee the protection of both populations and the surrounding ecosystem. This entails doing routine checks on the water to ensure that it is free of contaminants (Khan et al., 2019). Educate the general public about the significance of good hospital waste disposal and the health concerns that are connected with incorrect treatment by implementing programs that educate people about these topics. In other words, this entails educating people about the reasons why it is essential for sanitizing the water (Khan et al., 2019). In order to implement the most effective methods and approaches, it is important to collaborate with local government authorities, environmental organizations, and international organizations. The greatest approach to cleaning the water is to collaborate with other people to develop the most effective methods (Khan et al., 2019).

3.7 Future Directions and Conclusion

3.7.1 Emerging Trends in Wastewater Treatment and Contaminant Mitigation

The future of cleaning wastewater and removing contaminants looks promising, thanks to new trends and technologies. Here are some exciting developments: Integrated Membrane Separation and Photocatalysis: This technology focuses on breaking down pollutants completely without creating harmful residues. Nanotechnology: Nanotechnology offers new ways to clean water, removing a wide range of pollutants. Thermal Hydrolysis: This method makes it easier to separate waste and process it, seeing sludge as a valuable energy source rather than just waste. Internet of Things (IoT): IoT can improve wastewater treatment by connecting data to a network, making processes more efficient. Waste Valorization: This approach turns waste into valuable products, encouraging better treatment of wastewater. Sustainable Effluent Treatment: Strategies like making treatment processes more intense and efficient are being developed for a sustainable and cost-effective future. In conclusion, the future of cleaning wastewater looks bright. New technologies and strategies not only tackle contamination but also promote sustainability. Continued research and development are crucial to meeting the increasing demands for wastewater treatment in an effective and eco-friendly way.

3.8 Summary of Key Findings and Implications for Environmental and Public Health

Inadequate management of wastewater from hospitals may have severe repercussions, not only for the overall health of the people but also for the environment. In the absence of appropriate treatment, wastewater from hospitals has the potential to damage water bodies, land, and air, in addition to causing a variety of environmental and health problems. Wastewater from hospitals often includes hazardous compounds such as medicines, microorganisms, and toxic chemicals (Ajala et al., 2022). Contaminants like these may be harmful to aquatic life and impair the quality of water if they are dumped into bodies of water without being treated. Moreover, they have the potential to pollute sources of drinking water, which poses a threat to human health. Pathogens, such as bacteria and viruses that are found in hospital wastewater have the potential to transmit illnesses if they penetrate the environment. Environmentally contaminated bodies of water have the potential to act as infectious agents for both people and animals. In addition, the release of drugs, particularly antibiotics, into the environment may have a role in the creation of bacteria that are resistant to antibiotics. Antibiotics may become less effective in treating infections as a result of this, which poses a substantial risk to the public's health. It is also possible for the pollution that comes from hospital wastewater to disturb ecosystems, which may result in the extinction of aquatic species and the loss of biodiversity. There is the potential for this to have long-term repercussions for the ecosystem and the services that it offers to people. People who live in close proximity to hospital wastewater treatment facilities that are not properly maintained may be exposed to pollutants via contaminants that are present in the air, soil, or water. This exposure has the potential to cause a variety of health concerns, such as respiratory troubles, infections of the skin, and gastrointestinal ailments. Not only may insufficient handling of waste from hospitals result in regulation breaches, but it can also expose healthcare institutions to the possibility of incurring penalties and other legal implications. It is essential to comply with environmental standards in order to safeguard both the environment and the health of the general people. It is vital to abide by regulatory requirements and deploy treatment methods that are effective in order to reduce the likelihood of these dangers occurring. The proper management of wastewater from hospitals is essential in order to forestall the deterioration of the environment and to protect the health of the general population (Ajala et al., 2022).

3.8.1 Concluding Remarks on the Necessity of Sustained Efforts for Sustainable Hospital Wastewater Management

Managing wastewater from hospitals is essential for both the environment and the health of the general population. There is a significant amount of wastewater produced by hospitals, which, if not properly handled, may be harmful to both the environment and individuals. Pharmaceuticals, biological materials, and other potentially hazardous compounds are common components of effluent from hospitals. Enhancing treatment may be accomplished by the use of cutting-edge technologies such as membrane separation, photocatalysis, nanotechnology, and thermal hydrolysis. Utilizing the Internet of Things for monitoring and control may also assist in improving wastewater management. The process of transforming trash into valuable resources, known as waste valorization, has the potential to make wastewater management more environmentally friendly. In order for these techniques to be successful, we need continual research, investments in technology, and regulations that foster their implementation. All parties involved, including politicians, healthcare professionals, researchers, and members of the community, must collaborate in order to achieve sustainable hospital wastewater management. This is a vital component for a healthy future.

References

Abdeljalil, A., Nabil, S., & Rachid, M. (2021). Towards a guideline of a spill management: Industrial sites as a case study. *International Journal of Ecosystems & Ecology Sciences,* 12.

Ahmad, N., Ahmed, S., Vambol, V., & Vambol, S. (2021). Treatment of drug residues (emerging contaminants) in hospital effluent by the combination of biological and physiochemical treatment process: A review. *Frontiers in Engineering and Built Environment, 1,* 1–13.

Ajala, O. J., Tijani, J. O., Salau, R. B., Abdulkareem, A. S., & Aremu, O. 2022. A review of emerging micro-pollutants in hospital wastewater: Environmental fate and remediation options. *Results in Engineering,* 100671.

Akin, B. S. (2016). Contaminant properties of hospital clinical laboratory wastewater: A physiochemical and microbiological assessment. *Journal of Environmental Protection, 7,* 635.

Alazaiza, M. Y., Albahnasawi, A., Eyvaz, M., Nassani, D. E., Abu Amr, S. S., Abujazar, M. S. S., & Al-Maskari, O. (2023). Electrochemical-based advanced oxidation for hospital wastewater treatment. *Desalination and Water Treatment, 300,* 44–56.

Alimba, C. G., Ajiboye, R. D., & Fagbenro, O. S. (2017). Dietary ascorbic acid reduced micronucleus and nuclear abnormalities in *Clarias gariepinus* (Burchell 1822) exposed to hospital effluent. *Fish Physiology and Biochemistry, 43,* 1325–1335.

Asfaw, T., Negash, L., Kahsay, A., & Weldu, Y. (2017). Antibiotic resistant bacteria from treated and untreated hospital wastewater at Ayder Referral Hospital, Mekelle, North Ethiopia. *Advances in Microbiology, 7,* 871–886.

Azar, A. M., Jelogir, A. G., Bidhendi, G. N., Mehrdadi, N., Zaredar, N., & Poshtegal, M. K. (2010). Investigation of optimal method for hospital wastewater treatment. *Journal of Food, Agriculture and Environment, 8,* 1199–1202.

Baghal Asghari, F., Dehghani, M. H., Dehghanzadeh, R., Farajzadeh, D., Shanehbandi, D., Mahvi, A. H., Yaghmaeian, K., & Rajabi, A. (2021a). Performance evaluation of ozonation for removal of antibiotic-resistant *Escherichia coli* and *Pseudomonas aeruginosa* and genes from hospital wastewater. *Scientific Reports, 11*, 24519.

Baghal Asghari, F., Dehghani, M. H., Dehghanzadeh, R., Farajzadeh, D., Yaghmaeian, K., Mahvi, A. H., & Rajabi, A. (2021b). Antibiotic resistance and antibiotic-resistance genes of *Pseudomonas* spp. and *Escherichia coli* isolated from untreated hospital wastewater. *Water Science and Technology, 84*, 172–181.

Bailey, E. S., Beetsch, N., Wait, D. A., Oza, H. H., Ronnie, N., & Sobsey, M. D. (2021). Methods, protocols, guidance and standards for performance evaluation for point-of-use water treatment technologies: History, current status, future needs and directions. *Water, 13*, 1094.

Bartley, J. M., Olmsted, R. N., & Haas, J. (2010). Current views of health care design and construction: Practical implications for safer, cleaner environments. *American Journal of Infection Control, 38*, S1–S12.

Behnam, B., Oishi, S. N., Uddin, S. M. N., Rafa, N., Nasiruddin, S. M., Mollah, A. M., & Hongzhi, M. (2020). Inadequacies in hospital waste and sewerage management in Chattogram, Bangladesh: Exploring environmental and occupational health hazards. *Sustainability, 12*, 9077.

Bekele, T., & Alamnie, G. (2022). Treatment of antibiotic-resistant bacteria by nanoparticles: Current approaches and prospects. *Annals of Advances in Chemistry, 6*, 001–009.

Bekele, T. M., Alamnie, G. A., Girma, A., & Mebratie, G. B. (2023). Nanoparticle therapy for antibiotic-resistant bacteria: Current methods and prospects. *Bioinspired, Biomimetic and Nanobiomaterials,* 12, 153–162.

Bhalla, G. S., Bandyopadhyay, K., & Sahai, K. (2019). Keeping in pace with the new biomedical waste management rules: What we need to know! *Medical Journal Armed Forces India/MJAFI, 75*(3), 240–245. https://doi.org/10.1016/j.mjafi.2018.12.003

Bhandari, G., Chaudhary, P., Gangola, S., Gupta, S., Gupta, A., Rafatullah, M., & Chen, S. (2023). A review on hospital wastewater treatment technologies: Current management practices and future prospects. *Journal of Water Process Engineering, 56*, 104516.

Boutin, S., Scherrer, M., Späth, I., Kocer, K., Heeg, K., & Nurjadi, D. (2024). Cross-contamination of carbapenem-resistant Gram-negative bacteria between patients and the hospital environment in the first year of a newly built surgical ward. *Journal of Hospital Infection, 144*, 118–127.

Bowler, P., Murphy, C., & Wolcott, R. (2020). Biofilm exacerbates antibiotic resistance: Is this a current oversight in antimicrobial stewardship? *Antimicrobial Resistance & Infection Control, 9*, 1–5.

Cabral, J., & Ag, R. (2019). Blue light disinfection in hospital infection control: Advantages, drawbacks, and pitfalls. *Antibiotics,* 8, 58.

Casas, M. E., Chhetri, R. K., Ooi, G., Hansen, K. M., Litty, K., Christensson, M., Kragelund, C., Andersen, H. R., & Bester, K. (2015). Biodegradation of pharmaceuticals in hospital wastewater by a hybrid biofilm and activated sludge system (Hybas). *Science of the Total Environment,* 530, 383–392.

Chandrappa, R., & Das, D. B. (2024). Biomedical waste. In *Solid waste management: Principles and practice.* Springer.

Chia, P. Y., Sengupta, S., Kukreja, A., Sl Ponnampalavanar, S., Ng, O. T., & Marimuthu, K. (2020). The role of hospital environment in transmissions of multidrug-resistant gram-negative organisms. *Antimicrobial Resistance & Infection Control, 9*, 1–11.

Cohen, R., Paikin, S., Rokney, A., Rubin-Blum, M., & Astrahan, P. (2020). Multidrug-resistant enterobacteriaceae in coastal water: An emerging threat. *Antimicrobial Resistance & Infection Control, 9*, 1–9.

Copple, R. C. (1989). *Hospital infectious waste disposal system design.*

Cusack, R., Little, E., & Martin-Loeches, I. (2024). Practical lessons on antimicrobial therapy for critically Ill patients. *Antibiotics, 13*, 162.

Dandachi, I., Chaddad, A., Hanna, J., Matta, J., & Daoud, Z. (2019). Understanding the epidemiology of multi-drug resistant gram-negative bacilli in the Middle East using a one health approach. *Frontiers in Microbiology, 10*, 1941.

Daoud, Z., Farah, J., Sokhn, E. S., El Kfoury, K., Dahdouh, E., Masri, K., Afif, C., Abdel-Massih, R. M., & Matar, G. M. (2018). Multidrug-resistant Enterobacteriaceae in Lebanese hospital wastewater: Implication in the one health concept. *Microbial Drug Resistance, 24*, 166–174.

Daouk, S., Chèvre, N., Vernaz, N., Widmer, C., Daali, Y., & Fleury-Souverain, S. (2016). Dynamics of active pharmaceutical ingredients loads in a Swiss university hospital wastewaters and prediction of the related environmental risk for the aquatic ecosystems. *Science of the Total Environment, 547*, 244–253.

Dargahi, A., Vosoughi, M., Normohammadi, A., Sedigh, A., Gholizadeh, H., Sadeghi, H., Karami, C., & Jeddi, F. (2022). Investigating SARS-CoV-2 RNA in five municipal wastewater treatment plants, hospital wastewater and wastewater collection networks during the COVID-19 pandemic in Ardabil Province, Iran. *Applied Water Science, 12*, 256.

De Almeida Cabral, J. M. (2019). *Light therapy in hospital infection control-advantages, drawbacks and pitfalls*.

Dehal, A., Vaidya, A. N., & Kumar, A. R. (2021). Biomedical waste generation and management during COVID-19 pandemic in India: Challenges and possible management strategies. *Environmental Science and Pollution Research International, 29*(10), 14830–14845. https://doi.org/10.1007/s11356-021-16736-8

Dinh, Q., Moreau-Guigon, E., Labadie, P., Alliot, F., Teil, M.-J., Blanchard, M., Eurin, J., & Chevreuil, M. (2017). Fate of antibiotics from hospital and domestic sources in a sewage network. *Science of the Total Environment, 575*, 758–766.

Domínguez, D. C., & Meza-Rodriguez, S. M. (2019). Development of antimicrobial resistance: Future challenges. *Pharmaceuticals and personal care products: Waste management and treatment technology*. Elsevier.

Emmanuel, E., Perrodin, Y., Keck, G., Blanchard, J.-M., & Vermande, P. (2002). Effects of hospital wastewater on aquatic ecosystem. In Proceedings of the XXVIII Congreso Interamericano de Ingenieria Sanitaria y Ambiental. Cancun, México, 2002. Citeseer, pp. 27–31.

Emmanuel, E., Perrodin, Y., Keck, G., Blanchard, J.-M., & Vermande, P. (2005). Ecotoxicological risk assessment of hospital wastewater: A proposed framework for raw effluents discharging into urban sewer network. *Journal of Hazardous Materials, 117*, 1–11.

Emmanuel, E., Pierre, M. G., & Perrodin, Y. (2009). Groundwater contamination by microbiological and chemical substances released from hospital wastewater: Health risk assessment for drinking water consumers. *Environment International, 35*, 718–726.

Escher, B. I., Baumgartner, R., Koller, M., Treyer, K., Lienert, J., & Mcardell, C. S. (2011). Environmental toxicology and risk assessment of pharmaceuticals from hospital wastewater. *Water Research, 45*, 75–92.

Ewing, L. J., Almgren, H. H., & Culp, R. L. (1978). *Effects of thermal treatment of sludge on municipal wastewater treatment costs*, Environmental Protection Agency, Office of Research and Development, Municipal Environmental Research Laboratory.

Eze, C., Oje, O., & Onwurah, I. (2016). Environmental risk assessment of hospital wastewater in Enugu, Nigeria. *Pollution Research, 35*, 55–61.

Fatta-Kassinos, D., Meric, S., & Nikolaou, A. (2011). Pharmaceutical residues in environmental waters and wastewater: Current state of knowledge and future research. *Analytical and Bioanalytical Chemistry, 399*, 251–275.

Frascaroli, G., Reid, D., Hunter, C., Roberts, J., Helwig, K., Spencer, J., & Escudero, A. (2021). Pharmaceuticals in wastewater treatment plants: A systematic review on the substances of greatest concern responsible for the development of antimicrobial resistance. *Applied Sciences, 11*, 6670.

Frédéric, O., & Yves, P. (2014). Pharmaceuticals in hospital wastewater: Their ecotoxicity and contribution to the environmental hazard of the effluent. *Chemosphere, 115*, 31–39.

Fuentefria, D. B., Ferreira, A. E., & Corção, G. (2011). Antibiotic-resistant *Pseudomonas aeruginosa* from hospital wastewater and superficial water: Are they genetically related? *Journal of Environmental Management, 92*, 250–255.

Gadipelly, C., Pérez-González, A., Yadav, G. D., Ortiz, I., Ibáñez, R., Rathod, V. K., & Marathe, K. V. (2014). Pharmaceutical industry wastewater: Review of the technologies for water treatment and reuse. *Industrial & Engineering Chemistry Research, 53*, 11571–11592.

Garbacz, K., Jarzembowski, T., & Rodriguez-Morales, A. J. (2023). *Bacterial infectious diseases annual volume 2023*, BoD–Books on Demand.

Gautam, A. K., Kumar, S., & Sabumon, P. (2007). Preliminary study of physico-chemical treatment options for hospital wastewater. *Journal of Environmental Management, 83*, 298–306.

Guide, A. A. (2010). *Guide to the elimination of multidrug-resistant Acinetobacter baumannii transmission in healthcare settings.*

Gupta, P., Mathur, N., Bhatnagar, P., Nagar, P., & Srivastava, S. (2009). Genotoxicity evaluation of hospital wastewaters. *Ecotoxicology and Environmental Safety, 72*, 1925–1932.

Gurjar, V., Khurana, A., Shaily, G., Kaushik, S., & Singh, N. (2022). Current challenges in treating MDR. In *Current developments in the detection and control of multi drug resistance* (p. 13).

Hawken, S. E., & Snitkin, E. S. (2019). Genomic epidemiology of multidrug-resistant Gram-negative organisms. *Annals of the New York Academy of Sciences, 1435*, 39–56.

Henrici De Angelis, L., D'Andrea, M., Di Pilato, V., Demattè, E., Brovarone, F., Nardini, P., Carretto, E., & Rossolini, G. (2017). First report of multi-drug resistant, VEB-6 producing Proteus mirabilis isolates from Italy. In *Abstract Book Congresso.*

Hocaoglu, S. M., Celebi, M. D., Basturk, I., & Partal, R. (2021). Treatment-based hospital wastewater characterization and fractionation of pollutants. *Journal of Water Process Engineering, 43*, 102205.

Karungamye, P., Rugaika, A., Mtei, K., & Machunda, R. (2023). Antibiotic resistance patterns of *Escherichia coli, Klebsiella pneumoniae*, and *Pseudomonas aeruginosa* isolated from hospital wastewater. *Applied Microbiology, 3*, 867–882.

Kaur, R., Yadav, B., & Tyagi, R. (2020). Microbiology of hospital wastewater. In *Current developments in biotechnology and bioengineering.* Elsevier.

Kehl, K., Schallenberg, A., Szekat, C., Albert, C., Sib, E., Exner, M., Zacharias, N., Schreiber, C., Parčina, M., & Bierbaum, G. (2022). Dissemination of carbapenem resistant bacteria from hospital wastewater into the environment. *Science of the Total Environment, 806*, 151339.

Kelambakkam, C. (2020). *Molecular methods for detecting biofilm producing multidrug resistant clinical isolates.*

Kemprai, L., Hussain, P., Kader, N. A., Nabil, S., Abedin, R. U. A., Roy, N. K., & Rahman, J. (2021). A study on some aspects of bacteriological qualities of hospital wastewater in and around Guwahati city. *The Pharma Innovation Journal, 10*, 1179–1183.

Kerneis, S., Woerther, P., Lepeule, R., Burdet, C., Decousser, J., Ruppé, É., Barbier, F., Armand-Lefèvre, L., Angebault, C., & Barbier, F. (2019). Controlling the diffusion of multidrug-resistant organisms in intensive care units. *Seminars in Respiratory and Critical Care Medicine*, 558–568.

Khan, F. A., Söderquist, B., & Jass, J. (2019). Prevalence and diversity of antibiotic resistance genes in Swedish aquatic environments impacted by household and hospital wastewater. *Frontiers in Microbiology, 10*, 688.

Khan, A. H., Khan, N. A., Ahmed, S., Dhingra, A., Singh, C. P., Khan, S. U., Mohammadi, A. A., Changani, F., Yousefi, M., & Alam, S. (2020a). Application of advanced oxidation processes followed by different treatment technologies for hospital wastewater treatment. *Journal of Cleaner Production, 269*, 122411.

Khan, H. K., Rehman, M. Y. A., & Malik, R. N. (2020b). Fate and toxicity of pharmaceuticals in water environment: An insight on their occurrence in South Asia. *Journal of Environmental Management, 271*, 111030.

Khan, N. A., Ahmed, S., Farooqi, I. H., Ali, I., Vambol, V., Changani, F., Yousefi, M., Vambol, S., Khan, S. U., & Khan, A. H. (2020c). Occurrence, sources and conventional treatment techniques for various antibiotics present in hospital wastewaters: A critical review. *TrAC Trends in Analytical Chemistry, 129*, 115921.

Khan, N. A., Khan, S. U., Ahmed, S., Farooqi, I. H., Yousefi, M., Mohammadi, A. A., & Changani, F. (2020d). Recent trends in disposal and treatment technologies of emerging-pollutants—A critical review. *TrAC Trends in Analytical Chemistry, 122*, 115744.

Khan, N. A., Ullah Khan, S., Ahmed, S., Farooqi, I. H., Hussain, A., Vambol, S., & Vambol, V. (2020e). Smart ways of hospital wastewater management, regulatory standards and conventional treatment techniques: A short review. *Smart and Sustainable Built Environment, 9*, 727–736.

Khan, A., Aziz, H., Khan, N., Hasan, M. A., Ahmed, S., Farooqi, I., Dhingra, A., Vambol, V., Changani, F., & Yousefi, M. (2021a). Impact, disease outbreak and the eco-hazards associated with pharmaceutical residues: A critical review. *International Journal of Environmental Science and Technology*, 1–12.

Khan, M. T., Shah, I. A., Ihsanullah, I., Naushad, M., Ali, S., Shah, S. H. A., & Mohammad, A. W. (2021b). Hospital wastewater as a source of environmental contamination: An overview of management practices, environmental risks, and treatment processes. *Journal of Water Process Engineering, 41*, 101990.

Khan, N. A., Vambol, V., Vambol, S., Bolibrukh, B., Sillanpaa, M., Changani, F., Esrafili, A., & Yousefi, M. (2021c). Hospital effluent guidelines and legislation scenario around the globe: A critical review. *Journal of Environmental Chemical Engineering, 9*, 105874.

Khan, A. H., Aziz, H. A., Palaniandy, P., Naushad, M., Cevik, E., & Zahmatkesh, S. (2023). Pharmaceutical residues in the ecosystem: Antibiotic resistance, health impacts, and removal techniques. *Chemosphere, 339*, 139647.

Klai, N., & Sellamuthu, B. (2020). Bacteriophages isolated from hospital wastewater and its role in controlling drug-resistant pathogens. In *Current developments in biotechnology and bioengineering*. Elsevier.

Kümmerer, K. (2001). Drugs in the environment: Emission of drugs, diagnostic aids and disinfectants into wastewater by hospitals in relation to other sources—a review. *Chemosphere, 45*, 957–969.

Lamba, M., Graham, D. W., & Ahammad, S. (2017). Hospital wastewater releases of carbapenem-resistance pathogens and genes in urban India. *Environmental Science & Technology, 51*, 13906–13912.

Le, T.-H., Ng, C., Chen, H., Yi, X. Z., Koh, T. H., Barkham, T. M. S., Zhou, Z., & Gin, K.Y.-H. (2016). Occurrences and characterization of antibiotic-resistant bacteria and genetic determinants of hospital wastewater in a tropical country. *Antimicrobial Agents and Chemotherapy, 60*, 7449–7456.

Le Corre, K. S., Ort, C., Kateley, D., Allen, B., Escher, B. I., & Keller, J. (2012). Consumption-based approach for assessing the contribution of hospitals towards the load of pharmaceutical residues in municipal wastewater. *Environment international, 45*, 99–111.

Li, T., Zhang, Y., Fu, Y., Li, X., & Chui, W. (2022). Recent progress in phage therapy for controlling multidrug-resistant *Acinetobacter baumannii* including in human and poultry.

Lin, A.Y.-C., & Tsai, Y.-T. (2009). Occurrence of pharmaceuticals in Taiwan's surface waters: Impact of waste streams from hospitals and pharmaceutical production facilities. *Science of the Total Environment, 407*, 3793–3802.

Liu, Q., Zhou, Y., Chen, L., & Zheng, X. (2010). Application of MBR for hospital wastewater treatment in China. *Desalination, 250*, 605–608.

Loganathan, P., Vigneswaran, S., Kandasamy, J., Cuprys, A. K., Maletskyi, Z., & Ratnaweera, H. (2023). Treatment trends and combined methods in removing pharmaceuticals and personal care products from wastewater—A review. *Membranes, 13*, 158.

Mackuľak, T., Cverenkárová, K., Vojs Staňová, A., Fehér, M., Tamáš, M., Škulcová, A. B., Gál, M., Naumowicz, M., Špalková, V., & Bírošová, L. (2021). Hospital wastewater—Source of specific micropollutants, antibiotic-resistant microorganisms, viruses, and their elimination. *Antibiotics, 10*, 1070.

Magalhães, M. J. T. L., Pontes, G., Serra, P. T., Balieiro, A., Castro, D., Pieri, F. A., Crainey, J. L., Nogueira, P. A., & Orlandi, P. P. (2016). Multidrug resistant *Pseudomonas aeruginosa* survey in a stream receiving effluents from ineffective wastewater hospital plants. *BMC Microbiology, 16*, 1–8.

Majumder, A., Otter, P., Röher, D., Bhatnagar, A., Khalil, N., Gupta, A. K., Bresciani, R., & Arias, C. A. (2024). Combination of advanced biological systems and photocatalysis for the treatment of real hospital wastewater spiked with carbamazepine: A pilot-scale study. *Journal of Environmental Management, 351*, 119672.

Manyele, S. V., & Tanzania, V. (2004). Effects of improper hospital-waste management on occupational health and safety. *African Newsletter on Occupational Health and Safety, 14*, 30–33.

Marschall, J., Snyders, R. E., Sax, H., Newland, J. G., Guimarães, T., & Kwon, J. H. (2023). Perspectives on research needs in healthcare epidemiology and antimicrobial stewardship: What's on the horizon—Part I. *Antimicrobial Stewardship & Healthcare Epidemiology, 3*, e199.

Mataraci-Kara, E., Ataman, M., Yilmaz, G., & Ozbek-Celik, B. (2020). Evaluation of antifungal and disinfectant-resistant Candida species isolated from hospital wastewater. *Archives of Microbiology, 202*, 2543–2550.

Mathur, V., Dwivedi, S., Hassan, M., & Misra, R. (2011). Knowledge, attitude, and practices about biomedical waste management among healthcare personnel: A cross-sectional study. *Indian Journal of Community Medicine, 36*, 143–145.

Mendoza, A., Aceña, J., Pérez, S., De Alda, M. L., Barceló, D., Gil, A., & Valcárcel, Y. (2015). Pharmaceuticals and iodinated contrast media in a hospital wastewater: A case study to analyse their presence and characterise their environmental risk and hazard. *Environmental Research, 140*, 225–241.

Mir, M. A. (2022). *Human pathogenic microbes: Diseases and concerns*. Academic Press.

Mirghani, R., Saba, T., Khaliq, H., Mitchell, J., Do, L., Chambi, L., Diaz, K., Kennedy, T., Alkassab, K., & Huynh, T. (2022). Biofilms: Formation, drug resistance and alternatives to conventional approaches. *AIMS Microbiology, 8*, 239.

Mir-Tutusaus, J. A., Parladé, E., Llorca, M., Villagrasa, M., Barceló, D., Rodriguez-Mozaz, S., Martinez-Alonso, M., Gaju, N., Caminal, G., & Sarrà, M. (2017). Pharmaceuticals removal and microbial community assessment in a continuous fungal treatment of non-sterile real hospital wastewater after a coagulation-flocculation pretreatment. *Water Research, 116*, 65–75.

Moccia, G., De Caro, F., Pironti, C., Boccia, G., Capunzo, M., Borrelli, A., & Motta, O. (2020). Development and improvement of an effective method for air and surfaces disinfection with ozone gas as a decontaminating agent. *Medicina, 56*, 578.

Moges, F., Endris, M., Belyhun, Y., & Worku, W. (2014). Isolation and characterization of multiple drug resistance bacterial pathogens from waste water in hospital and non-hospital environments, Northwest Ethiopia. *BMC Research Notes, 7*, 1–6.

Mullot, J.-U., Karolak, S., Fontova, A., Huart, B., & Levi, Y. (2009). Development and validation of a sensitive and selective method using GC/MS-MS for quantification of 5-fluorouracil in hospital wastewater. *Analytical and Bioanalytical Chemistry, 394*, 2203–2212.

Murugaiyan, J., Kumar, P. A., Rao, G. S., Iskandar, K., Hawser, S., Hays, J. P., Mohsen, Y., Adukkadukkam, S., Awuah, W. A., & Jose, R. A. M. (2022). Progress in alternative strategies to combat antimicrobial resistance: Focus on antibiotics. *Antibiotics, 11*, 200.

National Research Council, Division on Earth, Life Studies, Board on Chemical Sciences, Committee on Prudent Practices in the Laboratory and An Update. (2011). *Prudent practices in the laboratory: Handling and management of chemical hazards, updated version.*

Ngigi, A. N., Magu, M. M., & Muendo, B. M. (2020). Occurrence of antibiotics residues in hospital wastewater, wastewater treatment plant, and in surface water in Nairobi County, Kenya. *Environmental Monitoring and Assessment, 192*, 18.

Nyaga, M. N., Nyagah, D. M., & Njagi, A. (2020). *Pharmaceutical waste: Overview, management, and impact of improper disposal.*

Oliveira, T. S., Murphy, M., Mendola, N., Wong, V., Carlson, D., & Waring, L. (2015). Characterization of pharmaceuticals and personal care products in hospital effluent and waste water influent/effluent by direct-injection LC-MS-MS. *Science of the Total Environment, 518*, 459–478.

Parida, V. K., Sikarwar, D., Majumder, A., & Gupta, A. K. (2022). An assessment of hospital wastewater and biomedical waste generation, existing legislations, risk assessment, treatment processes, and scenario during COVID-19. *Journal of Environmental Management, 308*, 114609.

Pariente, M., Segura, Y., Álvarez-Torrellas, S., Casas, J., De Pedro, Z., Diaz, E., García, J., López-Muñoz, M., Marugán, J., & Mohedano, A. (2022). Critical review of technologies for the on-site treatment of hospital wastewater: From conventional to combined advanced processes. *Journal of Environmental Management, 320*, 115769.

Pérez-Alvarez, I., Islas-Flores, H., Gómez-Oliván, L. M., Barceló, D., De Alda, M. L., Solsona, S. P., Sánchez-Aceves, L., Sanjuan-Reyes, N., & Galar-Martínez, M. (2018). Determination of metals and pharmaceutical compounds released in hospital wastewater from Toluca, Mexico, and evaluation of their toxic impact. *Environmental Pollution, 240*, 330–341.

Pirsaheb, M., Mohamadi, M., Mansouri, A. M., Zinatizadeh, A. A. L., Sumathi, S., & Sharafi, K. (2015). Process modeling and optimization of biological removal of carbon, nitrogen and phosphorus from hospital wastewater in a continuous feeding & intermittent discharge (CFID) bioreactor. *Korean Journal of Chemical Engineering, 32*, 1340–1353.

Prado, T., Silva, D. M., Guilayn, W. C., Rose, T. L., Gaspar, A. M. C., & Miagostovich, M. P. (2011). Quantification and molecular characterization of enteric viruses detected in effluents from two hospital wastewater treatment plants. *Water Research, 45*, 1287–1297.

Queen, A., Yusuf, M., Qadir, A., & Aamir, S. (2022). Health risk associated with hospital effluents. *Hospital Wastewater Treatment: Global Scenario and Case Studies.*

Ramdani, N., Benouis, K., Lousdad, A., Hamou, A., & Boufadi, M. Y. (2018). Physicochemical and bacteriological characterization of hospital effluents and their impact on the environment. *Chemistry International, 4*, 102–108.

Rankin, D. A., Walters, M. S., Caicedo, L., Gable, P., Moulton-Meissner, H. A., Chan, A., Burks, A., Edwards, K., Mcallister, G., & Kent, A. (2024). Concurrent transmission of multiple carbapenemases in a long-term acute-care hospital. *Infection Control & Hospital Epidemiology*, 1–10.

Robertson, L., Hermansen, L., & Gjerde, B. (2006). Occurrence of Cryptosporidium oocysts and giardia cysts in sewage in Norway. *Applied and Environmental Microbiology, 72*, 5297–5303.

Rodríguez-Rodríguez, C. E., Ramírez-Morales, D., Masis-Mora, M., Montiel-Mora, J. R., Soto-Garita, C., Araya-Valverde, E., Cambronero-Heinrichs, J. C., Sànchez-Melsió, A., Briceño-Guevara, S., & Mendez-Rivera, M. (2023). Occurrence and risk assessment of pharmaceuticals in hospital wastewater in Costa Rica. *Chemosphere, 339*, 139746.

Samiksha, Bisht, K. K., Gupta, A., Kant, R., & Kumar, R. (2024). Chemical management of industrial hazardous and non-hazardous waste. In *Integrated waste management: A sustainable approach from waste to wealth.* Springer.

Santos, L. H., Gros, M., Rodriguez-Mozaz, S., Delerue-Matos, C., Pena, A., Barceló, D., & Montenegro, M. C. B. (2013). Contribution of hospital effluents to the load of pharmaceuticals in urban wastewaters: Identification of ecologically relevant pharmaceuticals. *Science of the Total Environment, 461*, 302–316.

Schmiege, D., Zacharias, N., Sib, E., Falkenberg, T., Moebus, S., Evers, M., & Kistemann, T. (2021). Prevalence of multidrug-resistant and extended-spectrum beta-lactamase-producing *Escherichia coli* in urban community wastewater. *Science of the Total Environment, 785*, 147269.

Sehulster, L., Chinn, R. Y., Arduino, M. J., Carpenter, J., Donlan, R., Ashford, D., Besser, R., Fields, B., Mcneil, M., & Whitney, C. (2003). Guidelines for environmental infection control in health-care facilities. In *Morbidity and mortality weekly report recommendations and reports RR* (p. 52).

Shehabi, A. A., Haider, A. A., & Fayyad, M. K. (2011). Frequency of antimicrobial resistance markers among *Pseudomonas aeruginosa* and *Escherichia coli* isolates from municipal sewage effluent water and patients in Jordan. *The International Arabic Journal of Antimicrobial Agents, 1*.

Sim, W.-J., Lee, J.-W., Lee, E.-S., Shin, S.-K., Hwang, S.-R., & Oh, J.-E. (2011). Occurrence and distribution of pharmaceuticals in wastewater from households, livestock farms, hospitals and pharmaceutical manufactures. *Chemosphere, 82*, 179–186.

Siri, Y., Bumyut, A., Precha, N., Sirikanchana, K., Haramoto, E., & Makkaew, P. (2024). Multidrug antibiotic resistance in hospital wastewater as a reflection of antibiotic prescription and infection cases. *Science of the Total Environment, 908*, 168453.

Songca, S. P. (2022). Applications of nanozymology in the detection and identification of viral, bacterial and fungal pathogens. *International Journal of Molecular Sciences, 23*, 4638.

Svebrant, S., Spörndly, R., Lindberg, R. H., Sköldstam, T. O., Larsson, J., Öhagen, P., Lindström, H., & Järhult, J. D. (2021). On-site pilot testing of hospital wastewater ozonation to reduce pharmaceutical residues and antibiotic-resistant bacteria. *Antibiotics, 10*(6), 684. https://doi.org/10.3390/antibiotics10060684

Tickner, J., & Jacobs, M. (2016). Improving the identification, evaluation, adoption and development of safer alternatives: Needs and opportunities to enhance substitution efforts within the context of REACH. *Environmental Health Perspectives, 124*, 265–280.

Tsakona, M., Anagnostopoulou, E., & Gidarakos, E. (2007). Hospital waste management and toxicity evaluation: A case study. *Waste Management, 27*, 912–920.

Uhrbrand, K., Schultz, A. C., Koivisto, A., Nielsen, U., & Madsen, A. (2017). Assessment of airborne bacteria and noroviruses in air emission from a new highly-advanced hospital wastewater treatment plant. *Water Research, 112*, 110–119.

Ulvi, A., Aydin, S., & Aydin, M. E. (2022). Fate of selected pharmaceuticals in hospital and municipal wastewater effluent: Occurrence, removal, and environmental risk assessment. *Environmental Science and Pollution Research, 29*, 75609–75625.

Varela, A. R., André, S., Nunes, O. C., & Manaia, C. M. (2014). Insights into the relationship between antimicrobial residues and bacterial populations in a hospital-urban wastewater treatment plant system. *Water Research, 54*, 327–336.

Verlicchi, P. (2018). Hospital wastewaters. In *The handbook of environmental chemistry*. Springer International Publishing. https://doi.org/10.1007/978-3-319-62178-4

Verlicchi, P., Galletti, A., Petrovic, M., & Barceló, D. (2010). Hospital effluents as a source of emerging pollutants: An overview of micropollutants and sustainable treatment options. *Journal of Hydrology, 389*, 416–428.

Verlicchi, P., Galletti, A., & Al Aukidy, M. (2013). Hospital wastewaters: Quali-quantitative characterization and for strategies for their treatment and disposal. In *Wastewater reuse and management* (pp. 225–251).

Verlicchi, P., Al Aukidy, M., & Zambello, E. (2015). What have we learned from worldwide experiences on the management and treatment of hospital effluent?—An overview and a discussion on perspectives. *Science of the Total Environment, 514*, 467–491.

Wang, Q., Wang, P., & Yang, Q. (2018). Occurrence and diversity of antibiotic resistance in untreated hospital wastewater. *Science of the Total Environment, 621*, 990–999.

Wang, J., Wang, S., Chen, C., Hu, J., He, S., Zhou, Y., Zhu, H., Wang, X., Hu, D., & Lin, J. (2022). Treatment of hospital wastewater by electron beam technology: Removal of COD, pathogenic bacteria and viruses. *Chemosphere, 308*, 136265.

White, P. A., Netzler, N. E., & Hansman, G. S. (2016). *Foodborne viral pathogens*. CRC Press.

Wiest, L., Chonova, T., Bergé, A., Baudot, R., Bessueille-Barbier, F., Ayouni-Derouiche, L., & Vulliet, E. (2018). Two-year survey of specific hospital wastewater treatment and its impact on pharmaceutical discharges. *Environmental Science and Pollution Research, 25*, 9207–9218.

Yadav, B., Pandey, A. K., Kumar, L. R., Kaur, R., Yellapu, S. K., Sellamuthu, B., Tyagi, R., & Drogui, P. (2020). Introduction to wastewater microbiology: Special emphasis on hospital wastewater. In *Current developments in biotechnology and bioengineering*. Elsevier.

Yamashiro, S., Foco, M. L. R., Pineda, C. O., José, J., Nour, E. A. A., Siqueira-Castro, I. C. V., & Franco, R. M. B. (2019). Giardia spp. and Cryptosporidium spp. removal efficiency of a combined fixed-film system treating domestic wastewater receiving hospital effluent. *Environmental Science and Pollution Research, 26*, 22756–22771.

Yan, H. (2021). *Investigation of the role of glycan binding on human norovirus using mass spectrometry.* Staats-und Universitätsbibliothek Hamburg Carl von Ossietzky.

Yan, S., Zhang, X. L., Tyagi, R., & Drogui, P. (2020). Guidelines for hospital wastewater discharge. In *Current developments in biotechnology and bioengineering.* Elsevier.

Yilmaz, G., Kaya, Y., Vergili, I., Beril Gönder, Z., Özhan, G., Ozbek Celik, B., Altinkum, S. M., Bagdatli, Y., Boergers, A., & Tuerk, J. (2017). Characterization and toxicity of hospital wastewaters in Turkey. *Environmental Monitoring and Assessment, 189*, 1–19.

Zhang, S., Huang, J., Zhao, Z., Cao, Y., & Li, B. (2020a). Hospital wastewater as a reservoir for antibiotic resistance genes: A meta-analysis. *Frontiers in Public Health, 8*, 574968.

Zhang, X., Yan, S., Chen, J., Tyagi, R., & Li, J. (2020b). Physical, chemical, and biological impact (hazard) of hospital wastewater on environment: Presence of pharmaceuticals, pathogens, and antibiotic-resistance genes. In *Current developments in biotechnology and bioengineering.* Elsevier.

Zhang, Y., Lin, Y., Galgano, S., Houdijk, J., Quan Xie, W., Jin, Y., Lin, J., Song, W., Fu, Y., & Li, X. (2022). Recent progress in phage therapy for controlling multidrug-resistant *Acinetobacter baumannii* including in human and poultry.

Chapter 4
Emerging Contaminants in Wastewater and Their Treatment Plans: A Major Focus on Radioactive Components Emitted from Medical Waste

Rahul Maddheshiya, Soniya Dhiman, and Himanshu Gupta

Abstract Emerging contaminants (ECs) have become a significant concern globally in recent years due to potential risks posed by them to the environment and human health. Antibiotics, hormones, pharmaceuticals and personal care products are widely used for medical and other purposes, have led to rapid emergence of environmental pollutants. Most concerned ECs also involve the radioactive wastes released from the nuclear medicine hospitals and cancer hospitals. These contaminants are made for the betterment of life but they are of great concern now-a-days, since they are slowly absorbed by soils or released directly into water bodies. Consequently, drinking water sources become contaminated, health problems arise and water quality declines. The presence of such pollution in aquatic environments, especially in surface water, poses a serious concern. The study examines a number of associated concerns of ECs in aquatic environments, including strategies for EC removal. Despite their advantages and disadvantages, ECs treatment procedures encompass various techniques, including oxidation, physico-chemical, and biological processes. Despite the abundance of scientific research on many areas of monitoring and analyzing contaminants, mainly radioisotopes from medical wastes, it is believed that a comprehensive assessment of these studies is urgently needed. This chapter provides an overview of the case studies related with the discharge of the radioactive isotopes.

Keywords Emerging contaminants (ECs) · Radioactive contaminants · Medical waste · Health impacts · Treatment methods

---

R. Maddheshiya
Department of Zoology, University of Allahabad, Prayagraj, Uttar Pradesh 211002, India

S. Dhiman (✉)
Helmholtz-Zentrum Dresden-Rossendorf, Helmholtz Institute Freiberg for Resource Technology, Bautzner Landstraße 400, Dresden 01328, Germany
e-mail: s.dhiman@hzdr.de

H. Gupta
Department of Chemistry, School of Sciences, IFTM University, Lodhipur Rajput, Moradabad, Uttar Pradesh 244102, India

H. Gupta et al. (eds.), *Emerging Contaminants in Water and Wastewater*, Environmental Science and Engineering, https://doi.org/10.1007/978-3-031-82579-8_4

4.1 Introduction

Due to the possible threats that emerging contaminants (ECs) bring to human health as well as environment, ECs have received increased attention recently on a global scale. ECs, also referred as micropollutants, are generated through a range of methods utilizing both natural and artificial materials. It has recently been demonstrated that drinking water, groundwater, food supplies, municipal wastewater, and surface water may all include a distinct class of chemical substances known as new ECs (Arman et al., 2021). These pollutants, often referred to as ECs, are chemical compounds' mixtures that are frequently encountered in the environment, especially in aquatic ecosystem and soil. But it's just lately been determined that they're serious water pollutants. Endocrine-disrupting chemicals (EDCs), hormones, pharmaceutical residues, microbial pathogens, radioisotopes, flame retardants, industrial additives, pesticides, nanomaterials, and personal care products (PCPs), polycyclic aromatic hydrocarbons are some examples of essential and commonly used ECs in our modern lives (Arman et al., 2021; Fatimazahra et al., 2023; Gupta, 2016; Gupta & Gupta, 2015; Mazhar et al., 2024; Rani et al., 2022).

As per the US Geological Survey, EC is "any compound of natural origin or engineered or any microbe that is not commonly found in the surrounding, although it may cause detrimental effects on environmental and human health". The term "ECs" refers to substances that may bioaccumulate, persist in the environment as well as poses threats to human health (Dulio et al., 2018; Gupta and Dhiman, 2023; Kaur et al., 2023). These substances may cause neurodevelopmental delays, stunted growth, decreased fertility and reproductive wellness, inhibit wildlife populations, decline aquatic environments, and possibly interfere with human immune systems (Rodriguez-Narvaez et al., 2017). This may be crucial to emphasize that most ECs are not entirely new or recently entered toxins in environmental segments. Thus, the term "emerging" refers to both the contaminants and the new problems which arise. As a result, ECs are also referred as "chemicals/contaminants of emerging concern" (Arman et al., 2021).

Due to pollution from economic growth, urbanization, agriculture, and health care, the quality of water resources has decreased (Patel et al., 2019). A large number of researches were performed to evaluate the sources and levels of contaminants in water (Lim et al., 2017). The variability in concentrations as well as lack of frequent monitoring programs, the understanding is still lacking regarding transformation products, metabolites and drinking water treatment (Arman et al., 2021; Gupta, 2018; Gupta & Gupta, 2016, 2018). Radionuclides or radioactive elements are in vast employed in the treatment of cancer treatment but due to less degree of degrading mechanisms or lack of disposal infrastructure they are directly or indirectly disposed in the water causing its contamination (Puspita et al., 2023).

Future researches must focus to enhance risk-based screening models and frameworks. Rodriguez-Narvaez et al. (2017) thoroughly examined few treatment strategies, including phase-change, biological and advanced oxidation processes, for remediation of water from ECs. Furthermore, Abaroa-Pérez et al. (2018) assessed the relationship between microplastics and contaminants. Microplastics are categorized as ECs. Khan et al. (2020) reported that medications are a significant source of environmental contamination. Hospital effluent discharge was taken into consideration, and they focused on treatment methods like activated sludge, membrane biological reactors, sequencing batch reactors, UV/H_2O_2, ozone treatment, upflow anaerobic sludge blank, carbon nanotubes and activated carbon. Roy et al. (2021) emphasized on photodegradation and their combination with nanomaterials, considering antibiotics as the only type emerging water contaminant.

Despite the abundance of scientific research on many areas of monitoring and analyzing contaminants, mainly radioisotopes from medical wastes, it is believed that a comprehensive assessment of these studies is urgently needed. Therefore, this chapter focuses on occurrence of ECs in water, their distribution, analytical techniques, toxicological impacts, and treatment methods. This chapter also explains the case studies related with the discharge of the radioactive isotopes.

4.2 Occurrence of Emerging Contaminants in Water Resources

ECs were found to exist in groundwater as well as surface water. As surface water gets effluents directly from wastewater treatment plants (WWTP) with shorter residual time compared to groundwater, it consists of greater level of ECs. The occurrence of specific medications, PAHs, phenols and PCPs in surface waters was found to vary greatly. The usage patterns and doses in various areas, as well as the efficiency of the WWTP system, are likely the main causes of the variations (Kumar & Gupta, 2020; Kumar et al., 2023; Munthe et al., 2017). The movement, fate, and presence of ECs in the environment are all influenced by numerous circumstances. These include latitude and longitude, as well as physico-chemical properties of water and surrounding environment.

Moreover, nature of source affects the level of exposure and the characteristics of the substance (Stefanakis & Becker, 2015). The sources and potential pathways for ECs discharged into the atmosphere and their dissolution into different recipients (drinking, surface, and ground water) are shown in Fig. 4.1.

4.2.1 Personal Care Products (PPCPs) and Pharmaceuticals, Antiseptics, Hormones, Perfluoronated Compounds (PFCs) Pesticides

Chemicals found in healthcare items and cosmetics (such as over-the-counter medicines, dietary supplements, and prescription medications) are included in PCPs. After their discharge from sewage treatment plants or leached farmed manure, these

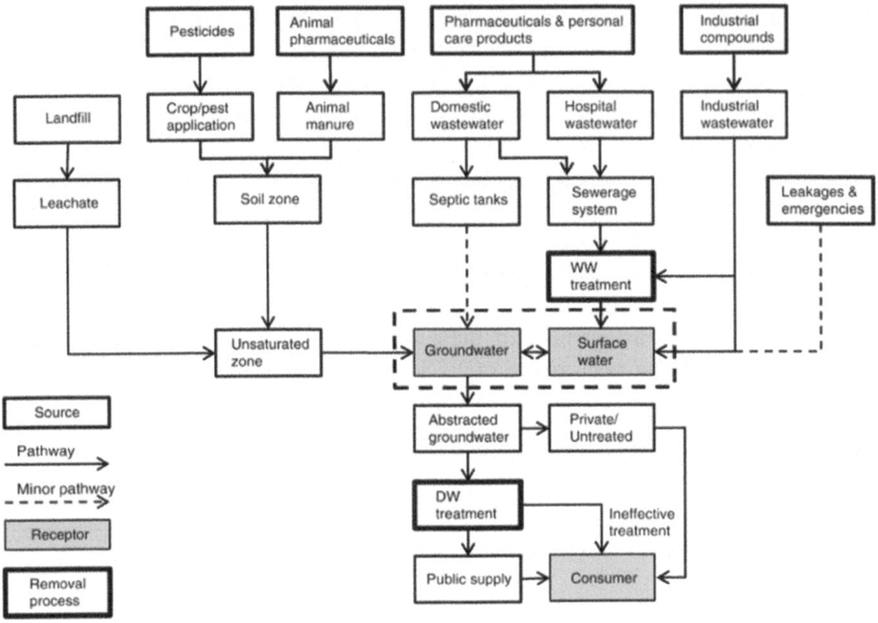

Fig. 4.1 Sources and possible pathways of ECs in the aquatic environment. Reprinted from original article by Stuart et al. (2012), after taking permission from Elsevier using RightsLink

contaminants find their way into soil and community water systems (Parra-Saldivar et al., 2020). A significant participation of the pharmaceutical industry in WWTP activities was reported (Palli et al., 2019). Commonly used antiseptic is triclosan that may be found in clothes, bedding, plastic, fabric, appliances, and soaps. In the environment, triclosan may breakdown into methyl triclosan or specific dioxins, the most common ones are 2,8-dichlorodibenzo-p-dioxin and 2,4-dichlorophenol. It leads to contamination in water (Gwenzi et al., 2018).

Steroid hormones, both synthetic and natural, including estrogens, androgens, and gestagens, are among the many prescription, and nonprescription medications (Wöhler et al., 2020). Due to their ability to modify physiology and activate vital regulatory functions in body fluids, these medicines have been used for various purposes in both humans as well as animals. The main source of steroid estrogens was agricultural residues and conjugated estrogens from dairy production might make up as much as 22% of the overall estrogen load (Tremblay et al., 2018). According to Maddheshiya and Singh (2021a, 2021b, 2021c) and Maddheshiya (2021) pesticides contamination is a also a major emerging contaminants in aquatic environment.

Paints, textiles, food packaging, adhesives, electronics, polishes, stain repellents, and waxes are just a few products that use PFCs. Perfluorocarbon sulfonic acids and perfluorinated carboxylic acids and are most prevalent, whereas, the most widely known are perfluoro octane sulfonate and perfluorooctanoic acid (Bai et al., 2018). Drinking and purified water sources include PFOA because of discharges from industrial facilities, WWTPs, and non-point sources (Sznajder-Katarzynska et al., 2019).

4.2.2 Radioactive Wastes Substances

In view of non-renewable characteristics of fossil fuel and difficulty in obtaining renewable energy sources, nuclear energy is becoming more and more recognized as a viable option. As new nuclear technology has been developed globally, a significant amount of radioactive discards has been generated by a range of processes (Table 4.1). (Zakrzewska-Trznadel, 2013; Zhang et al., 2019). This is significant to report that the environment and public health are seriously threatened by these type of emerging radioactive pollutions, as evidenced by significant nuclear incidents in history, which have raised serious concerns at the highest levels to the government. Since dissolved radionuclides have mobility, these may find their way into groundwater and rivers as well as other aquatic habitats without being properly treated (Zhang et al., 2019). This will unavoidably raise the possibility of radioactive exposure in humans.

Most importantly, algae, crops, microorganisms, plankton or plants can absorb dissolved radioactive nuclides through aqueous media, and they would then further accumulate and concentrate through cascading food chains. It might be seriously

Table 4.1 Showing different sources of radioactive waste

Waste category	Common sources of waste	Common radionuclides
High-level radioactive waste (LLRW)	Partially used fuel from nuclear reactors; liquid waste from the reprocessing of used fuel	^{90}Sr, half-life: 29.78 y ^{137}Cs, half-life: 30.07 y
Transuranic waste	Weapons-production factory waste included mixed transuranic waste	^{238}Pu, half-life: 87.7 y ^{241}Am, half-life: 432.7 y
Mixed waste	Weapons-production waste and some research wastes	^{239}Pu, half-life: 24,100 y ^{241}Pu, half-life: 14.4 y
Naturally occurring Radioactive material (NORM) waste	Scale buildup on pipe walls that carry petroleum products	^{226}Ra, half-life: 1599 y ^{228}Ra, half-life: 5.76 y
Uranium or thorium mill tailings waste	Production only at the rare earth extraction milling location	^{230}Th, half-life: 75,400 y
Low-level radioactive waste (LLRW)	Nuclear power plant industrial waste; paper, plastic, and glass waste for research, medicine, and academia	^{3}H, half-life:12.32 y ^{60}Co, half-life: 5.27 y
Very low activity waste	From medical procedures	^{131}I, half-life: 8.027 d

Reprinted from the original article by Zhang et al. (2019) after permission from Elsevier through RightsLink detrimental to the human health. The data indicates that steady build-up of radionuclides within the human body would result in an extended period of exposure to high radioactivity levels, raising the risk of anemia, leukemia, metabolic diseases, and ultimately death (Kinly, 2005). This is evident to report that the removal of radioactive contaminants from wastewater has grown to be a crucial technological, social as well as political challenge (Zhang et al., 2019).

4.3 Case Studies Related to Special Insight on the Radioactive Wastes from the Medical Centers

4.3.1 Case Study 1: Urine and Excrement from People Receiving Large Doses of Radioisotopes (Khan et al., 2010)

Patients receiving high therapeutic doses of radioisotopes (e.g., Iodine-131 for thyroid carcinoma) were segregated in isolation units till their emission levels of radiation decrease to levels below which the public can tolerate the lowest levels (3 m Roentgens/hour at one meter distance). Patients' excreta and from high-dose isolation wards (such as Iodine-111) are flushed, and the quickest route feasible is taken by PVC pipes to delay tanks, which are specialized storage tanks where the materials are held in reserve until they are released into the sewage system. The delay tank may be placed having little foot traffic. The number of patients admitted each day determines the tank's capacity. A hospital would need two 6000-L delay tanks if it were to admit two patients at the same time. This capacity was calculated under the supposition that each patient needs, on average, 100 L water each day. Each patient uses three thousand litres a month, for a total of six thousand litres used.

After one month, the tank is filled, at which point the gate valve is opened to the next tank. The full tank remains sealed for the month that it takes for the second tank to fill. As such, radioactive waste is kept in each tank for a period of two months, long enough for Iodine-131 to decay to very low levels (Delay and Decay). Before the tank's discharge is released in public sewer system, samples should be taken to evaluate the activity; the sample's concentration shouldn't be higher than 1.2 microcuries per liter. Khan et al. (2010) reported that no hospital is permitted to release more than 37 G Bq (1 Curie) of liquid radioactive waste into the community's sewer system in one calendar year.

4.3.2 *Case Study 2: Radioisotope Releases for Nuclear Medicine in Spanish Hospitals (Krawczyk et al., 2013)*

The two general hospitals with radiation therapy services in central Granada (Clínico and Ruiz de Alda) were picked for this investigation. For a variety of uses, both institutions employ non-encapsulated radioactive isotopes, which are typically liquid. Non-encapsulated radioisotopes are used in the Ruiz de Alda hospital in both in- and out-patient diagnosis and treatment. The majority of the radioactive material generated during patient stays is found in the excreta of hospitalized patients. Patients receiving I^{131} cancer treatments spend four to seven days in specialized chambers where pee excrement is gathered in specially designed decomposition storage tanks. The Ruiz de Alda has two of these tanks, with the second being used after the first fills up. Eventually, this procedure results in the emission of waste around once every three months. This system's main goal is for the isotopes' radioactivity to decrease so that, before effluents are released into public sewers, their radioactive concentrations are below the set limits. Radioactive residue-containing excreta effluents from Clínico Hospital's nuclear health division, an outpatient support, drain directly into the sewer system.

There were no $^{90}Sr/^{90}Y$ or ^{14}C activity detected in the two hospitals' wastewater. It is possible to observe activity levels of man-made radionuclides and gamma emitters, including ^{131}I, ^{99m}Tc, ^{67}Ga, and ^{111}In, and radioactive tracers, including ^{3}H and ^{14}C, that are frequently utilized in research. The following is an upward trend in the percentages of gamma radiation in the discharges: Ruiz de Alda (61.85%) < Clínico 1 (3.77%) < Clínico 2 (34.85%). The radionuclide activity concentrations in the samples collected at the Clínico 2 outflow are listed in the following order: ^{131}I < ^{67}Ga < ^{99m}Tc < ^{111}In. The fact was that ^{99m}Tc and ^{67}Ga were found ten times lower at Clínico 1 point of control than at Clínico 2. It was confirmed by presence of 3 H at Clínico 1, a radioisotope that is commonly used as a tracer in biological and medical research. The exceedingly low ^{14}C values found in the wastewater from both hospitals are demonstrated in Fig. 4.2 (Krawczyk et al., 2013).

All three outputs had high amounts of ^{99m}Tc, which is compatible with the radionuclide's primary use in hospitals as a tracer due to its radiological characteristics and adaptability for illness diagnosis. Clínico 1 < Clínico 2 < Ruiz de Alda was the range of activity it covered. The samples obtained at Ruiz de Alda did not contain ^{67} Ga or ^{111}In, but they did have the highest activity of ^{131}I (59.29 Bq/L1) of any sample analyzed. This could be because ^{131}I is used for both therapeutic and diagnostic purposes at this facility (Krawczyk et al., 2013).

At control point of Ruiz de Alda, total specific activity of all radionuclides was below upper limits allowed by Spanish law on Intake for adults, considering constant yearly intake of 600 L of water per person. However, sewerage outlet from Clínico 2 had a little greater activity concentration of ^{99m}Tc. However, Ruiz Alda's result (0.797) fell inside these limitations.

It is impossible to overestimate the significance of storage vessels in hospitals with nuclear health units as these reduce amount of radioactive discharge into the

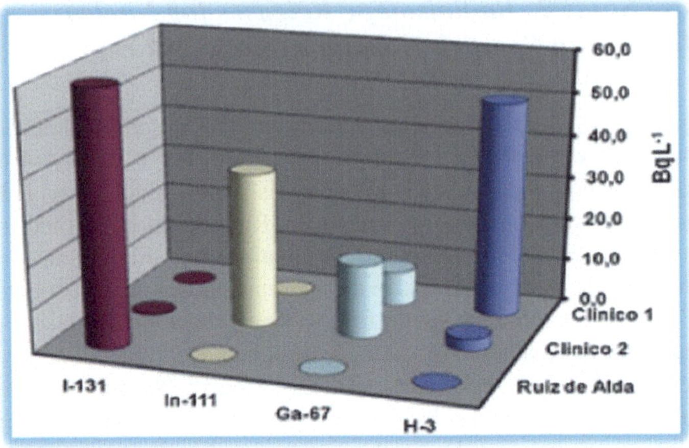

Fig. 4.2 The concentrations of the radioisotopes used in nuclear medicine, with the exception of ^{99m}Tc, in the hospital discharges. Reprinted from original article by Krawczyk et al. (2013) after permission from Elsevier using RightsLink

wastewater. It is illustrated by the ^{99m}Tc, whose activity reduces to about one-third following its storage in Ruiz de Alda tank, thereby reducing its environmental impact (Krawczyk et al., 2013).

4.3.3 Case Study 3: Radiological Examination of a Wastewater Treatment Facility Connected to Radioiodine Therapy at a West Java, Indonesia, Hospital (Puspita et al., 2023)

This study was carried out in West Java, Indonesia, by NM unit of referral hospital. Building 1, an outdated building that the unit utilized for its service operations, no longer had the waste pipe system blueprint. Meanwhile, NM service was housed in the basement of an entirely novel, six-story building at the second location. The investigation's primary focus was on how radioiodine therapy patients' excreta were managed in both locations. Liquid waste from building 1's NM installation did not travel to centralized WWTP of hospital, in contrast to building 2's NM installation, which was momentarily settled before entering the WWTP. Radioiodine concentrations in the main sewage pit may arise from patients' waste entering decomposition tank C. Decomposition tank B takes waste from hot lab and the four uptake rooms, while tank A receives trash via isolation and post-injection rooms.

The minimum detectable quantity of I-131 was found at 2.6 × 104 Bq/m^3. As part of the I-131 therapy, the fluid waste contents in three decay tanks (A–C) were measured every twelve hours for three days. With tanks B and C and the remarks

A1-4, I-131 therapy of patient results from the first 12 h of treatment to the 48th hour. Since tanks A, B, and C receive a direct flow, tank D can have a high I-131 concentration. But the concentration of radioactivity at Point H is higher than what is considered reasonable in the surrounding air. So, steps must be taken to lower the I-131 concentration levels. The percentage of I-131 removal produced a high efficiency with the present waste disposal system for NM patients (Puspita et al., 2023).

The concentration was found to be higher than the ambient radioactivity standard at point H, the hospital wastewater treatment facility's outflow point. The surrounding settlements relied on the river for irrigation water, and the water company used it as a source. Consequently, materials containing I^{131} were ingested by the community both directly and indirectly. Based on data from the nuclear medicine department's assessments of I^{131} concentrations in four decay tanks, assessment of risks was conducted considering "single" 24-h discharge of I^{131} by hospital's WWTP effluents. This highlights significance of I^{131} monitoring before the discharge of fully treated effluent from medical WWTPs.

4.4 Treatment Techniques to Remove Emerging Contaminants

EC removal from waters, but occurrence of disinfection by-products significantly impact quality of drinking water supply. Many treatment techniques, including oxidation, physico-chemical, and biological techniques, were be employed to remove ECs. However, many of these techniques have drawbacks, including high maintenance costs, secondary pollutants (Arman et al., 2021; Grassi et al., 2012).

4.4.1 Physico-Chemical Treatment

Physico-chemical methods are generally applied in traditional treatment of water for elimination of pathogens, reduction of turbidity, odor issues and taste control. Such procedures can be additionally beneficial for reducing the concentration of ECs in water for drinking purposes (Benner et al., 2013). The physico-chemical process includes membrane filtration, flocculation, coagulation and activated carbon treatment techniques. Coagulation-flocculation is common technique used to treat water. Recent research on halo-benzoquinone, one of the newly discovered DBPs, indicates that the precursor ingredients in drinking water sources could not be entirely eliminated by coagulation (Diemert et al., 2013). Another study found that the only way to achievement of the maximum rate of removal as 51% and 59%, respectively, was to reduce the precursors of THMs and halogenic acetic acids under improved coagulation by poly aluminum chloride (Zhao et al., 2013). Drinking water and industry wastewater can be treated and purified through activated carbon adsorption as a workable solution (Jusoh et al., 2011; Nazifa et al., 2018).

4.4.2 Biological Treatment

Since the early 1900s, sewage has been treated with microbial biomass to break down organics, contaminants, and nutrients. However, the use of this type of biological method to treat drinking water is not highly applied. New techniques are currently being applied to improve viability, feasibility as well as application of biological drinking water treatment techniques. Among these developments are the subsequent ones: (1) the growing complexity and expense of managing water treatment residuals; (2) need for methods which efficiently eliminate pollutants instead of just concentrating; (3) emergence of new pollutants, like perchlorate, which is highly susceptible to biological degradation; (4) invention of membrane-supported water treatment equipment; (5) regulations that restrict DBPs formation (Jensen et al., 2014; Prest et al., 2016). Compared to more sophisticated treatment techniques, activated sludge (CAS) has a lower capital cost because of its lower operational requirement. However, CAS is not as environmentally harmful as chlorination, it still has a number of disadvantages, such as a propensity to generate larger volumes of sludge, a high energy requirement, and the potential for bulk sludge, foam, and color generated by secondary clarifiers (Arman et al., 2021; Hussain & Dubey, 2014).

Membrane bioreactors (MBRs) are an additional efficient biological treatment technique for mass organics removal. MBRs can be applied in conjunction to CAS or as hybrid systems (Mazzei et al., 2017). Research has shown that MBR can effectively remove both chemical and biological oxygen demand at high rates in production plants (Hamon et al., 2018). MBRs have been shown to be successful in eliminating estrogens from wastewater; they have been shown to achieve moderate removal efficiencies of up to 67% for 17β-estradiol and up to 91% for estrone (Katibi et al., 2021). On the other hand, no appreciable clearance of 17-estradiol-glucuronides, estrone-3-glucuronide, or estrone-3-sulfate was seen.

A more recent development in biological treatment for getting rid of ECs is enzyme-based therapy. The reports have showed that diclofenac cytotoxicity may be totally eliminated under the right circumstances when laccase treatment is paired with the natural redox mediator syringaldehyde (Bokare et al., 2010; Sathishkumar et al., 2014). Laccase-poly (lactic-co-glycolic acid) nanofibers were found as an efficient way for removal of diclofenac from water sources and detoxify them when syringaldehyde is present (Sathishkumar et al., 2012). According to recent research, combining an enzymatic cascade reaction with cross-linked enzyme aggregates (combi-CLEA) can remove pharmaceutically active chemicals at a rate that is higher than 80% (Touahar et al., 2014). Enzymes are environmentally friendly, thus treating ECs with them has a bright future ahead of it.

4.4.3 Adsorption Method

Adsorption method is highly applied and effective in the treatment of water, because it is easy to design, yields no unwanted byproducts, and is not affected by potentially harmful substancess (Singh et al., 2018, 2020, 2022; Tong et al., 2010). Even though activated carbons are the best option for eliminating different ECs, their usage is occasionally restricted because of their expensive cost. Activated carbon may be regenerated for further use even though it has been used up. However, carbon will be depleted during regeneration; the resultant product may not have as much adsorption capacity as freshly made activated carbon (Gupta et al., 2009). Eighty different endocrine-disrupting compounds (EDCs) were tested utilizing bench-scale water treatment plant (WTP) simulation model and natural water spiked with thirty medicinal products. Leusch et al. (2019) found that, aside from oxidation via chlorination, ozonation and powdered activated carbon (PAC), no significant compound removal was observed. Subsequently, Bolong et al. (2009) and Abd El-Gawad and Abd ElAziz (2018) confirmed that granular activated carbon regeneration or an appropriate PAC dose is necessary for achievement of higher rates of removal.

4.4.4 Reverse Osmosis (RO) and Nanofiltration (NF)

RO and NF have found significant concern in the recent times. Membrane filtration techniques successfully remove the majority of inorganic and organic contaminants. These are currently widely utilized to treat wastewater and drinking water (Touati et al., 2018). The removal of ECs appears to be greatly promising when using RO and NF membranes (Ouyang et al., 2019; Semião et al., 2013). Although RO membrane is still efficient compared to NF for removing ECs, it is less recommended as it requires more energy. Utilizing RO and NF membranes in treatment of water, efficiently removes PCPs and EDCs, based on reports in literature (Amin et al., 2014; Aziz & Ojumu, 2020). Moreover, studies have showed that application of nanofiltration in aqueous media can efficiently remove pesticides (duringon, atrazine, and simazine metabolite) (Musbah et al., 2013). Atrazine exhibits maximum retention rate and diuron exhibits lowest rate among several treatment procedures; NF membranes outperform dual medium filtration, particle coagulation-flocculation, and sedimentation (Plakas & Karabelas, 2012). In another study, nanofiltration membranes were utilized to remove DDT and other organic chlorinated contaminants through drinking water. The findings demonstrated that DDT may successfully extract from water at a rate of up to 95% by utilizing an NF membrane (Pang et al., 2010).

4.4.5 *Ozonation*

Due to higher hydroxyl radical generation as well as photon-initiated carbon-halogen bond cleavage, ozonation individually is less successful than integrating advanced oxidation processes (AOPs) based on ozone, like Fenton-type, photochemical, O_3/UV and O_3/H_2O_2 based processes. AOPs are recommended to treat recalcitrant substances (Wang & Zhuan, 2020). Many nations are concerned about the potential for ionizing radiation to break down hazardous organic pollutants such as antibiotics, nitrophenols, and chlorophenols in aqueous solutions (Alkhuraiji & Alkhuraiji, 2019; Arman et al., 2021; Liu & Wang, 2013; Shaoqing et al., 2010).

4.5 Radioactive Waste Management

Presently, there are two management systems which are used practically to treat radioactive wastes by patients undergoing radionuclide therapy:

4.5.1 *Delay and Decay*

The most common approach for securely managing radioactive waste materials is delay and decay, which is employed by several businesses, institutions, research centers, and hospitals. It is primarily utilized for I-131 radionuclide abatement. Due to its intricate nature, the repair procedure requires money from a highly specialized funding organization. The dose of the integrated radionuclide determines the effectiveness of I-131 ablation therapy. Radionuclide leftovers resulting from integration have been shown in multiple cases to cause loss of taste, neck pain, sialoadenitis, and gastritis (Banerjee, 2013).

4.5.2 *Biochroma*

A novel and patented biological treatment method featuring an adsorption phase at the end is called BioChroma. This method is based on comprehensive optimization of wastewater treatment principles that were created and constructed in the late 1970s and principle of chromatography. However, this approach was never employed again because of its high cost and labour requirements.

The activated carbon employed in the adsorption stage needed to be changed regularly, which posed a significant challenge to cost effectiveness. As an alternative to these intricate and costly systems, BioChroma was created, utilizing and optimizing the core components of chromatography-based technology. The wastewater produced in radiological therapy ward is collected by the BioChroma system, which is fed by specialized pumps equipped to device which shreds solid debris and homogenizes effluent. The holding tanks, which have an aeration system, serve as prelude to biological treatment stage and aid in preventing anaerobic processes and the disagreeable odors that are sometimes associated with them. Larger particles that could impair the function of activated sludge within biological reactor were isolated during preclarification stage of the wastewater before it reaches the biological treatment plant.

This biological reactor can be thought of as a miniature wastewater treatment facility. It has a filter to remove any suspended materials and a secondary clarity stage. The downstream adsorption filter system can be shielded from undesirable blockages and ultimate organic contamination is minimized. The aforementioned final adsorption filter line is made up of selective ion exchangers and activated carbon filters that remove any dissolved radioactive components from wastewater is collected in storage tank, where it is continuously monitored, and finally released into sewage system (Banerjee, 2013).

Merits of Biochroma Processing Technology

1. Space requirement is lowered by 50% and fully automated therefore cost saving.
2. More flexible continuous operation, providing bulk capacity to absorb abrupt peaks and more comfort for the final user.
3. Due to the absence of anaerobic reaction no unpleasant odour.
4. No risk of leakage or cross contamination.
5. This system is easy to install.

4.5.3 Dilute and Disperse

Articles with low activity solids that have an activity of less than 1.35 microcurie (50 KBq) or a total package concentration of less than 135 microcurie/m^3 (5 MBq/m^3) can be disposed of as regular hospital waste. Items of this type consist of tissue paper, cotton swabs, vials, syringes, and more. Analogously, radioactive liquid waste that has activity levels lower than microcurie may be dumped into sanitary sewer system after sufficiently flushing with water. It is imperative that the stipulated limitations be adhered to when it comes to the overall discharge of liquid radioactive material into the sanitary sewage system (Khan et al., 2010).

4.5.4 Concentrate and Contain

When disposing the radioactive waste, this method was occasionally employed for materials having lengthy half-lives (more than one month) and extremely high activity levels. Their longer storage period makes the delay and decay technique of disposal impracticable, especially if available space is restricted. After being gathered in appropriately labeled and constructed containers, radioactive waste is interred in places designated for such purposes by the relevant authorities. Such radioactive waste is uncommon in the day-to-day operations of hospitals, and as a result, this way of disposing of radioactive waste is not frequently employed (Khan et al., 2010).

4.5.5 Ion Exchange Method

Numerous naturally occurring materials, such as clays, perovskites, zeolites, and others, show adsorption and ion exchange capabilities for different types of radionuclides. The cesium adsorption in radioactive wastewater was investigated using modified zeolite made from fly ash (El-Naggar et al., 2008), which had the maximum cesium uptake of 64.1% as well as capacity of cation exchange as 4.624 meq/g. Additionally, bentonites and montmorillonite were used to adsorb soluble cesium. Furthermore, it was shown that the synthetic $SrTiO_3$ could adsorb europium (Eu^{3+}). As a result, $SrTiO_3$ might be utilized in the form of permeable barriers to clean radioactive wastewater (Zhang et al., 2019).

4.5.6 Evaporation Technique

A plant with a 100,000 gallons-per-month capacity was constructed as an example of an evaporator, demonstrating the effectiveness of this tried-and-true technique for decreasing radioactive wastewater volumes. A recently implemented forced-circulation evaporator type was used in a radioactive wastewater treatment plant. The evaporator's top injected steam was accompanied by a heat exchanger that pumped feed water into the unit, and the bottom of the unit discharged condensate. After settling for approximately 26 min, the unit-step response curve's maximum overshoot was determined to have decreased to 25%. Based on evaporation, it looks that radioactive wastewater could have a decontamination factor (DF) of up to 104–106. In fact, according to Zhang et al. (2019), evaporation is a practical solution for managing radioactive wastewater that has a high radioactivity level and a limited volume.

4.6 Conclusion

Environmental preservation is a global problem due to the effects of various contaminants, especially wastewaters, contaminating the environment and human health. It is challenging to design sustainable detection and cost-effective, risk evaluation, and removal programs for all of these components because of the rapid expansion of ECs in the environment. Various techniques and tactics as ion exchange, evaporation mechanism, biochroma, delay tactic, reverse osmosis, ozonation have been employed in the removal of medically radioactive components and various other harmful compounds from waste water. The present study also discusses various case studies related to the radioactive contamination from the wastewater of hospitals. This chapter provides information upon the importance of the storage tanks in the hospitals using radio-immunotherapy because they help in the reduction of the radioactive components in the sewage.

References

Abaroa-Pérez, B., Sánchez-Almeida, G., Hernández-Brito, J. J., & Vega-Moreno, D. (2018). In situ miniaturised solid phase extraction (m-SPE) for organic pollutants in seawater samples. *Journal of Analytical Methods in Chemistry, 2018*, 7437031.

Abd El-Gawad, S. A., & Abd ElAziz, H. (2018). Effective removal of chemical oxygen demand and phosphates from aqueous medium using entrapped activated carbon in alginate. *MOJ Biology and Medicine, 3*, 227–236.

Alkhuraiji, T. S., & Alkhuraiji, W. S. (2019). Detailed study of water radiolysis-based degradation of chloroorganic pollutants in aqueous solutions. *Journal of Hazardous Materials, 368*, 569–577.

Amin, M. T., Alazba, A. A., & Manzoor, U. (2014). A review of removal of pollutants from water/wastewater using different types of nanomaterials. *Advances in Materials Science and Engineering, 2014*, 825910.

Arman, N. Z., Salmiati, S., Aris, A., Salim, M. R., Nazifa, T. H., Muhamad, M. S., & Marpongahtun, M. (2021). A review on emerging pollutants in the water environment: Existences, health effects and treatment processes. *Water, 13*, 3258. https://doi.org/10.3390/w13223258

Aziz, M., & Ojumu, T. (2020). Exclusion of estrogenic and androgenic steroid hormones from municipal membrane bioreactor wastewater using UF/NF/RO membranes for water reuse application. *Membranes, 10*, 37.

Bai, X., Lutz, A., Carroll, R., Keteles, K., Dahlin, K., Murphy, M., & Nguyen, D. (2018). Occurrence, distribution, and seasonality of emerging contaminants in urban watersheds. *Chemosphere, 200*, 133–142.

Banerjee, S. (2013). Radioactive and hospital waste management: A review. *International Journal of Latest Trends in Engineering and Technology, 3*(1). https://www.researchgate.net/publication/349725019

Benner, J., Helbling, D. E., Kohler, H.-P.E., Wittebol, J., Kaiser, E., Prasse, C., Ternes, T. A., Albers, C. N., Aamand, J., Horemans, B., et al. (2013). Is biological treatment a viable alternative for micropollutant removal in drinking water treatment processes? *Water Research, 47*, 5955–5976.

Bokare, V., Murugesan, K., Kim, Y.-M., Jeon, J.-R., Kim, E.-J., & Chang, Y. S. (2010). Degradation of triclosan by an integrated nano-bio redox process. *Bioresource Technology, 101*, 6354–6360.

Bolong, N., Ismail, A. F., Salim, M. R., & Matsuura, T. (2009). A review of the effects of emerging contaminants in wastewater and options for their removal. *Desalination, 239*, 229–246.

Diemert, S., Wang, W., Andrews, R. C., & Li, X.-F. (2013). Removal of halo-benzoquinone (emerging disinfection by-product) precursor material from three surface waters using coagulation. *Water Research, 47*, 1773–1782.

Dulio, V., van Bavel, B., Brorström-Lundén, E., Harmsen, J., Hollender, J., Schlabach, M., Slobodnik, J., Thomas, K., & Koschorreck, J. (2018). Emerging pollutants in the EU: 10 years of NORMAN in support of environmental policies and regulations. *Environmental Sciences Europe, 30*, 1–13.

El-Naggar, M. R., El-Kamash A. M., El-Dessouky M. I., Ghonaim A. K. (2008) Two-step method for preparation of NaA-X zeolite blend from fly ash for removal of cesium ions. *Journal of Hazardous Materials, 154*(1-3), 963–972. https://doi.org/10.1016/j.jhazmat.2007.10.115

Fatimazahra, S., Latifa, M., Laila, S., & Monsif, K. (2023). Review of hospital effluents: Special emphasis on characterization, impact, and treatment of pollutants and antibiotic resistance. *Environmental Monitoring and Assessment, 195*, 393. https://doi.org/10.1007/s10661-023-11002-5

Grassi, M., Kaykioglu, G., Belgiorno, V., & Lofrano, G. (2012). Removal of emerging contaminants from water and wastewater by adsorption process. In *Springer briefs in molecular science* (pp. 15–37). Springer.

Gupta, H. (2016). Photocatalytic degradation of phenanthrene in the presence of akaganeite nano-rods and the identification of degradation products. *RSC Advances, 6*(114), 112721–112727.

Gupta, H. (2018). Anthracene removal from water onto activated carbon derived from vehicular tyre. *Separation Science and Technology, 53*(4), 613–625.

Gupta, H., & Dhiman, S. (2023). Synthetic nanoparticle-based remediation of soils contaminated with polycyclic aromatic hydrocarbons. In U. Shanker, C. M. Hussain, & M. Rani (Eds.), *Handbook of green and sustainable nanotechnology*. Springer. https://doi.org/10.1007/978-3-030-69023-6_102-1

Gupta, B., & Gupta, H. (2015). Iron oxide mediated degradation of mutagen pyrene and determination of degradation products. *International Journal of Environmental Science and Development, 6*(12), 908.

Gupta, H., & Gupta, B. (2016). Adsorption of polycyclic aromatic hydrocarbons on banana peel activated carbon. *Desalination and Water Treatment, 57*(20), 9498–9509.

Gupta, H., & Gupta, B. (2018). Vehicular tire as potential adsorbent for the removal of polycyclic aromatic hydrocarbons. *Polycyclic Aromatic Compounds, 38*(4), 354–368.

Gupta, V. K., Carrott, P. J. M., Ribeiro Carrott, M. M. L., & Suhas. (2009). Low-cost adsorbents: growing approach to wastewater treatment—A review. *Critical Reviews in Environmental Science and Technology, 39*, 783–842.

Gwenzi, W., Mangori, L., Danha, C., Chaukura, N., Dunjana, N., & Sanganyado, E. (2018). Sources, behaviour, and environmental and human health risks of high-technology rare earth elements as emerging contaminants. *Science of the Total Environment, 636*, 299–313.

Hamon, P., Moulin, P., Ercolei, L., & Marrot, B. (2018). Oncological ward wastewater treatment by membrane bioreactor: Acclimation feasibility and pharmaceuticals removal performances. *Journal of Water Process Engineering, 21*, 9–26.

Hussain, A., & Dubey, S. K. (2014). Specific methanogenic activity test for anaerobic treatment of phenolic wastewater. *Desalination and Water Treatment, 52*, 7015–7025.

Jensen, V. B., Darby, J. L., Seidel, C., & Gorman, C. (2014). Nitrate in potable water supplies: Alternative management strategies. *Critical Reviews in Environment Science and Technology, 44*, 2203–2286.

Jusoh, A., Hartini, W. J. H., Ali, N., & Endut, A. (2011). Study on the removal of pesticide in agricultural run off by granular activated carbon. *Bioresource Technology, 102*, 5312–5318.

Katibi, K. K., Yunos, K. F., Che Man, H., Aris, A. Z., Bin Mohd Nor, M. Z., Binti Azis, R. S. (2021). Recent advances in the rejection of endocrine-disrupting compounds from water using membrane and membrane bioreactor technologies: A review. *Polymers, 13*, 392.

Kaur, A., Gupta, H., Dhiman, S. (2023). Biogenic metallic nanoparticles: Synthesis and applications using medicinal plants. In U. Shanker, C. M. Hussain, M. Rani (Eds.), *Handbook of green and sustainable nanotechnology*. Springer.

Khan, S., Syed, A. T., Ahmad, R., Rather, T. A., Ajaz, M., & Jan, F. A. (2010). Radioactive waste management in a hospital. *International Journal of Health Sciences, Qassim University, 4*(1), 39–46.

Khan, N. A., Khan, S. U., Ahmed, S., Farooqi, I. H., Yousefi, M., Mohammadi, A. A., & Changani, F. (2020). Recent trends in disposal and treatment technologies of emerging-pollutants—A critical review. *Trends in Analytical Chemistry, 122*, 115744.

Kinly, D. III. (2005). Chernobyl's legacy: health, environmental and socio-economic impacts and recommendations to the governments of Belarus, the Russian Federation and Ukraine. The Chernobyl Forum.

Krawczyk, E., Pinero-Garcia, F., & Ferro- Garcia, M. A. (2013). Discharges of nuclear medicine radioisotopes in Spanish hospitals. *Journal of Environmental Radioactivity, 116*, 93–98. https://doi.org/10.1016/j.jenvrad.2012.08.011

Kumar, A., & Gupta, H. (2020). Activated carbon from sawdust for naphthalene removal from contaminated water. *Environmental Technology & Innovation, 20*, 101080.

Kumar, N., Kumar, B., Gupta, H., & Gupta, A. (2023). Development and evaluation of cellulose/graphene-oxide based composite for removing phenol from aqueous solutions. *Polymers, 15*(3), 572.

Leusch, F. D. L., Neale, P. A., Busetti, F., Card, M., Humpage, A., Orbell, J. D., Ridgway, H. F., Stewart, M. B., van de Merwe, J. P., & Escher, B. I. (2019). Transformation of endocrine disrupting chemicals, pharmaceutical and personal care products during drinking water disinfection. *Science of the Total Environment, 657*, 1480–1490.

Lim, F., Ong, S., & Hu, J. (2017). Recent advances in the use of chemical markers for tracing wastewater contamination in aquatic environment: A review. *Water, 9*, 143.

Liu, Y., & Wang, J. (2013). Degradation of sulfamethazine by gamma irradiation in the presence of hydrogen peroxide. *Journal of Hazardous Materials, 250–251*, 99–105.

Maddheshiya, R. (2021). Effect of a novel juvenoid fenoxycarb on the pupal-adult transformation in the blowfly, *Chrysomya megacephala* (Fabricius, 1794) (Diptera: Calliphoridae). *Parasitology Research, 120*, 2351–2356. https://doi.org/10.1007/s00436-021-07205-9

Maddheshiya, R., Singh, K. P. (2021a). Efficacy of fenoxycarb on pupal-adult transformation of *Sarcophaga ruficornis* (Diptera: Sarcophagidae). *Journal of Experimental Zoology, India, 24*(1), 353–360. https://connectjournals.com/03895.2021.24.353

Maddheshiya, R.; Singh, K. P. (2021b). Effect of juvenile hormone analogue, fenoxycarb on post embryonic development of blowfly, *Chrysomya megacephala*. *Journal of Experimental Zoology, India, 24*(1), 405–4013. https://connectjournals.com/03895.2021.24.405

Maddheshiya, R., & Singh, K. P. (2021c). Effect of *Lantana camara* essential oil on embryogenesis and postembryonic development of the blowfly, *Chrysomya megacephala* (Diptera: Calliphoridae) following egg treatment. *Invertebrate Reproduction & Development, 66*(1), 16–22. https://doi.org/10.1080/07924259.2021.2016508

Mazhar, M. A., Madhav, S., Ahmed, S., Kumar, P., & Springer, C. (2024). Drinking water chlorination and disinfection by-products: Formation, history, and regulations. In *Drinking water disinfection by-products: Sources, fate and remediation* (pp. 21–34). Springer Nature Switzerland.

Mazzei, R., Piacentini, E., Gebreyohannes, A. Y., & Giorno, L. (2017). Membrane bioreactors in food, pharmaceutical and biofuel applications: State of the art, progresses and perspectives. *Current Organic Chemistry, 21*, 1671–1701.

Munthe, J., Brorström-Lundén, E., Rahmberg, M., Posthuma, L., Altenburger, R., Brack, W., Bunke, D., Engelen, G., Gawlik, B. M., van Gils, J., et al. (2017). An expanded conceptual framework for solution-focused management of chemical pollution in European waters. *Environmental Sciences Europe, 29*, 13.

Musbah, I., Cicéron, D., Saboni, A., & Alexandrova, S. (2013). Retention of pesticides and metabolites by nanofiltration by effects of size and dipole moment. *Desalination, 313*, 51–56.

Nazifa, T. H., Habba, N., Aris, A., & Hadibarata, T. (2018). Adsorption of procion Red MX-5B and crystal violet dyes from aqueous solution onto corncob activated carbon: Adsorption of Procion Red MX-5B and crystal violet dyes. *Journal of the Chinese Chemical Society, 65*, 259–270.

Ouyang, Z., Huang, Z., Tang, X., Xiong, C., Tang, M., & Lu, Y. (2019). A dually charged nanofiltration membrane by pH-responsive polydopamine for pharmaceuticals and personal care products removal. *Separation and Purification Technology, 211*, 90–97.

Palli, L., Spina, F., Varese, G. C., Vincenzi, M., Aragno, M., Arcangeli, G., Mucci, N., Santianni, D., Caffaz, S., & Gori, R. (2019). Occurrence of selected pharmaceuticals in wastewater treatment plants of tuscany: An effect-based approach to evaluate the potential environmental impact. *International Journal of Hygiene and Environmental Health, 222*, 717–725.

Pang, W., Gao, N., & Xia, S. (2010). Removal of DDT in drinking water using nanofiltration process. *Desalination, 250*, 553–556.

Parra-Saldivar, R., Castillo-Zacarías, C., Bilal, M., Iqbal, H. M. N., & Barceló, D. (2020). Sources of pharmaceuticals in water. In *The handbook of environmental chemistry* (pp. 33–47). Springer.

Patel, M., Kumar, R., Kishor, K., Mlsna, T., Pittman, C. U., Jr., & Mohan, D. (2019). Pharmaceuticals of emerging concern in aquatic systems: Chemistry, occurrence, effects, and removal methods. *Chemical Reviews, 119*, 3510–3673.

Plakas, K. V., & Karabelas, A. J. (2012). Removal of pesticides from water by NF and RO membranes—A review. *Desalination, 287*, 255–265.

Prest, E. I., Hammes, F., van Loosdrecht, M. C. M., & Vrouwenvelder, J. S. (2016). Biological stability of drinking water: Controlling factors, methods, and challenges. *Frontiers in Microbiology, 7*, 45.

Puspita, T. D., Moersidik, S. S., Pratama, M. A. (2023). Radiological study of a wastewater treatment plant associated with radioiodine therapy at a hospital in West Java, Indonesia. *Journal of Radiological Protection, 43*(3). https://doi.org/10.1088/1361-6498/ace6ad

Rani, L., Srivastav, A. L., Kaushal, J., Grewal, A. S., & Madhav, S. (2022). Heavy metal contamination in the river ecosystem. In *Ecological significance of river ecosystems* (pp. 37–50). Elsevier.

Rodriguez-Narvaez, O. M., Peralta-Hernandez, J. M., Goonetilleke, A., & Bandala, E. R. (2017). Treatment technologies for emerging contaminants in water: A review. *Chemical Engineering Journal, 323*, 361–380.

Roy, N., Alex, S. A., Chandrasekaran, N., Mukherjee, A., & Kannabiran, K. (2021). A comprehensive update on antibiotics as an emerging water pollutant and their removal using nano-structured photocatalysts. *Journal of Environmental Chemical Engineering, 9*, 104796.

Sathishkumar, P., Chae, J.-C., Unnithan, A. R., Palvannan, T., Kim, H. Y., Lee, K.-J., Cho, M., Kamala-Kannan, S., & Oh, B.-T. (2012). Laccase-poly(lactic-co-glycolic acid) (PLGA) Nanofiber: Highly stable, reusable, and efficacious for the transformation of diclofenac. *Enzyme and Microbial Technology, 51*, 113–118.

Sathishkumar, P., Mythili, A., Hadibarata, T., Jayakumar, R., Kanthimathi, M. S., Palvannan, T., Ponraj, M., Salim, M. R., & Mohd Yusoff, A. R. (2014). Laccase mediated diclofenac transformation and cytotoxicity assessment on mouse fibroblast 3T3-L1 preadipocytes. *RSC Advances, 4*, 11689.

Semião, A. J. C., Foucher, M., & Schäfer, A. I. (2013). Removal of adsorbing estrogenic micropollutants by nanofiltration membranes: Part B—Model development. *Journal of Membrane Science, 431*, 257–266.

Shaoqing, Y., Jun, H., & Jianlong, W. (2010). Radiation-induced catalytic degradation of p-nitrophenol (PNP) in the presence of TiO_2 nanoparticles. *Radiation Physics and Chemistry, 79*, 1039–1046.

Singh, S., Parveen, N., & Gupta, H. (2018). Adsorptive decontamination of rhodamine-B from water using banana peel powder: A biosorbent. *Environmental Technology & Innovation, 12*, 189–195.

Singh, S., Kumar, A., & Gupta, H. (2020). Activated banana peel carbon: A potential adsorbent for Rhodamine B decontamination from aqueous system. *Applied Water Science, 10*(8), 1–8.

Singh, S., Gupta, H., Dhiman, S., & Sahu, N. K. (2022). Decontamination of cationic dye brilliant green from the aqueous media. *Applied Water Science, 12*(4), 61.

Stefanakis, A. I., & Becker, J. A. (2015). A review of emerging contaminants in water: Classification, sources, and potential risks. In *Practice, progress, and proficiency in sustainability* (pp. 55–80). IGI Global.

Stuart, M., Lapworth, D., Crane, E., & Hart, A. (2012). Review of risk from potential emerging contaminants in UK groundwater. *Science of the Total Environment, 416*, 1–21.

Sznajder-Katarzynska, K., Surma, M., & Cieslik, I. (2019). A review of perfluoroalkyl acids (PFAAs) in terms of sources, applications, human exposure, dietary intake, toxicity, legal regulation, and methods of determination. *Journal of Chemistry, 2019*, 2717528.

Tong, D. S., Zhou, C. H. C., Lu, Y., Yu, H., Zhang, G. F., & Yu, W. H. (2010). Adsorption of acid red g dye on octadecyl trimethylammonium montmorillonite. *Applied Clay Science, 50*, 427–431.

Touahar, I. E., Haroune, L., Ba, S., Bellenger, J.-P., & Cabana, H. (2014). Characterization of combined cross-linked enzyme aggregates from laccase, versatile peroxidase and glucose oxidase, and their utilization for the elimination of pharmaceuticals. *Science of the Total Environment, 481*, 90–99.

Touati, K., Gzara, L., Mahfoudhi, S., Bourezgui, S., Hafiane, A., & Elfil, H. (2018). Treatment of coastal well water using ultrafiltration-nanofiltration-reverse osmosis to produce isotonic solutions and drinking water: Fouling behavior and energy efficiency. *Journal of Cleaner Production, 200*, 1053–1064.

Tremblay, L. A., Gadd, J. B., & Northcott, G. L. (2018). Steroid estrogens and estrogenic activity are ubiquitous in dairy farm watersheds regardless of effluent management practices. *Agriculture, Ecosystems & Environment, 253*, 48–54.

Wang, J., & Zhuan, R. (2020). Degradation of antibiotics by advanced oxidation processes: An overview. *Science of the Total Environment, 701*, 135023.

Wöhler, L., Niebaum, G., Krol, M., & Hoekstra, A. Y. (2020). The grey water footprint of human and veterinary pharmaceuticals. *Water Research X, 7*, 100044.

Zakrzewska-Trznadel, G. (2013). Advances in membrane technologies for the 860 treatment of liquid radioactive waste. *Desalination, 321*, 119–130.

Zhang, X., Gu, P., & Liu, Y. (2019). Decontamination of radioactive wastewater: State of the art and challenges forward. *Chemosphere, 215*, 543–553. https://doi.org/10.1016/j.chemosphere.2018.10.029

Zhao, Y., Xiao, F., Wang, D., Yan, M., & Bi, Z. (2013). Disinfection byproduct precursor removal by enhanced coagulation and their distribution in chemical fractions. *Journal of Environmental Sciences, 25*, 2207–2213.

Chapter 5
Endocrine Disrupting Chemicals and Antibiotics in Wastewater

Jahnavi Alwala, Aayasha Negi, Ramchander Merugu, K. V. N. Rajeswari, and S. Suma

Abstract Humanistic hustle has popularised a huge number of contaminants in the emerging world of environment. Presently, the most important micropollutants considered worldwide were hormone disrupting chemicals and antibiotics remarkably affects the living flora and fauna. In the present world of system, the onset and destiny of endocrine disrupting chemicals in this environment has a top solicitude for public health. In the comprehensive and investigative destiny of this situation in environment has a grid being air, water, soil, and biota computes the clear link among the hazards and its effect on humans and primitive world. Endocrine disrupting chemicals were endowed on the surface, ground and drinking water which are found in the greater extent in association with industrial pharmaceuticals effluents, heavy metals, fluorinated compounds, plastic materials, fertilizers, pesticides, herbicides and fungicides. These sewerage discharges the major sources of the therapeutic and illicit drugs, fertilizers, anti-bacterial agents for the production of hormones intended to usage of the agriculture, as personal care products and the most hazardous substances which assist in undergoing various metabolic transformations on human beings, plants and microbes for the release of hormones like hydroxy pregnenolone, dihydroxy testosterone, pregnenolone, phytoestrogens and thyroid hormones. Endocrine disruptors and antibiotics were an emerging pollutant commenced in wastewater interferes with the endocrine system and inhibits the synthesis of biomolecules in humans. Ecological, synthetic decay with few physical, chemical and biological treatments may leads to reformed articles which may or may not have chemical and biological activities. The risk associated activities and substances which are assessed are very important to detect and remove these EDCs continuously for the growth and demand of water necessity by great recycling and reusing methods of waste water. They are known to cause various human disorders which are chronic

J. Alwala · K. V. N. Rajeswari · S. Suma (✉)
Department of Biochemistry, Chaitanya (Deemed to be University), Hyderabad, India
e-mail: dr.ssuma20@gmail.com

A. Negi
Department of Chemistry, IFTM University, Moradabad, Uttar Pradesh, India

R. Merugu
Department of Biochemistry, Mahatma Gandhi University, Nalgonda, India

H. Gupta et al. (eds.), *Emerging Contaminants in Water and Wastewater*, Environmental Science and Engineering, https://doi.org/10.1007/978-3-031-82579-8_5

in nature. The antibiotics which are present in the waste water are ciprofloxacin, norfloxacin, tetracycline, vancomycin and β-lactams. The present chapter describes the various endocrine disruptors and antibiotics which are present in waste water for the existence of current research and literature.

Keywords Endocrine system · Antibiotics · Pollutants · Waste water · Humans

5.1 Introduction

Water is an essential and scarce natural resource. As an assertion for water increase, an effective existence of emerging contaminants in wastewater thrives as a remark for aquatic environments. In spite of that, biological transformation and adsorption towards the biological mass are the most crucial systems for eliminating emerging contaminants (ECs) (Singh et al., 2018; Madhav et al., 2024). Emerging contaminants introduce various chemical components like endocrine disrupting chemicals (EDCs), fire preventives, therapeutics, exclusive care, household cleaning, lawn care, and agricultural products. These components will unanimously be biomagnifying the food lattice for different hazards to the human hormone system with the status of biological cycles in the environment (Drewes et al., 2001). The rigorous principles involved implementing the regulation of concentrations in various, agricultural waters and in the environment for the general effects of limited information interaction and for their actual toxicological impact. The chemicals that were an emerging contaminant passed through certain principles (Noguera-Oviedo et al., 2016; Khan et al., 2023) by various compounds with unfavourable effects on public well-being. Different emerging contaminants have specific adverse effects to the environment (Gupta, 2018; Gupta & Gupta, 2016, 2018; Singh et al., 2020, 2022).

The most prospective residential proximity and factor for the agricultural activity was often describes the abnormalities for the development and epidemiological studies on underweight infants (Xiang et al., 2000), foetal demise (Bell et al., 2001), and preadolescence cancers (Reynolds et al., 2002). On contrary to the high prevalence in cryptorchidism and hypospadias (Kristensen et al., 1997; Carbone et al., 2006) were in the places of large-scale farming and fertilizers usage on children and female workers in garden (Weidner et al., 1998). Recently the most prominent relationship between cryptorchidism and persistent pesticide concentration has been reported in maternal breast milk (Skakkebaek et al., 2001, 2002; Damgaard et al., 2006).

Emerging Contaminants like Pharmaceuticals, Personal care products, Pesticides as an Endocrine disrupting compounds (EDCs) compounds were described (Shaheen et al., 2022; Kim et al., 2016) based on their availability of motive. Universally accepted anthropogenic chemicals energetically contaminates the natural waters quite affectively in the aquatic ecosystem for high toxicity and persistence (Diamanti-Kandarakis et al., 2009). EDCs may be defined as "an exogenous (non-natural) chemical, or mixture of chemicals, that interferes with many aspects of hormone action."

Hormones are naturally secreted chemicals produced by the endocrine cells within endocrine glands present all over the body.

Endocrine secreting hormones regulates the development of secretions frames blood, bones, brain, and other target tissues. As per the biological survey, an individual develops the diverging changes and requirements of each organ and tissues. Endocrine hormones regulate the body's response on the lower concentrations for the regulation of reproductive functions, developmental activities of the body and nutritional requirements for the hunger, starvation and obesity.

Antibiotics, the valuable successful remakes with huge mutations of human well-being in this whole worldwide as an extensive usage for improvement of human well-being to kill or inhibiting the growth of bacteria. The term "Antibiotic" refers to "antibiosis," which biologically means "antilife," which defines by every microorganism for toxicity to microorganisms, inhibition and growth and killing of various microorganisms at lower concentrations. Antibiotics are known to be called as an antimicrobial drug which may be used for the treatment of flora and fauna infections and with a wide-spread diseases in animal farming. In this contemporary world, there were a wide-spectrum antibiotics which are universally and adequately in use.

The potential EDCs and antibiotics their modes of action with adverse effects and their analysis on the human health, biological detection and chemical evaluation methods, for experimental findings and survey on endocrine disruptors in humans. The main objective for this present study in this chapter is regarding the various types of Endocrine disrupting hormones and antibiotics with their mode of action at target site or tissues on the biological systems. The adverse effects of the EDCs and antibiotics present in the waste water and also in all the bioaccumulated compounds inside the human bodies and the hormonal imbalances with effects like mental retardation, improper foetal growth, cardiovascular diseases, diabetes mellitus, kidney dysfunction and also cancers. This chapter interpret all the information about the hazardous eluents of EDCs and antibiotics in wastewater and also from the various sources of contamination. These compounds have to be activated by various treatments for the complete exhaustion from the various bioremediation, biodegradation, decomposition and physio-chemical methods of the bio transformed materials inside the human bodies. In spite of intriguing, studies on endocrine dysfunctions brought by EDCs are challenging for the investigations on adverse effects of EDCs on humans. It's been a vast difficulty to assess the impact of human exposure on EDCs as a hazardous effect on the development inherently and revealed later stages in few people. Despite undoubtedly these investigations represents a new intriguing field of science research.

Human Endocrine System

Human endocrine system with array of glands secretes hormones and distributes throughout the body. Hormones were naturally secreted chemicals which are secreted by endocrine cells inside the specific endocrine gland and targets at specific site in blood to reach a targeted organs for the proper functioning (shown in Table 5.1). Accordingly at binding site of the hormonal receptors at specified receptors, triggers a response for production of various hormone in metabolism for the behavioural response at its target site. Some endocrine glands produce a single hormone, while others produce multiple endocrine hormones. For example, parathyroid gland produces a single hormone (parathyroid hormone), pituitary gland produces eight or more hormones like prolactin and growth hormone. Prolactin helps in secreting breast milk which synthesises and releases milk in lactating women for infants by pituitary glands. By disparity with all the other endocrine hormones, growth hormone is synthesized throughout life which is important for growth and development in childhood to build and maintain the proper muscles in the human body and the skeleton in adulthood. The pancreas are the best examples for non-endocrine functioning for the secretion of insulin and circulation into the bloodstream which is necessary for normal regulation of blood sugar and digestive enzymes produces the enzymes in the digestive tract may not release into the blood and not the similar to endocrine hormones. Evidently, these endocrine systems with their functions makes complex and diverse at each gland and secretes hormone and plays the unique roles in healthy and well-being. As the major role played by this endocrine system of the many important biological and physiological functions impairs diseases leading to health disorders and even may leads to death. Various hormonal disorders like infertility, growth disturbances, sleep disorders and many acute and chronic diseases which are released in appropriate quantities may change environment to enable the healthy life.

The chemical content has its three-dimensional nature of all endocrine hormone is unique and has its receptor for the localization of the target cells. EDCs can interfere with all of these endocrine signalling pathways and functioning may have the consequences on endocrine functioning. In addition to these interferences, EDCs which are not naturally secreted hormones for ability and affective multiple hormones signalling pathways and it is likely to disrupt the variety of EDC of multiple and single endocrine functions, with a widespread consequence on the biological processes that are controlled by endocrine glands. The most privileged improvement of EDC related information is crucial for global regulatory measures by standardised and streamline manner (usage, disposal and treatment) to ensure better protection for citizens and the environment.

Naturally manufactured chemicals which includes the Endocrine Disruptor Compounds (EDCs) to reach the aquatic environment at concentrations ranges from ng/L to g/L for the most common resources of water micropollutants. Extreme toxicity recognizes widely as a severe threat to the environment and humanity (Gogoi et al., 2018). The most prevalent household detergents, disinfectants, plastics and pesticides contain chemical ingredients which enters the body for the developmental, reproductive and disrupts the hormones for causing adverse effects with in the endocrine system, brings out the destruction of humans and wildlife.

Endocrine disrupting agents leads to tumours of cancerous specific for the organ level, genetical defects during developmental stages of foetus and infants like reproductive system, brain, immune system etc. (Vieira et al., 2021). EDCs were the substances present in the environment interferes the hormone biosynthesis which causes a diverse effect from normal homeostatic regulation (Birch et al., 2015). These compounds are widely used in industrial applications and human exposure was widespread which occurs via ingestion, inhalation and dermal contacts. Endocrine

Table 5.1 Major endocrine hormones

S. No.	Endocrine gland	Tissue or target	Hormone secreted	Functions
1	Pineal gland	Brain	Melatonin	Circadian rhythm and regulates body temperature
2	Hypothalamus	Brain	Pituitary hormones	Stimulates or inhibits the pituitary hormones
3	Pituitary gland	Anterior pituitary gland	Growth hormone	Growth and development
			TSH	Activities and growth of thyroid gland
			ACTH	Regulates the activities of adrenal gland
			FSH	Stimulates the production of oestrogen in females Stimulates the production of spermatozoa
			LH	Oestrogen and progesterone in females Testosterone in males
			LTH/prolactin	Prolactin secretion in lactating women
		Posterior pituitary gland	Oxytocin	Contracts uterus during delivery Secretion of milk from mammary glands
			ADH/vasopressin	Decreases urine output Increases blood pressure
4	Thyroid gland	Thymus/lobes of neck region	T3 and T4	Increases the oxygen consumption Complete body regulation and activities
5	Parathyroid gland	Interior lobes of thyroid gland	Parathormone	Regulates the calcium levels in the body

(continued)

Table 5.1 (continued)

S. No.	Endocrine gland	Tissue or target	Hormone secreted	Functions
6	Adrenal gland	Adrenal cortex	Glucocorticoids and mineralocorticoids	Influences the carbohydrate metabolism and protein synthesis, etc. (glucocorticoids) aldosterone regulates water and electrolyte balance (mineralocorticoids)
		Adrenal medulla	Epinephrine Non-epinephrine Dopamine	Increases the heart rate
7	Pancreas	Under the stomach	Glucagon Insulin	Regulates the blood sugar levels
8	Ovaries	Female gonads	Oestrogen	Control of ovulation and secondary sexual characteristics
			Progesterone	Formation of uterine lining for receiving embryo
9	Testes	Male gonads	Testosterone	Development of sperm production and secondary sexual characteristics
10	Thymus	Thymus	Thymosin	Development, production and maturity of WBCs

disruptors mimic their adverse defects on natural hormones in human body and also, they are considered as the major health effect growing day by day on human health.

EDCs from various resources makes the exposure to humans in several ways like breathing, eating, drinking and through the skin. In the Fig. 5.1, shows a source of EDCs with their concern on environment affecting and on which components, biotransformation and human exposure to EDCs on consumption of food variables and drinking water including behaviour and adverse effects with these EDCs on human body.

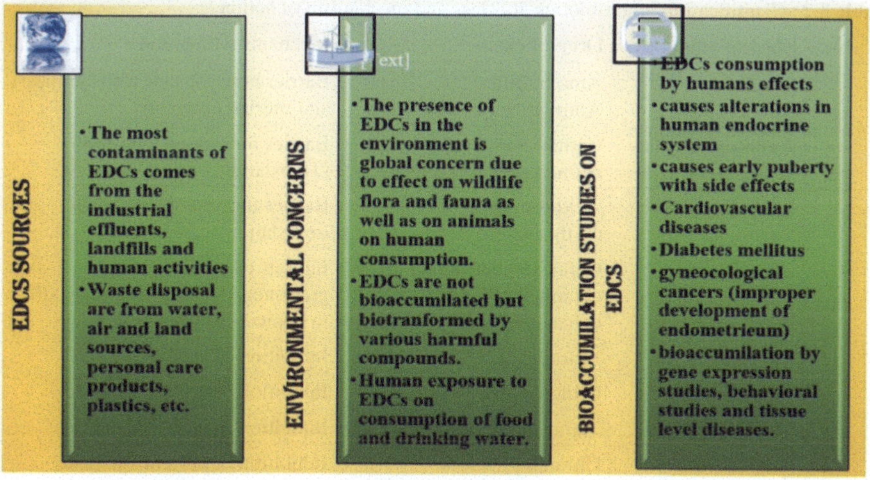

Fig. 5.1 EDCs sources and occurrence in the environment

5.2 Antibiotics

Antibiotics are considered as the alkylating agents which releases in huge varieties into natural ecosystems at lower concentrations as stable antibiotic precursors and metabolic compounds which are conventional in water treatment systems and ecosystem.

Antibiotics are classified into many aspects based on the mode of action, chemical structure and entry into the body. The direct discharge of antibiotics into the waste water sludges by pharmaceutical industries and hospitals and not for effecting the wastewater treatment plants and landfills. Existence of these antibiotics in waste water is considered as a pollution (Peng et al., 2009) and is used as the sludge for the fertilizers as primitive source of antibiotics in agricultural soil and water irrigation systems (Park et al., 2007) and it might cause allergic and toxic reactions for the development of antibiotic resistance in human beings (Reddersen et al., 2002).

Industrial effluents are known to have the human welfare for the economy in modern commodities for the adverse effects and detects the worldwide contamination in the natural resources. Antibiotics play an emerging role for biological abnormalities which were estimated to be as remark for the biological life. These antibiotics which are present in the environment at higher concentrations make the remarkable cause and lower concentrations causes the effect on flora and fauna.

The growth and release of these antibiotics and its genes into the ecosystems gains the huge privilege extraordinarily and for impacting on an ecosystem attains the active pace of the interactions with these compounds for environmental significance. Antibiotics are further classified to emphasize their apt usage as a tool to be better

Table 5.2 Classification of antibiotics, its class and mechanism of action

S. No.	Class of antibiotic	Drug of class	Mechanism of action
1	β-lactam	Amoxicillin Ampicillin	Barrier between cell wall synthesis and murine assembly
2	Amino-glycosides	Gentamicin, kanamycin, neomycin	Barrier for protein synthesis (Translation)
3	Macrolides	Azithromycin, erythromycin	Barrier for protein synthesis (translation)
4	Cephalosporins	Cefalexin, cetirizine, cefamandole, cefmetazole, cefcapenepivoxil, cefixime	Inhibits the growth of gram-negative bacteria for binding to penicillin and its proteins
5	Monobactams	Aztreonam	Inhibition of cell wall
6	Lincosamides	Clindamycin	Inhibition of protein synthesis
7	Tetracyclines	Tetracyclines, doxycycline	Inhibition of protein synthesis
8	Amphenicols	Chloramphenicol	Inhibition of protein synthesis
9	Fluoroquinolones	Ciprofloxacin, norfloxacin, levofloxacin	Inhibition of DNA replication
10	Glycopeptides	Vancomycin	Inhibition of murine synthesis and assembly
11	Trimethoprim	Trimethoprim	Inhibition of beta-lactamase
12	Carbapenems	Biapenem, doripenem	Peptidoglycan inhibition
13	Rifamycin	Rifaximin, rifamycin, rifampicin	Bacterial DNA polymerase inhibition
14	Nitro-derivatives	Fusidanos, phosphonates, pleuromulins, diamino pyridines	Inhibition of DNA, RNA and protein synthesis

supportive antibiotics into this completely polluted world (shown in Table55..2) which was represented in detail (Adzitey et al., 2015).

5.3 Endocrine Association with Human body and Their Adverse Effects

EDCs are mixed chemicals that can interfere with the hormonal framework of the endocrine system of the human body. Some EDCs act like "hormone mimics" and cover the human body for hormonal balance. Few EDCs access or decline the matched levels of hormones in the bloodstream with adverse effects and secretions in our body. Few EDCs vary with the sensitivity of our human hormone system. EDCs may disrupt various hormones in the human body that are adversely affecting the human health for the fallout alterations and parameters in the human body like traits of sperm and fertility, abnormal functions of sex organs, endometriosis, early puberty retarded, nervous system functioning, immune functioning, certain cancer detection and effects on the human body, respiratory problems, metabolic related issues, diabetes mellitus, obese related issues, cardiovascular problems, growth, neurological and mental disabilities with speech and hearing disorders, etc. A potential connectivity between these substances which are encountered per day and medical conditions related from cancer and low levels of IQ in children can be complicated for the powerful knowledge levels setting. Endocrine disruptors or disrupting chemicals interfere with the action of hormones in the human body focusing on the whole rather than the other chemicals that are adopted or affected by the polluted environment. The adverse effects of this pollution may be particularly classified as human carcinogens. They are more effective in agricultural commodities who consumes more among infants and children in rural areas and very low levels in adult population in urban areas.

5.4 Adverse Effects of Endocrine Disruption According to US Environmental Protection Agency

In the past few decades, there has been a spreading of information about the hazardous effects on humans. Enduring the evidence of these chemicals causes the effects on wildlife limits the evidence and potentiality of chemicals on humans with their adverse effects, which may be environmental exposure levels. Current standardized testing methods do not provide a competent data for the identification of endocrine disruptors (EDs) or to assess their risks to humans and wildlife.

5.5 Disruption of the Endocrine System

Environmental contaminants disrupt the endocrine system, which leads to adverse effects on health consequences, as their scientific research on human beings, animals, aquatics and wildlife epidemiology shows the drastic changes.

It is important for the improved empathetic, specified concentrations of chemicals in the environment cause adverse effects. Numerous varieties of scientific studies explain epidemiology, mammalian, and ecological toxicology, making it necessary to determine the ambiguity on endocrine disruption, which undergoes modulation for the governing agencies, industries and academic records.

Disruption of the endocrine system occurs in different ways. Some chemical compounds mimic naturally for the specific hormones trifles the human body restoring up to the stimulus on a growth hormone that results in increased muscle mass or restoring at unseemly produces the insulin.

5.6 Case Studies on Endocrine Disruption on Living World

The most devastating consequences on the exposure of endocrine disruptor on humans, animals as the potent cause for the drug diethylstilbesterol (DES), as a synthetic oestrogen. Erroneously the doctors prescribed DES to millions of pregnant women for the blockage of spontaneous abortion and promote foetal growth. During the developmental stages of children at puberty and vaginal cancer.

The regulatory drugs and various chemicals have been improved by Congress as it was established as an endocrine disruptors screening program with high significant values.

Humans, domestic animals, aquatic species and wildlife species have shown evidence for growth and exhibited adverse health defects from exposure to environmental chemicals which interacts with endocrine system.

Lower levels of endocrine disruptors which effects on exposure and evaluation on the aquatic organisms in the waste water systems. Higher levels of various contaminants report the adverse effects on humans.

Over the last four decades, several conflicted reports have concerned for the decrease the quality of sperm fertility in humans has been reported and increased in cancers like breast, prostate and testicular cells. The endocrine effects the environmental factors spectacularly cause the specific possibility for these endocrine related chemicals. Endocrine functioning with minute disturbances as consent of particular function and highly sensitive stages like development, pregnancy and lactation leads to the profound and lasting effects of life cycle.

5.7 Endocrine Disruptors and Antibiotics Functions by Various Commodities in Action

Mimicking naturally secreted hormones of the human endocrine system such as oestrogen, testosterone for the various chemical reactions mechanisms of hormone balance. Blockage of secondary hormone receptors and their cells action. The composite organisation, anabolism, catabolism and the whole metabolic processes of various hormones with their mechanisms alters the concentrations of naturally synthesised hormones of the various glands of endocrine system. The provoke linking attention on the endocrine disrupting hormones deficit on the metabolic disorders of diabetes, cardiovascular disease, hyperactivity disorders, Parkinson's and Alzheimer's diseases and acute disorders like obesity, early puberty, infertility and other reproductive disorders, human cancers including various metabolic dysfunctions.

The most adversely affecting endocrine disrupting hormones and familiar for us in our daily life are lead on shampoo bottles, lids of coffee cups and ingredients of our friendly pet's lawn fertilizers, etc. The most important endocrine disrupting hormones and antibiotics which are present in wastewater are as follows.

5.8 Description of Various Compounds with Sources, Adverse Causes and Preventive Measures

1. **BPA-Bisphenols**: These are the most commonly used materials in our daily lives like all types of hard and soft plastic bottles and sports water bottles, metal canned food containers, baby pacifiers and toys with foreign made trinket toys, printed receipts from banks and grocery stores and labelled polycarbonate plastics (Konieczna et al., 2015). BPA mimics the estragon hormone in human endocrine system. These BPA compounds are about 94% in our bodies. So according the survey we need to prefer the "BPA free labelled" bottles or containers for the usage or else we can use glass bottles always.
2. **Perchlorates**: These have combination of chemicals like chlorine and oxygen which is found in waste water, fireworks, rocket fuels and perchlorate salts in food processing industries (Srinivasan et al., 2009). Perchlorates causes the declined levels of iodine in thyroid, metabolism, brain functioning and internal organs functioning also. The most prevalent risk factors are avoided by having filtered drinking water (under reverse osmosis system), using iodized salts and also buying organic foods for our daily usage makes the disruption of these EDHs.
3. **Dioxins**: These are the organic pollutants in the environment and they are specific in the waste water and industrial and manufacturing wastes. They are chemically known as persistent environmental toxic pollutants quickly settle in water, soil. Dioxins mostly causes damage to tissue levels for all the systems in living organisms like immune system, cancer, diabetes, delays sex hormones signals and damages the reproductive processes like decline the sperm quality and count (Birnbaum et al., 1994). The most effective way to avoid dioxins are properly cooked animal food and raw plant food is advisable for the human life style.
4. **Atrazine**: These are herbicides used in the agricultural fields, roadsides and railroads and considered as environmental fate. These compounds are washed away from soil to steams upon affects the drinking water sources and wildlife too. These compounds on direct exposure cause skin damage, during pregnancy (Mole et al., 2015; Huang et al., 2016). Atrazine disrupts the hormone synthesis for ovaries and testicles delaying the puberty, causes breast and prostate cancer. These atrazine compounds are prevented by drinking water (good filtration system), avoiding pesticides and herbicides and usage of organic foods.
5. **Phthalates**: These chemicals are found in plasticizers like transparent bottles, detergents, deodorants, nail polishes, body lotions, aftershave lotion, and polyvinyl toys and floors, in some consumer goods like dibutyl phthalate, dimethylphthalalte, diethyl phthalate, diethyl hexyl phthalate. These compounds disrupt the endocrine system of the human body and blocks the testosterone and leads to the abnormality in the development of foetus, birth defects like hypospadias, early puberty in girls and insulin resistant. These chemicals usage is prevented by using plastics labelled with phthalate free,

avoiding perfumes, fragrances, cosmetics, PVC toys and consumer goods, avoid using the food products of animals and plants largely sprayed with these compounds and prefer organic foods.

6. **Chlorine compounds**: Chlorine is a toxic green gas and found in the bleaching detergents, insecticides, dyes, paints and also in the swimming pools and hot tubs, paper products chlorine is mostly used. This chlorine mostly affects the interference with production of testosterone levels, highly levels of it causes the heart failure and pulmonary oedema and with mild levels causes kin irritation, nausea, breathing disorders (Nickmilder et al., 2011; Grandjean et al., 2014; Everett et al., 2011). Avoid using the swimming pools with chlorine and water also with chlorine levels.
7. **Fluoride compounds**: Fluoride compounds are most dangerous found in leaching from pipelines of drinking water, tooth pastes and mouth wash bottles, Teflon pans mechanically processed foods, dental care products. These fluoride compounds cause disruption in endocrine system by displacing the iodine levels in the cells which leads to thyroid dysfunction like grave's disease. Fluoride is neurotoxin which affects the child development (James Deal et al., 2013; Fängström et al., 2005; Hong et al., 2014). Fluoride can be avoided by adding more content in water pipelines and using Teflon coated vessels for cooking, tooth pastes, mouth washing liquids and use the reverse osmosis systems for drinking water filtration.
8. **Fire retardants**: These fire retardants are used in every office, homes and schools as well for the sake of the carpeting, furniture and complete interior designing works, car parts, polyurethane foam. These compounds are most dangerous endocrine disrupting compounds which affects mostly thyroid dysfunction, decline in IQ, brain damages, cancer and reproductive problems (Eskenazi et al., 1999). So, these are avoided by decreasing the usage of polyurethane foams and furnished architecture in all work places.
9. **Lead compounds**: These compounds found in the paints, old pipelines with paints, wood works, antique toys and furniture, gasoline, imported candies, furniture. These compounds cause huge damage endocrine system like sex hormones and body tissues with mood disorders like depression and anxiety, high blood pressure with uncontrolled stress, brain damage, miscarriages and decreased IQ. These compounds can be avoided by eating processed foods and filtered water through a reverse osmosis filtration system.
10. **Arsenic compounds**: Arsenic compounds are found in the environment and contaminants in waste water as inorganic compounds in many areas, including pesticides, rat poison, glass treatments, cigarettes, mining activities and burning fossil fuels (Murata et al., 2007). This arsenic exposure or intake indirectly from different food sources leads to harmful diseases like cancer, respiratory, neuro, immune, gastrointestinal and reproductive dysfunction for the human body. It also disrupts the endocrine system which relates to the glucocorticoid system resulting in the uncontrolled weight gain, diabetes and developmental problems. Arsenic can be avoided by using water purifier system which can filter out arsenic and having grains in daily diet.

11. **Organophosphate Pesticides**: These compounds are found in insecticides and pesticides; they are confirmed as the pest control products in the agricultural fields and garden. These compounds destroy nervous and endocrine systems and causes the cancers like pancreatic, thyroid and adrenal glands (Afrifa et al., 2018). Organophosphate pests are avoided completely in our daily life as pest control products.
12. **Mercury**: Mercury is a toxic chemical present in underground water, seafood, fluorescent bulbs and batteries. This mercury disrupts the endocrine system and dangerous to pregnant women, brain damage and delayed the baby development (Rice et al., 2014). Endocrine system disrupts away the hormones like adrenalin, insulin and testosterone and also the disorders of menstruation, ovulation, mood defects, suicidal thoughts (pituitary problems), thyroid and diabetes. Mercury should be avoided by farmed fish and also seafood.
13. **Perfluorinated chemicals PFCs**: PFCs are found in nonstick cookware, interior furniture, paper wrappers and plates, carpets cleaners and beauty products. These compounds cause the defects like decline in sperm count, fertility problems in women and risk factor (Kjeldsen et al., 2013). PFCs can be avoided by nonstick and Teflon coated cookware and stain-resistant, water repellents, non-sticky coatings and paintings.
14. **Glycol Ethers**: Glycol ethers are found in cleaning products of floor, oven, glass, liquid, paints, varnish for wood stain and lacquers for wood and metal finishing, dyes, perfumes for fragrance and cosmetic naturally and synthetic products. These ethers on long term exposure cause decreased sperm count for men and frequent miscarriages and undeveloped foetus in women with anaemia and autoimmune disorders. But sudden exposure of the compounds leads to liver and kidney damage. These compounds can be avoided by using vinegar, baking soda and hydrogen peroxide.

5.9 Antibiotics in Wastewater

Antibiotic-based pollution is being found in soil/water resources/sewage (Kemper et al., 2008; Chang et al., 2015; Granados-Chinchilla et al., 2017; Mohanan et al., 2022) which has a harmful threats for human health and ecosystems as thus it causes bacterial resistance and makes treatment against any infection ineffective (Mahamallik et al., 2015; Daghrir et al., 2013; Paramonov et al., 2022; Kim et al., 2011). Antibiotics are the considered as the marginally influencing compounds to be present in the waste water for the organic matter degradation. These antibiotics cause destruction of digestive flora and fauna, bacterial bloating as a common one. There are multiple antibiotics which have the metabolites belonging to five various classes like Tetracyclines, β-lactams, quinolones, macrolides and sulphonamides commonly found in activated sludge, sewage, digested sludge, hospital waste and effluents as pseudo-persistent pollutants (Chen et al., 2015; Burke et al., 2016) which are discharged into water channels was shown in Fig. 5.2. Among them three classes

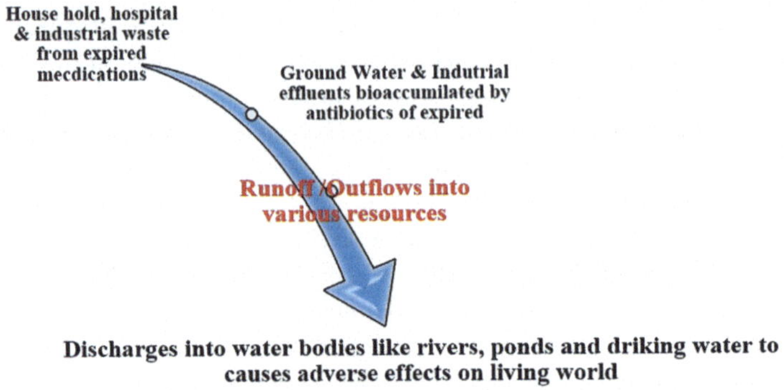

Fig. 5.2 Antibiotics contamination through outflows of various wastewater sources

are majorly present in wastewater systems. They are sulphonamides, tetracyclines and β-lactams and vey finite quantities of ciprofloxacin, norfloxacin, vancomycin.

5.10 Adverse Effects and Importance of Antibiotics Present in Wastewater

1. **Sulphonamides**: Sulphonamides inhibit bacterial folic acid synthesis, have broad-spectrum variability against staining of the different groups of bacteria for usage and treatment of various infections (Rodriguez-Mozaz et al., 2020; Wang et al., 2019; Seydel et al., 1968; Supuran et al., 2017). Sulphonamides which were known as sulpha drugs are a group of synthesised medicines used to treat bacterial infections, Crohn's disease, diabetes, gout, fluid retention, high BP, rheumatic arthritis. It has side effects like gastrointestinal, headaches, itchy skin and rashes, candidiasis (Supuran et al., 2003; Giles et al., 2019; Hruska et al., 2012; Khan et al., 2018). It also has adverse effects which comes from the wastewater sludges, tap water and also from the improperly filter drinking water allergies, hematologic, renal, and gastrointestinal effects. These can be avoided by drinking filtered water and synthetic drugs (Khan et al., 2019; Yousef et al., 2018; Zawodniak et al., 2010).
2. **Tetracyclines**: Tetracycline, broad-spectrum antibiotic in the living world, has activity against various bacterial infections (Wang et al., 2014; Biošić et al., 2017; Begum et al., 2021) and makes it effective in human therapy (Wang et al., 2020; Grossman et al., 2016) and veterinary medicine. Tetracycline which are capable of accumulating the food chain and causes toxicity to microbial world in waste water, restores the advanced developments and spread of antibiotic resistance genes for creating threats in drinking and irrigation water and these antibiotics disrupts the microbial flora in the human intestine.

3. **β-lactams**: These are known as beta-lactam antibiotics which are a very crucial division of antibiotics that contains a ring. Their mechanism of cell wall biosynthesis which operates against the various types of bacteria. The most common adverse effects of these b-lactams antibiotics which comes from drinking tap water and river include fever, nausea, vomiting, diarrhoea and rashes and also urticaria. Some of the severe adverse effects are erythema, dermatitis, angioedema and pseudomembranous colitis (Cai et al., 2007), painful inflammation and allergic reactions with anaphylaxis rarely.
4. **Ciprofloxacin**: Ciprofloxacin is a fluoroquinolone antibiotic which helps in usage for the treatment of variety of bacterial infections like cystitis, urinary tract infections (UTIs), prostrate infections, lung infections like bronchitis and pneumonia, skin infections, bone and joint infections and gonorrhoea, plague and anthrax viral infections shows the effectiveness and some shows resistance towards it. The most effected adverse effects are nausea, vomiting, diarrhoea and liver functioning disorders. These all dysfunctions can be avoided by drinking tap water and unfiltered water from the river.
5. **Norfloxacin**: Norfloxacin is an active class of medicine in the field of antibiotics. These are active against bacteria like enteric pathogens and penicillin resistant gonorrhoeae. This norfloxacin has severe side effects like pancreatitis, stomach pain, back pain and abnormal weight gain and also kidney problems, sweeling of mouth, face and throat. Thus, these effects will be generally caused due to the improperly filtered water (reverse osmosis filtered system).
6. **Vancomycin**: Vancomycin are glycopeptide antibiotics used for the treatment of bacterial infections like dermatitis, circulatory infections like swelling in endocardium and myocardium, bone and joint infections, and meningitis. So, these antibiotics are mainly meant for possible for sake of direct resistance. Though these antibiotics are indirectly affecting humans from the environment by having direct tap water and unfiltered water.

Antibiotics play a crucial role in human and animal healthcare which helps in preventing, alleviating, and curing diseases. In spite of all these, execution makes the significant segment for these drugs were discharged and stable into water bodies. The ultimate fate of antibiotics remains largely unknown and research is in progress.

5.11 Analysis and Detection Methods of EDCs

These EDCs can be are the traditionally analysed by using techniques high-performance liquid chromatography, mass spectrometry, ultra-performance liquid chromatography with mass spectrometry, MALDI-TOF and gas chromatography and time-of-flight in combination with mass spectrometry (Castillo-Zacarías et al., 2021; Chabchoubi et al., 2021) (Fig. 5.3).

These techniques are highly sensitive for the detection and reproducibility for various methods of sample preparation with varied volumes of the samples, long

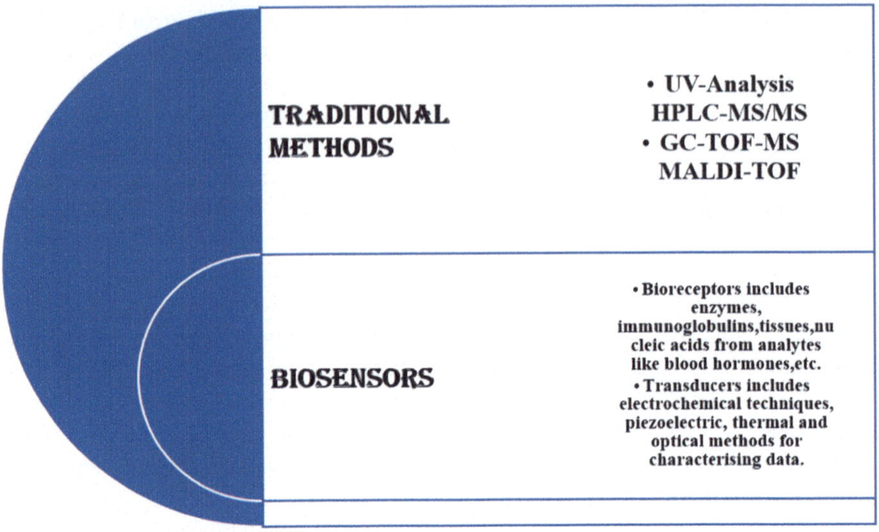

Fig. 5.3 Detection methods for EDCs

term analysis, elongated consumables, and equipment are required promptly. Electrochemical sensors reported a new rapid procedure and application economically. Biosensors are the advanced and developed by using cells with specific molecules, single stranded oligonucleotides targeted at specific site, antibodies, DNA, and model organisms which were used for the application (Jaffrezic-Renault et al., 2020; Kaya et al., 2020; Lu et al., 2020). In recent studies has developed an electrochemical method for the detection, quantification and an electrochemical sensor for the development of covalent organic frameworks charges and discharges with specified detection limits (Pang et al., 2020). EDCs especially focused on analogy and oestrogens.

Consequences for the removal of Endocrine Disruptors in Waste Water Treatment Plants may be summarized. These disruptors can be specifically degraded by high efficiency methods like basic treatment stages like primary, secondary and tertiary treatments (Awad et al., 2019; Zamri et al., 2021) shown in Fig. 5.4.

5.12 Analysis and Detection Methods of Antibiotics

Most of the Antibiotics persistently reaches the water bodies (Singer et al., 2016; Carvalho & Santos, 2016; Segura et al., 2015) unaltered as active pharmaceutical ingredients which can transform into metabolites through natural bioaccumulation. Based on those natural processes in the environment, various technologies are used for treating the antibiotics in wastewater by physical, chemical and biological processes which was time-consuming, ineffective, or inadequate for emerging

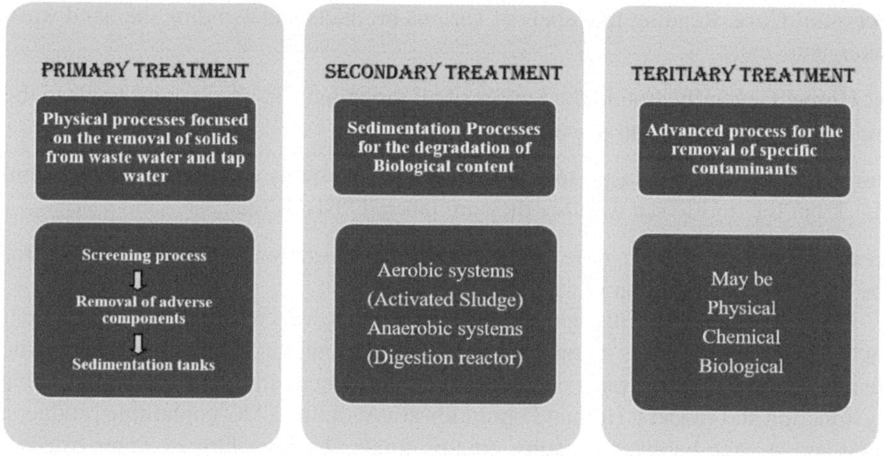

Fig. 5.4 High efficiency methods for waste water treatment plans

contaminants (Guo et al., 2016). The most innovative approaches are advanced oxidation processes as an expensive and energy-intensive which produces reactive and unstable oxidants may not effectively degrade the refractory contaminants. The key processes which include in the treating of antibiotic pollutants are electrochemical techniques like oxidation, reduction, coagulation, electron-Fenton, photoelectron-Fenton and Sono-electrochemical.

5.13 Best Practices to Avoid Disruption of Endocrine System by EDCs and Antibiotics Competence

Endocrine-disrupting chemicals (EDCs) and antibiotics can have significant impacts on our health, and it's essential to take steps to minimize exposure. Some best practices to avoid disruption of the endocrine system by notified 14 endocrine disruptors and antibiotics can be avoided by practicing and eliminating the harmful compounds from current lifestyle. Some of the health defects which are not fully proven takes them to a precautionary wise becomes more familiar with EDCs to the families who are exposed frequently. The following is a list of precautionary steps to be taken to minimize EDC exposures.

Food stuffs and water: Consulting the safeguard materials to consume, cooking fat to drain, thorough washing of veggies and fruits.

Physical activity: Quality of oxygen levels should be measured within customized places.

Personal Care: Reading the labels of various products and avoiding the hand wash detergents, etc.

At Home: Generally around the houses of all individuals in all the areas to clean the wells for replacement and discarding.

For Children: Avoid plastic toys for the usage of infants and formula bottles and check out the toys to see whether they are labeled "BPA-Free".

Transportation: Choosing the options in the transportation to transit the routes for the limiting time variation in traffic.

Thus, these endocrine disruptors which are present everywhere in the environment including drinking, Waste water, industrial wastes, nontoxic repellents leading to the disruption of the endocrine systems with consequences like thyroid, reproductive dysfunction and cancers. Human Population are Avoiding EDC-containing products, food and storage habits and eating organic foods, drinking filtered water (reverse osmosis system), Avoiding the plastics, cosmetics and beauty products, Keep the environment clean by reducing the chemicals in residential places and choose HEPA filters in industries, Intake of animal fats and seafoods should be decreased, Wash your hands often.

The challenging potentiality regarding the health risks from EDCs will be assessed by their endpoints and those risk factors for cancer risk will be quantified. Some assays are commonly used studying EDCs may be labeled, which are not in regulatory practice. Various compounds are neurotoxic for a few compounds like organophosphate pesticides and their regulation at the point of departure (cholinesterase inhibition) and the initiating point regulatory limit. EDCs remain in the environment for many years to build our bodies and for the constant exposure. Research reveals numerous EDCs are most commonly tested for adults as well as newborns.

5.14 Conclusion

The main prototype for these hazardous eluents originated in wastewater from pharmaceutical industries, agricultural products, transformations. Bioremediation, biochemical degradation, biochemical decomposition and various physio-chemical biological treatments provides the transformed materials. The importance of detecting and degradation of Endocrine disrupting Compounds continues to expand and increase the necessity of water for the recycling and reuse of wastewater and their wastewater treatments. The multiple storage of these EDCs is treated taking into consideration as concern for the human accumulation. The complete removal these EDCs in the wastewater treatment technique will be done in unit level representation.

In the present future, advanced technologies will increase the availability of more evaluated risks will be recognized. The regulatory agencies are likely to evaluate the potential risk for regulation and the biased activity of samples for individual EDCs.

Analytically, current research erroneously requires the multi-compound assay for targeting in favor of a biological activity assessment.

An apt condition for the environmental changes predicts the indications on hazardous EDCs and antibiotics in wastewater are increasing in the upcoming decades. A very vast usage of widespread of policy-based actions manages, minimize and eliminate the usage of chemicals in protection against human health in the society. The specified challenges in the existing world are describing, definite existing properties and threat for EDCs accounts for the exposed effects till the late life. The cumulative exposure of these chemicals in this real-world reflects by updating the testing paradigms. The strategic requirement for the health-care products focuses on the prevention of EDCs and antibiotics in the clinical care. The critical steps for EDCs and antibiotics universally framework is the prevention of these compounds in environment for the human health. Thus, present study explains the endocrine disruptors and antibiotics present in the wastewater which are most life-threatening affects and they are causing huge adverse effects and how they should be avoided was explained clearly. The present research rapidly helps in the growing environmental changes drastically makes an urge threat for the pollution to affect the human health impacts on endocrine system and antibiotics usage for the equitable justice.

Endocrine disruption and the public

As the most advanced research in past decade develops the Summit on Environmental Challenges to Reproductive Health and Fertility at the University of California, San Francisco which recommends the future research on health care, policy, community action, and occupational health (Koesukwiwat et al., 2007). The provided and approved recommendations of this summit enhances the collaborations from various organisations and researchers regarding the directed neonatal exposures in the National Children's Health Study which leverages the specific laboratory experimental data into the National Health and Nutrition Examination Survey studies for the developing biomarkers of exposure and disease while increases the funding effects of chemicals on the epigenome, developmental programming, transgenerational effects, and cross-talk among endocrine systems and metabolic and immune systems.

API compounds, which are important for human welfare in economically advanced communities, have been detected worldwide in the natural environment. Antibiotics play a key role in evolving biological defects as a remarkable nature for the biota.

5.15 Future Aspects of Current Research

The most upcoming adverse effects of direct exposures to EDCs and their diseases on humans result from basic research and epidemiological studies to make it clear about screening, exposures, and targeting groups for the risk at high priority. In addition to this enhanced research, an important and effective approach is the prevention of diseases.

Based on the updated observations and research, it concludes as follows: (1) Advanced methods limit the lab and pilot scale investigations with huge system efficiency. (2) Applications for steroid hormones act as a single pollutant. (4) Membrane systems help in the removal of hormones with greater efficiency by transferring brine before discharging waste water and focus on the complete destruction of hormones without the transfer. (5) Advanced waste water treatment methods like AOP, membrane adsorption, execution of complex processes, and maintenance require huge investment (6) Advanced treatment techniques focus on increasing the exist with updated research methods. (7) The detection of these hormones in waste water develops new methods for the beneficiary task of WWTP and their operations and methodologies for regulatory administration for standardization and concentration limits. (8) Most of these methodologies for impact steroid hormones like oestrogen which are needed to understand the environmental risks of the environment.

The presence of Endocrine disrupting Hormones was observed in humans, aquatic life, and plants, and their effects. According to the state-of-the-art, waste water treatment methods have removed the contaminants presented in this study. Despite the advancements in this study available for treatment approaches there are gaps that need to be addressed in terms of cost-effectiveness, sustainability, and environmental safety.

References

Adzitey, F. (2015). Antibiotic classes and antibiotic susceptibility of bacterial isolates from selected poultry; a mini review. *World Veterinary Journal, 5*(3), 36–41.

Afrifa, J., Ogbordjor, W. D., & Duku-Takyi, R. (2018). Variation in thyroid hormone levels is associated with elevated blood mercury levels among artisanal small-scale miners in Ghana. *PLoS ONE, 13*(8), e0203335. https://doi.org/10.1371/journal.pone.0203335

Awad, H., Alalm, M. G., & El-Etriby, H. K. (2019). Environmental and cost life cycle assessment of different alternatives for improvement of wastewater treatment plants in developing countries. *Science of the Total Environment, 660*, 57–68.

Begum, S., Begum, T., Rahman, N., & Khan, R. A. (2021). A review on antibiotic resistance and way of combating antimicrobial resistance. *GSC Biological and Pharmaceutical Science, 14*, 87–97. https://doi.org/10.30574/gscbps.2021.14.2.0037

Bell, E. M., Hertz-Picciotto, I., & Beaumont, J. J. (2001). A case-control study of pesticides and fetal death due to congenital anomalies. *Epidemiology, 12*, 148–156.

Biošić, M., Mitrevski, M., & Babić, S. (2017). Environmental behavior of sulfadiazine, sulfamethazine, and their metabolites. *Environmental Science and Pollution Research, 24*(10), 9802–9812. https://doi.org/10.1007/s11356-017-8639-8

Birch, G. F., Drage, D. S., Thompson, K., Eaglesham, G., & Mueller, J. F. (2015). Emerging contaminants (pharmaceuticals, personal care products, a food additive and pesticides) in waters of Sydney Estuary, Australia. *Marine Pollution Bulletin, 97*, 56–66.

Birnbaum, L. S. (1994). Endocrine effects of prenatal exposure to PCBs, doxins, and other xenobiotics: Implications for policy and future research. *Environmental Health Perspectives, 102*, 676–6797895708.

Burke, V., Richter, D., Greskowiak, J., Mehrtens, A., Schulz, L., & Massmann, G. (2016). Occurrence of antibiotics in surface and groundwater of a drinking water catchment area in Germany. *Water Environment Research, 88*, 652659. https://doi.org/10.2175/106143016X14609975746604

Cai, Y., Wang, R., Pei, F., & Liang, B. B. (2007). Antibacterial activity of allicin alone and in combination with beta-lactams against *Staphylococcus* Spp. and *Pseudomonas aeruginosa. Journal of Antibiotics, 60*(5), 335–338. https://doi.org/10.1038/ja.2007.45

Carbone, P., Giordano, F., Nori, F., Mantovani, A., Taruscio, D., Lauria, L., & Figà-Talamanca, I. (2006). Cryptorchidism and hypospadias in the Sicilian district of Ragusa and the use of pesticides. *Reproductive Toxicology, 22*, 8–12.

Carvalho, I. T., & Santos, L. (2016). Antibiotics in the aquatic environments: A review of the European scenario. *Environment International, 94*, 736–757.

Castillo-Zacarías, C., Barocio, M. E., Hidalgo-Vázquez, E., Sosa-Hernández, J. E., Parra-Arroyo, L., López-Pacheco, I. Y., et al. (2021). Antidepressant drugs as emerging contaminants: Occurrence in urban and non-urban waters and analytical methods for their detection. *Science of the Total Environment, 757*, 143722.

Chang, Q., Wang, W., Regev-Yochay, G., Lipsitch, M., & Hanage, W. P. (2015). Antibiotics in agriculture and the risk to human health: How worried should we be? *Evolutionary Applications, 8*, 240–247. https://doi.org/10.1111/eva.12185

Chen, A., Chen, Y., Ding, C., Liang, H., & Yang, B. (2015). Effects of tetracycline on simultaneous biological wastewater nitrogen and phosphorus removal. *RSC Advances, 5*, 59326–59334. https://doi.org/10.1039/C5RA08434B

Daghrir, R., & Drogui, P. (2013). Tetracycline antibiotics in the environment: A review. *Environmental Chemistry Letters, 11*, 209–227. https://doi.org/10.1007/s10311-013-0404-8

Damgaard, I. N., Skakkebaek, N. E., Toppari, J., Virtanen, H. E., Shen, H., Schramm, K. W., Petersen, J. H., Jensen, T. K., & Main, K. M. (2006). Persistent pesticides in human breast milk and cryptorchidism. *Environmental Health Perspectives, 114*, 1133–1138.

Diamanti-Kandarakis, E., Bourguignon, J., Giudice, L. C., Hauser, R., Prins, G. S., Soto, A. M., Zoeller, R. T., & Gore, A. C. (2009). Endocrine-disrupting chemicals: An endocrine society scientific statement. *Endocrinology Review, 30*, 293–342 (2009).

Drewes, J. E., & Shore, L. S. (2001). Concerns about pharmaceuticals in water reuse, groundwater recharge, and animal waste. In *Pharmaceuticals and care products in the environment: Scientific and regulatory issues* (Vol. 791, pp. 206–228). American Chemical Society.

Eskenazi, B., Bradman, A., & Castorina, R. (1999). Exposures of children to organophosphate pesticides and their potential adverse health effects. *Environmental Health Perspectives., 107*(Suppl 3), 409–419.

European Centre for Ecotoxicology and Toxicology of Chemicals. *The toxicology of glycol ethers and its relevance to man.* Technical Report No. 95.

Everett, E. T. (2011). Fluoride's effects on the formation of teeth and bones, and the influence of genetics. *Journal of Dental Research., 90*(5), 552–560. https://doi.org/10.1177/0022034510384626

Fängström, B., Strid, A., Grandjean, P., Weihe, P., & Bergman, Å. (2005). A retrospective study of PBDEs and PCBs in human milk from the Faroe Islands. *Environmental Health, 4*, 12. https://doi.org/10.1186/1476-069X-4-12

Ferber, D. (2003). Antibiotic resistance. WHO advises kicking the livestock antibiotic habit. *Science, 301*, 1027.

Giles, A., Foushee, J., Lantz, E., & Gumina, G. (2019). Sulfonamide allergies. *Pharmacy, 7*(3), 1–12. https://doi.org/10.3390/pharmacy7030132

Gogoi, A., Mazumder, P., Tyagi, V. K., Tushara Chaminda, G. G., An, A. K., & Kumar, M. (2018). Occurrence and fate of emerging contaminants in water environment: A review. *Groundwater for Sustainable Development, 6*, 169–180.

Granados-Chinchilla, F., & Rodríguez, C. (2017). Tetracyclines in food and feedingstuffs: From regulation to analytical methods, bacterial resistance, and environmental and health implications. *Journal of Analytical Methods in Chemistry, 2017*, 1–25. https://doi.org/10.1155/2017/1315497

Grandjean, P., & Landrigan, P. J. (2014). Neuro behavioural effects of developmental toxicity. *The Lancet Neurology, 13*(3), 330–338. https://doi.org/10.1016/S1474-4422(13)70278-3

Grossman, T. H. (2016). Tetracycline antibiotics and resistance. *Cold Spring Harbor Perspectives in Medicine, 6*, a025387. https://doi.org/10.1101/cshperspect.a025387

Guo, X. Y., Hao, L. J., Qiu, P. Z., et al. (2016). Pollution characteristics of 23 veterinary antibiotics in livestock manure and manure-amended soils in Jiangsu province, China. *Journal of Environment Science Health B, 51*(6), 383–392.

Gupta, H. (2018). Anthracene removal from water onto activated carbon derived from vehicular tyre. *Separation Science and Technology, 53*(4), 613–625.

Gupta, H., & Gupta, B. (2016). Adsorption of polycyclic aromatic hydrocarbons on banana peel activated carbon. *Desalination and Water Treatment, 57*(20), 9498–9509.

Gupta, H., & Gupta, B. (2018). Vehicular tire as potential adsorbent for the removal of polycyclic aromatic hydrocarbons. *Polycyclic Aromatic Compounds, 38*(4), 354–368.

Hong, Y.-S., Song, K.-H., & Chung, J.-Y. (2014). Health effects of chronic arsenic exposure. *Journal of Preventive Medicine and Public Health., 47*(5), 245–252. https://doi.org/10.3961/jpmph.14.035

Hruska, K., & Franek, M. (2012). Sulfonamides in the environment: A review and a case report. *Review Article, 57*(1), 1–35. https://www.epa.gov/endocrine-disruption/overview-endocrine-system

Huang, P.-C., Tsai, C.-H., Liang, W.-Y., Li, S.-S., Huang, H.-B., & Kuo, P.-L. (2016). Early phthalates exposure in pregnant women is associated with alteration of thyroid hormones. *PLoS ONE, 11*(7), e0159398. https://doi.org/10.1371/journal.pone.0159398

Chabchoubi, I., Belkhamssa, N., Ksibi, M., & Hentati, O. (2021). Trends in the detection of pharmaceuticals and endocrine-disrupting compounds by field-effect transistors (FETs), *Trends Environment Analysis Chemistry, 30*, e00127.

Jaffrezic-Renault, N., Kou, J., Tan, D., & Guo, Z. (2020). New trends in the electrochemical detection of endocrine disruptors in complex media. *Analytical and Bioanalytical Chemistry, 412*(24), 5913–5923. Available from: https://doi.org/10.1007/s00216-020-02516-9

James Deal, J. D. (2013). Fluoride class action, "266 trillion atoms of lead," May 2, 2013.

Kaya, S. I., Cetinkaya, A., Bakirhan, N. K., & Ozkan, S. A. (2020). Trends in sensitive electrochemical sensors for endocrine disruptive compounds. *Trends Environmental Analysis Chemistry, 28*, e00106.

Kemper, N., Farber, H., Skutlarek, D., & Krieter, J. (2008). Analysis of antibiotic residues in liquid manure and leachate of dairy farms in northern Germany. *Agricultural Water Management, 95*, 1288–1292.

Khan, S. A., Maher, S., Naheed, N., Mahmood, T., Ayub, K., Farooq, A., Khan, S. B., & Shah, Z. (2018). Isolation, spectroscopic and density functional theory of two with anolide glycosides. *Journal of Molecular Structure*, 1–26. https://doi.org/10.1016/j.molstruc.2018.09.078

Khan, D. A., Knowles, S. R., & Shear, N. H. (2019). Sulfonamide hypersensitivity: Fact and fiction. *Journal of Allergy and Clinical Immunology: In Practice, 7*(7), 2116–2123. https://doi.org/10.1016/j.jaip.2019.05.034

Khan, W. U., Ahmed, S., Dhoble, Y., & Madhav, S. (2023). A critical review of hazardous waste generation from textile industries and associated ecological impacts. *Journal of the Indian Chemical Society, 100*(1), 100829.

Kim, M., & Zoh, K. (2016). Occurrence and removals of micropollutants in water environment. *Environmental Engineering Research, 21*, 319e332 (2016).

Kim, K. R., Owens, G., Kwon, S. I., So, K. H., Lee, D. B., & Ok, Y. S. (2011). Occurrence and environmental fate of veterinary antibiotics in the terrestrial environment. *Water, Air, and Soil Pollution, 214*, 163–174. https://doi.org/10.1007/s11270-010-0412-2

Kjeldsen, L. S., & Bonefeld-Jørgensen, E. C. (2013). Perfluorinated compounds affect the function of sex hormone receptors. *Environment Science Pollution Research International, 20*(11), 8031–8044. Published online June 14, 2013. https://doi.org/10.1007/s11356-013-1753-3

Koesukwiwat, U., Jayanta, S., & Leepipatpiboon, N. (2007). Validation of a liquid chromatography–mass spectrometry multi-residue method for the simultaneous determination of sulfonamides, tetracyclines, and pyrimethamine in milk. *Journal of Chromatography A, 1140*, 147–156.

Konieczna, A., Rutkowska, A., & Rachoń, D. (2015). Rocz Panstw Zakl Hig. *Health risk of exposure to Bisphenol A (BPA), 66*(1), 5–11.

Kristensen, P., Irgens, L. M., Andersen, A., Bye, A. S., & Sundheim, L. (1997). Birth defects among offspring of Norwegian farmers; 1967–1991. *Epidemiology, 8*, 537–544.

Lu, X., Sun, J., & Sun, X. (2020). Recent advances in biosensors for the detection of estrogens in the environment and food, TrAC. *Trends in Analytical Chemistry, 127*, 115882.

Madhav, S., Mishra, R., Kumari, A., Srivastav, A. L., Ahamad, A., Singh, P., Ahmed, S., Mishra, P. K., & Sillanpää, M. (2024). A review on sources identification of heavy metals in soil and remediation measures by phytoremediation-induced methods. *International Journal of Environmental Science and Technology, 21*(1), 1099–1120.

Mahamallik, P., Saha, S., & Pal, A. (2015). Tetracycline degradation in aquatic environment by highly porous MnO_2 nanosheet assembly. *Chemical Engineering Journal, 276*, 155–165. https://doi.org/10.1016/j.cej.2015.04.064

Mole, B. (2015). Atrazine's path to cancer possibly clarified, science news. January 21, 2015.

Monahan, C., Harris, S., Morris, D., & Cummins, E. (2022). A comparative risk ranking of antibiotic pollution from human and veterinary antibiotic usage—An Irish case study. *Science of the Total Environment, 826*, 154008. https://doi.org/10.1016/j.scitotenv.2022.154008

Murata, K., Dakeishi, M., Shimada, M., & Tohoku, S. H. (2007). Assessment of intrauterine methylmercury exposure affecting child development: messages From the newborn. *Journal of Experimental Medicine, 213*(3), 187–202.

Nickmilder, M., & Bernard, A. (2011). Associations between testicular hormones at adolescence and attendance at chlorinated swimming pools during childhood. *International Journal of Andrology, 34*(5pt2), e446–e458. https://doi.org/10.1111/j.1365-2605.2011.01174

Noguera-Oviedo, K., & Aga, D. S. (2016). Lessons learned from more than two decades of research on emerging contaminants in the environment. *Journal of Hazardous Materials, 316*, 242–251.

Pang, Y. H., Huang, Y. Y., Wang, L., Shen, X. F., & Wang, Y. Y. (2020). Determination of bisphenol A and bisphenol S by a covalent organic framework electrochemical sensor. *Environmental Pollution, 263*, 114616.

Paramonov, S., Zelikova, D., Sklyarova, L., & Alkhutova, I. (2022). Environmental risks of microcontamination of the environment with tetracycline. *Pharmacy Formulations, 4*, 76–88. https://doi.org/10.17816/phf106547

Park, J., Kim, M. H., Choi, K. et al. (2007). *Environmental risk assessment of pharmaceuticals: model application for estimating pharmaceutical exposures in the Han River Basin, Re-06* (p. 201). Korea Environment Institute.

Peng, X., Tang, C., Yu, Y., et al. (2009). Concentrations, transport, fate, and releases of polybrominated diphenyl ethers in sewage treatment plants in the Pearl River Delta, South China. *Environment International, 35*, 303–309.

Reddersen, K., Heberer, T., & Dünnbier, U. (2002). Identification and significance of phenazone drugs and their metabolites in ground and drinking water. *Chemosphere, 49*, 539–544.

Reynolds, P., Von Behren, J., Gunier, R. B., Goldberg, D. E., Hertz, A., & Harnly, M. E. (2002). Childhood cancer and agricultural pesticide use, an ecologic study in California. *Environmental Health Perspectives, 110*, 319–324.

Rice, K. M., Walker, E. M., Wu, M., Gillette, C., & Blough, E. R. (2014). Environmental mercury and its toxic effects. *Journal of Preventive Medicine and Public Health, 47*(2), 74–83. https://doi.org/10.3961/jpmph.2014.47.2.74

Rodriguez-Mozaz, S., Vaz-Moreira, I., Della Giustina, S. V., Llorca, M., Barcelo, D., Schubert, S., Berendonk, T. U., Michael-Kordatou, I., Fatta-Kassinos, D., Martinez, J. L., et al. (2020).

Antibiotic residues in final effluents of European wastewater treatment plants and their impact on the aquatic environment. *Environment International, 140*, 105733. https://doi.org/10.1016/j.envint.2020.105733

Segura, P. A., Takada, H., Correa, J. A., et al. (2015). Global occurrence of anti-infectives in contaminated surface waters: Impact of income inequality between countries. *Environment International, 80*, 1386–1398. https://doi.org/10.1016/j.envint.2015.04.001

Seydel, J. K. (1968). Sulfonamides, structure-activity relationship, and mode of action. *Journal of Pharmaceutical Sciences, 57*, 1455–1478.

Shaheen, J. F., Sizirici, B., & Yildiz, I. (2022). Fate, transport, and risk assessment of widely prescribed pharmaceuticals in terrestrial and aquatic systems: a review. *Emerging Contaminants*.

Singer, A. C., Shaw, H., Rhodes, V., et al. (2016). Review of antimicrobial resistance in the environment and its relevance to environmental regulators. *Frontiers in Microbiology, 7*, 1–22. https://doi.org/10.3389/fmicb.2016.01728

Singh, P., Singh, R., Borthakur, A., Madhav, S., Singh, V. K., Tiwary, D., Srivastava, V. C., & Mishra, P. K. (2018). Exploring temple floral refuse for biochar production as a closed loop perspective for environmental management. *Waste Management, 77*, 78–86.

Singh, S., Gupta, H., Dhiman, S., & Sahu, N. K. (2022). Decontamination of cationic dye brilliant green from the aqueous media. *Applied Water Science, 12*(4), 61. https://doi.org/10.1007/s13201-022-01596-5

Singh, S., Kumar, A., & Gupta, H. (2020). Activated banana peel carbon: A potential adsorbent for Rhodamine B decontamination from aqueous system. *Applied Water Science, 10*, 185. https://doi.org/10.1007/s13201-020-01274-4

Skakkebaek, N. N. (2002). Endocrine disrupters and testicular dysgenesis syndrome. *Hormone Research, 57*, 43.

Skakkebæk, N. E., Rajpert-De Meyts, E., & Main, K. M. (2001). Testicular dysgenesis syndrome, an increasingly common developmental disorder with environmental aspects. *Human Reproduction, 16*, 972–978.

Srinivasan, A., & Viraraghavan, T. (2009). Perchlorate: Health effects and technologies for its removal from water resources. *International Journal of Environmental Research and Public Health, 6*(4), 1418–1442. https://doi.org/10.3390/ijerph6041418

Supuran, C. T. (2017). Special issue: Sulfonamides. *Molecules, 22*(10), 1–5. https://doi.org/10.3390/molecules22101642

Supuran, C. T., Casini, A., & Scozzafava, A. (2003). Protease inhibitors of the sulfonamide type: Anticancer, antiinflammatory, and antiviral agents. *Medicinal Research Reviews, 23*(5), 535–558. https://doi.org/10.1002/med.10047

Vieira, W. T., de Farias, M. B., Spaolonzi, M. P., da Silva, M. G. C., & Vieira, M. G. A. (2021). Endocrine-disrupting compounds: occurrence, detection methods, effects and promising treatment pathways—A critical review. *Journal of Environmental Chemical Engineering, 9*, 104558.

Wang, B., Zhang, Y., Zhu, D., & Li, H. (2020). Assessment of bioavailability of biochar-sorbed tetracycline to *Escherichia coli* for activation of antibiotic resistance genes. *Environmental Science and Technology, 54*, 12920–12928. https://doi.org/10.1021/acs.est.9b07963

Wang, N., Yang, X., Jiao, S., Zhang, J., Ye, B., & Gao, S. (2014). Sulfonamide-resistant bacteria and their resistance genes in soils fertilized with manures from Jiangsu Province, Southeastern China Na. *PLoS ONE, 9*, 1–11. https://doi.org/10.1371/journal.pone.0112626

Wang, W., Zhang, W., Liang, H., & Gao, D. (2019). Occurrence and fate of typical antibiotics in wastewater treatment plants in Harbin, North-East China. *Frontier Environment Science Engineering, 13*, 34. https://doi.org/10.1007/s11783-019-1118-3

Weidner, I. S., Moller, H., Jensen, T. K., & Skakkebæk, N. E. (1998). Cryptorchidism and hypospadias in sons of gardeners and farmers. *Environmental Health Perspectives, 106*, 793–796.

Xiang, H., Nuckols, J. R., & Stallones, L. (2000). A geographic information assessment of birth weight and crop production patterns around mother's residence. *Environmental Research A, 82*, 160–167.

Yousef, F., Mansour, O., & Herbali, J. (2018). Sulfonamides: historical discovery development (structure-activity relationship notes). *IIIJ, 1*(1), 1–15.

Zamri, M. F. M. A., Bahru, R., Suja', F., Shamsuddin, A. H., Pramanik, S. K., & Fattah, I. M. R. (2021). Treatment strategies for enhancing the removal of endocrine-disrupting chemicals in water and wastewater systems. *Journal of Water Proceedings of Engineering, 41*, 102017.

Zawodniak, A., Lochmatter, P., Beeler, A., & Pichler, W. J. (2010). Cross-reactivity in drug hypersensitivity reactions to sulfasalazine and sulfamethoxazole. *International Archives of Allergy and Immunology, 153*(2), 152–156. https://doi.org/10.1159/000312632

Chapter 6
Water Disinfectant and Its Health Hazards to Every Form of Life

Garima Tomar and Vikas Kumar

Abstract The water disinfection process increases DBPs through chemical reactions of organic and inorganic substances present in the water. Developed or developing countries are currently grappling with secondary pollutants such as disinfectant by-products. Notably, DBPs are becoming an emerging concern by increasing the rate of eco-toxicological impact on a person's health and water ecosystem. This review article explores the presence of DBPs in the urban water cycle, its hazardous health impacts, various treatment methods for disinfectants, potable water treatment plants, excessive disinfectant removal techniques, and domestic, wastewater treatments. Internal and external factors are responsible for the effects of DBP in surface water systems. The impacts of DBP are observed in the urban water cycle and individuals who is associated with swimming associations, such as swimmers, pool workers, and sometimes children in swimming classes. This has raised serious concerns about respiratory and skin diseases in swimmers. Additionally, the review addresses the regulatory frameworks and standards for DBP levels in drinking water, highlighting the need for stringent monitoring and compliance to safeguard public health. The future perspective of this research involves developing predictive models for assessing the mitigation, significance, assessment, removal, and impact of DBPs on an aquatic ecosystem. These models also highlight the importance of DBP removal methods in terrestrial and atmospheric environments, emphasizing a comprehensive approach to managing DBPs in all environmental contexts.

Keywords DBPs · Water disinfectants · Wastewater treatment · Health hazards · Human life

G. Tomar (✉)
School of Allied Sciences, Dev Bhoomi Uttarakhand University, Dehradun 248007, U.K.
e-mail: garimatomar63@gmail.com

V. Kumar
School of Applied and Life Sciences, Uttaranchal University, Dehradun 248007, U.K.

H. Gupta et al. (eds.), *Emerging Contaminants in Water and Wastewater*, Environmental Science and Engineering, https://doi.org/10.1007/978-3-031-82579-8_6

6.1 Introduction

The World Health Organization and the United Nations have always expressed their concern for water (WHO, 2017). The quality characteristics of drinking water should be pure and drinkable for all living beings. It is necessary to supply water for human consumption and to maintain its characteristics. Various water treatment processes have been adopted to increase the abundance of potable water whether it is surface water or groundwater resources. This essential process is known as disinfection of water. It is critical for ensuring that all type of hydro resources is safe for human use (Lu et al., 2021). The surface water bodies, such as rivers and streams, often become polluted through the addition of wastewater, which introduces bacterial pathogens and chemical contaminants. Such type of water pollution's problem significantly increases the risk of waterborne diseases (Gupta, 2018; Khan et al., 2023; Kumar et al., 2014, 2017; Madhav et al., 2020; Zhang et al., 2021). The contaminants present in wastewater can react with disinfectants used in water treatment processes, leading to the formation of disinfection by-products (DBPs). These DBPs are chemical compounds that can have harmful effects on human health and the environment. Various researchers summarized the formation mechanism, occurrence, adverse effects, and reduction strategies for DBPs in drinking water (Ding et al., 2019; Mazhar et al., 2024; Saxena et al., 2024). Despite these efforts, there has been limited discussion on the abundance of DBP in domestic water and its transport and movement in the urban water cycle. Additionally, there is a lack of summaries or published research data showing concerns about the adoption of mitigation methods for the occurrence of DBP in drinking water (Alexandrou et al., 2018). The introduction of wastewater treatment techniques and reclaimed water has further contributed to the presence of DBPs in surface water bodies. These water conservation practices can introduce additional chemical reactions that form DBPs, complicating the challenge of ensuring safe drinking water. Adverse health effects are a growing concern with increasing concentration of DBPs. Research has linked exposure to DBPs with various health issues such as reproductive problems, increased risk of cancer, developmental disorders, etc. The impact of DBPs extends various problems, beyond human health, affecting aquatic ecosystems by disrupting the balance of these environments and harming aquatic life (Neale et al., 2017). Reclaimed water and wastewater techniques are added disinfectant by-products in surface water bodies. The objective of this study is:

- To evaluate the different types of DBPs, their releasing sources, and the health hazards, pose to life.
- To fill the gaps in current research by offering insights into potential mitigation strategies and focusing on a detailed analysis of DBPs in urban water cycles.
- To examine standards and regulatory frameworks for DBP levels in drinkable water, targeting the need for stringent monitoring and compliance to safeguard public health.

Future research is required to develop predictive models that can assess the significance, mitigation, removal, and impact of DBPs on aquatic ecosystems. These models will also highlight the importance of DBP removal methods in terrestrial and atmospheric ecosystems, emphasizing a comprehensive approach to managing DBPs across various environmental contexts. By addressing these issues, the study aims to contribute to the ongoing efforts to ensure safe and clean drinking water for all.

6.1.1 Disinfectant By-Products

Various water disinfectants are helpful in the reduction of waterborne disease epidemics (Postigo et al., 2021). Disinfectants inactivate pathogens, however, DBPs (disinfectant by-products) are used to disinfect water. It is the resultant reaction between water quality and its characteristics (Mohd et al., 2020). Various types of treated water contain DBPs similar to drinking water (Krasner et al., 2010); reclaimed wastewater (Peng et al., 2021), and treated leachates of landfills (Zeng et al., 2017) etc. contain DBPs before their discharge. The occurrence of disinfectant by-products in drinking water is of more concern than others, as drinking water is directly linked to biological safety (Cui et al., 2021; Wagner & Plewa, 2017). Polluted water-generated DBPs may be added into the freshwater bodies and reused potable drinking water (WHO, 2017). The introduced wastewater disinfectant by-products contribute to the decreasing rate of aquatic biodiversity (Luan et al., 2020). Additionally, the upstream water sources of wastewater treatment plants can degrade the water quality of downstream resources due to the high concentration of precursors. (Soller et al., 2019). In the context of drinking water, disinfectant products are also responsible for causing adverse effects on humans through their toxic properties including carcinogenic risks (Du et al., 2017). The urban water cycle amplifies DBP fluxes through wastewater treatment, collection, and supply to drinking water sources (Pal et al., 2014). Based on the toxic nature of DBPs, water treatment is required to maintain and move the urban water cycle and to have an adequate water supply and characteristics of the closed loop (Fu et al., 2017). The efficacy of freshwater aquatic systems and wastewater treatment processes will enhance human health protection from the adverse effects of water disinfectants and reduce DBP levels (Bond et al., 2015; Rong et al., 2022). Therefore, it is paramount to summarize and raise awareness about DBPs reduction in the water cycle, keeping in mind the importance of drinking water quality. By addressing the impacts and sources of DBPs, we can preserve the aquatic ecosystem and protect public health.

6.1.2 Impact of DBPs on Natural Aquatic Ecosystems

Aquatic organisms are very sensitive to the presence of any toxic element in the stream or river. Disinfectant by-products are responsible for posing threats to aquatic biodiversity, particularly near any outlet of a wastewater treatment plant into freshwater bodies (Cui et al., 2021). However, influent source water from these treatment plants can serve as freshwater downstream of the effluent treatment plant. Transport of wastewater effluent to receiving water bodies increases concentrations of disinfectant by-products and elevate the risk to aquatic life (Chen et al., 2010). Various removal techniques such as hydrolysis, photolysis, microbial degradation, volatilization, adsorption, evaporation, and reverse osmosis are adopted for disinfection by-products in the surface water ecosystem (Jin et al., 2012; Ding et al., 2019). The volatilization process is highly efficient for DBP removal in water systems due to the high efficiency of hydrogeological reactions of aquifers with Henry's law (Ding et al., 2018). According to Chen et al. (2014), DBP has a low absorption capacity which is responsible for making the adsorption process unimportant in the removal methods. However, the microbial degradation process is quite helpful in removing a variety of wastewater bio-degradable disinfectant by-products (Yu & Reckhow, 2015), similar to the microbial method, the hydrolysis process of wastewater treatment is also seen as very common (Richardson, 2011). Freshwater bodies and marine water bodies also receive chlorinated water after treatment (Abusallout & Hua, 2016). Typical reduction rates of I > Br > Cl have been observed in seawater, given the concentration of halophenolic disinfectant by-products from the photo-conversion process (Liu et al., 2019). Abusallout and Hua (2016) recorded a decreasing rate in total organic bromine, total organic iodine, and total organic chlorine through photo-degradation in the presence of UV irradiation. Lv et al. (2017) found that the cytotoxicity of reclaimed wastewater is effectively decreased by solar IR radiation. These various processes and techniques highlight the importance of effective DBP removal to protect both freshwater and marine ecosystems.

6.2 Occurrence of DBPs in Surface Water

Disinfectant by-products (DBPs) exacerbate problems in natural water flows by changing water quality characteristics after discharge. Whereas, the treated wastewater gets diluted after disposal into the water body as it is necessary to maintain the quality characteristics of the surface water body, especially in the dry season (Van Huy et al., 2011). Typically, surface water receives high concentrations and large amounts of DBPs through the discharge of disinfected treated wastewater. Volatile halogenated, non-halogenated, and N-nitrosodimethylamine (NDMA) were detected in the studies of various surface water bodies in China, Beijing, and Tokyo (Heng et al., 2021; Wang et al., 2016). Lei et al. (2021) analyzed that disinfectant concentrations of by-products are lower in surface water compared to tap water and wastewater-treated water. During trial testing and the SARS-CoV-2 pandemic, Wang et al. (2021) recorded that nitrosamines (NA) and DBPs concentrations were significantly higher in surface water. Notifying the issue of DBPs in surface water requires comprehensive strategies that include improving wastewater treatment processes, enhancing monitoring of water quality, and developing better DBP removal techniques. Such type of techniques as advanced oxidation processes, improved chemical dosing protocols, and membrane filtration can help reduce DBP levels. Additionally, regulatory measures and public awareness are essential to ensure pure water quality characteristics and provide protection to both environmental and human health.

6.2.1 Source of DBPs in the Urban Water Cycle

The urban water cycle is responsible for the transfer of energy, and contaminants and making an interconnection within other components such as source water, stormwater, sewage and wastewater-treated water, and drinking water (Pal et al., 2014). Sewage water treatment can be responsible for adding water into groundwater and nearby natural water resources which is an unplanned reuse of water (WHO, 2017). However, after AWTFs complete the process, the effluents from sewage treatment plants released into the water bodies are properly treated by advanced technologies (Soller et al., 2019). This treated water enters the drinking water supply, is potable for use directly or indirectly, and is added to environmental buffers (Huang et al., 2017). Source water affected by wastewater treatment plants or AWTF effluent can be distributed downstream as raw water and increase the production of potable water. (Xu et al., 2021). Subsequently, Later, the collected water after treatment will be transferred for individual use, and the water re-collected after domestic use into the sewer system by the SWTP, and again treated by the Effluent Water Treatment Plant (Pal et al., 2014). The DBP enters through wastewater treatment plants, water disinfection is processed to obtain disinfectant by-products, and through treated water into nearby surface water bodies (Li et al., 2021). As these DBPs circulate through the urban water cycle, they pose potential risks to both environmental and human health. Proper management and advanced treatment methods are essential to minimize the presence of DBPs and ensure the safety and quality of urban water supplies.

6.2.2 Addition of DBPs in Drinking Water

DBPs in potable water, however, are restricted for human direct use, water disinfectant by-products and their dangerous health effects associated with potable water DBPs (Richardson et al., 2007). Therefore, the disinfectant effects of drinking water studied are found to be maximum as compared to disinfectants of surface water. Drinking water has higher concentrations of haloacetic acids (HAAs) and trihalomethanes (THMs) than biological DBPs (Jasemizad et al., 2021; Mian et al.,

2018). DBPs might experience reductive dehalogenation, hydrolysis, and microbial degradation in DWDS (Postigo et al., 2020; Wright et al., 2017). Numerous processes are used to identify the level and variation in DBPs value in drinking water, which may contribute to record variability in DBP levels in distributed drinking water. The persistence and formation of DBPs are influenced by numerous factors including the type of water source, disinfectants used, and environmental conditions within the distribution system. Chloramine, Chlorine, and ozone are the common disinfectants, and each interacts differently with organic matter present in water, forming various DBPs (Xing-Fang et al., 2018). Over time, these by-products can transform or degrade, altering their composition and concentration (Ying et al., 2019; Uyak et al., 2014). Continuous monitoring and regulation are essential to mitigate the health risks posed by DBPs, ensuring safe drinking water quality (Tang et al., 2021). Alternative disinfection strategies and advanced treatment methods are being explored to minimize DBP formation while maintaining effective microbial control. Understanding the dynamics of DBP degradation and formation helps in developing the best way of water treatment and distribution practices.

6.2.3 Presence of DBPs in Reclaimed Wastewater

Wastewater treatment plants or sewage treatment plants are helping to resolve the water pollution problem by following some treatment methods, but in certain cases, it may enhance the sources of DBPs through some chemical disinfectant in the water resources (Li et al., 2021; Postigo et al., 2021). Chlorination process is a type of common wastewater treatment method which is also found for increasing the concentration of DBPs in chlorinated water than drinking water (Du et al., 2017). Watson et al. (2012) examined that THMs and NDMA are present in effluent water of wastewater treatment plants in Queensland (Australia) and Victoria. During SARS-CoV-2 pandemic, Li et al. (2015) was recorded that HAAs and THMs disinfectant by-products present in domestic or household wastewater which is recorded maximum in the wastewater samples (Postigo et al., 2020). Figure 6.1 represents the various additional sources of DBPs in surface water body. Li et al. (2015) detected haloacetic acids (HAAs) and THMs in domestic wastewater, and these are the most prevalent in wastewater samples (Postigo et al., 2020). These are certains factors which is responsible to influence the formation of DBPs in wastewater treatment plants, such as operational conditions of treatment process, presence of organic matter in the water and what type and amount of disinfectant used. THMs and HAAs are the form of DBPs that is increased by the use of chlorine in treatment process. These by-products of DBPs can caused potential health risks on life, and various type of environmental concerns when the water is reused or released back into natural water bodies. Strict regulatory guidelines and advanced treatment techniques are necessary, it may help to minimize the formation of DBPs and provide a surety of purity in treated wastewater. Understanding the mechanisms of DBP formation and implementing best practices

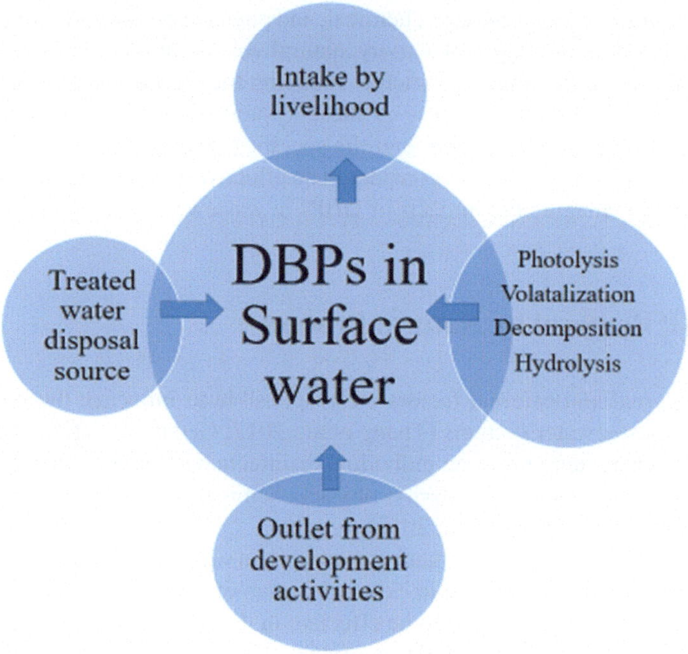

Fig. 6.1 DBPs occurrence, operational and treatment process in aquatic ecosystem

in water treatment can help reduce their prevalence and protect public health and the environment.

6.2.4 Constructed Wetland Systems

It has been observed by Keefe et al. (2004) that ecological buffers are commonly used in wastewater treatment by wetland systems. Volatilization and microbial degradation methods are considered to be the best methods for the removal of HAAs and THMs (Rostad et al., 2010; Chen et al., 2014). Rostad et al. (2000) HKs, TCNMs, HANs, HAAs, THMs, and TOHs have been reported to be reduced (35–100%) by evaporation methods in 2–3 days. Additionally, eleven major disinfectant co-products were investigated and successfully removed through constructed wetland systems with retention times of 5 days (Ged et al., 2015). Diversely, plants are also useful in TCM removal rather than by transpiration methods in surface wetland systems. (Seeger et al., 2011). Because of the diffusion process of volatilization, it is hindered by subsurface flow, and typically, TCM is removed by plant uptake rather than transpiration in wetland systems (Kumari & Gupta, 2018). However, plants are responsible for providing first-stage substrates and redox mediators that enhance anaerobic biodegradation in THMs, DBAN, and BCAN (Zhou et al., 2013). Wetland systems utilize a combination of biological, chemical, and physical processes to treat wastewater. The integration of microbial activity, natural volatilization processes, and plant uptake creates a multifaceted approach to reducing the high or major concentration of various DBPs. The effectiveness of these systems depends on factors such as water retention time, plant species, and the specific contaminants present. As such, the above-mentioned systems are considered an effective and sustainable method for mitigating and managing the impacts of DBPs in wastewater.

6.3 DBP Removal by Natural Process

Both the internal and external factors are responsible to influence the presence of DBPs in domestic water systems (Thang et al., 2017; Ged et al., 2015). The factors affecting the characteristics of an individual disinfectant are mainly disinfectant by-products, intrinsic, volatility, polarity, physicochemical properties, molecular size, and hydrophobicity (Zhang et al., 2018).

However, the wastewater treatment process, urban water cycle structure, and water quality characteristics are related to external factors of DBP. The processes of evaporation and adsorption are found to be effective in the removal techniques of THMs (Tak & Velanki, 2018). The evaporation process is most commonly used for potable water use from wastewater treatment in artificial wetland systems and open surface water systems (Bougeard et al., 2020; Zeng et al., 2016). Microbial degradation techniques and membrane filtration are widely and majorly used to reduce or remove HAAs from urban water, as it is biodegradable and present in ionic form (Meite et al., 2015; Huang et al., 2017). Overall, the DBPs removal process contains a combination of biological, chemical, and physical processes in a water system. Which process is more effective in DBP removal depends on the properties and characteristics of DBPs and the specific treatment conditions and environment. Optimizing and understanding these factors are optimal for improving water treatment methods and ensuring the quality and safety of domestic and urban water supplies.

6.3.1 Removal of DBPs Through Membrane Techniques

Advanced oxidation processes, membrane filtration, reverse osmosis, and microfiltration are used for water disinfectant removal (Zeng et al., 2016). Advanced water recycling facilities (AWRF) must be used for the treatment of potable water which is majorly affected by some pollutants that are found in effluent water-generated disinfectant products. Besides, chloramine and chlorine are used to manage membrane biofouling, which leads to an increase in DBP concentration in the feed water (Agus & Sedlak, 2010). Generally, the characteristics and physical and chemical properties of membrane and DBP have a significant impact on the operation process of membrane filtration and DBP removal. DBPs are causing environmental pollution as emerging pollutants in water bodies (Shiqing et al., 2015). As observed by the scientists, some disinfectant by-products are increasing (HAAs, THMs, HALs) and some (HNMs, HANs, HAMs) are showing a decreasing rate through drinking water and reclaimed wastewater. In addition, DBPs were described using heavy concentrations and high frequency in surface water, after being added to a wastewater treatment plant (Chen et al., 2021). This highlights the need for continuous monitoring and optimization of treatment methods to mitigate the environmental impact of DBPs and ensure the safety and quality of water.

6.3.2 Removal of Disinfectant By-Products Through Advanced Techniques

The hydrophobic components and volatile species are involved in the removal methods of THMs. Biodegradable DBPs such as HAAs are removed using the microbial degradation method, the hydrolysis process is most common in HAMs, HANs, and HALs, solar photolysis process is susceptible for N-DBPs (Uyak et al., 2014; McCurry et al., 2016). Solar light-sensitive cleavage of the halogen-carbon bond of disinfectant by-products leads to an increase in the number and size of substituted halogens. Biodegradation methods are more likely to be adopted by biologists to remove BCANs, DBANs, and THMs (Bougeard et al., 2020). The reverse osmosis process targets to HAAs i.e. negatively charged through MW compounds. UV254 is used for fluence-based photodegradation rate constants: IDBP > Br DBP > Cl DBP, and tertiary > secondary > primary halogenated DBP. The consciousness of the OH^- radical and functional group rank is given as follows: I > CHO > NO_2 > H > Br > CN > $CONH_2$ > CO_2. A large volume or equal to 50% removal of halogenated DBPs in the effluent water treatment plant has not been done by the UV process only (Kities et al., 2010). Some unique techniques and appropriate method implementation are mandatory to know for selecting the removal methods based on the specific properties of DBPs. Combining various methods such as solar photolysis, hydrolysis, advanced filtration techniques, and microbial degradation can enhance the overall efficiency of DBP removal and reduce their environmental impact. Continuous research and development in this field are crucial for improving water treatment practices and ensuring the safety of treated water.

6.3.3 Modern Techniques for DBP Removal

THMs and HAAs DBPs are removed by adsorption and microbial degradation methods in a granular activated carbon (GAC) treatment process. During this treatment method, the adsorption process occurred in the initial stage, and the biodegradation process was found after a long period of treatment (Chuang et al., 2021). The

DBP removal efficiency of the biodegradation method in terms of halogen number is opposite to that of the adsorption process (TCAA < DCAA < DBAA < MBAA < MCAA). Some modern techniques viz. Water quality, compound properties, operational parameters, and membrane characteristics affect the removal efficiency of DBPs by nanofiltration method (Wang et al., 2018; Li et al., 2017). The nanofiltration process represents the effectiveness of electrostatic repulsion on the sieve in HAAs removal. The filtration portion of the NF method is useful for non-ionized disinfectant by-products in DBP removal. Primary halogenated halo acetic acids are more environmentally friendly or biodegradable than halogenated and tertiary halogenated HAAs (Jin et al., 2012; Li et al., 2017).

The basic water quality characteristics of potable water or parameters as temperature, pH, TOC, and concentration or amount of disinfectant affect the biodegradable process of haloacetic acids. Reductive dehalogenation of elements represents a pattern of I > Br > Cl, which is the tendency for halogen removal in DBPs, and dehalogenation increases with the interchange of halogen by the methyl group (Ike et al., 2020). Physico-chemical parameters of water such as metal aging, DO, pH, and water temperature affect the disinfectant by-products in removal methods of photolysis, hydrolysis, abiotic dehalogenation, decarboxylation and thermal reactions in the non-rural water cycle. Domestic treatment shows a short-term and efficient solution to disinfectant by-product elimination. Various types of DBP i.e. low liquid solubility and polarity can be easily removed through an activated carbon filter (Chowdhury et al., 2011). Furthermore, DBPs having high hydrophobicity, large in size, and low polarity are highly rejected by RO filtration (Shah et al., 2012; Szczuka et al., 2017). The boiling process can increase DBP removal efficiency by adding hydrolysis and evaporation. However, the boiling process can change the concentration and level of haloacetic acids (Diana et al., 2019). So, it is very clear that only a single treatment method is not amenable to DBP removal, a combination of different types of methods is required to achieve control in DBPs. The adaptation of modern DBP removal methods is recommended for wastewater treatment methods.

6.4 Health Hazards of Disinfectant By-Products

Yang et al. (2022) observed in their study that DBPs disturb the menstrual cycle in women. The researchers assessed that high concentrations of disinfectants in by-products disturb estrous cyclicity and cause problems with ovarian steroidogenesis which directly affects the fertility of females. The presence of trichloroacetic and dichloroacetic acid in water is mainly linked to reclaimed wastewater disinfecting by-products and causes disturbances in a woman's reproductive system. Jianrong et al (2010) evaluated sewage treatment plants in Beijing. The researchers revealed that direct disposal of treated water resulted in maximum concentration of disinfectant byproducts in surface water resources. Yu et al. (2019) investigated the occurrence and transfer of DBPs in water resources (tap water, prepared water) in eastern China and revealed that disinfectant by-products such as HANS, TCNM, THM, and HAAS have the potential to increase cytotoxicity and genotoxicity in mammals. Yi-Xin et al (2014) found DBP effects on human health by examining human urine samples. The study found that disinfectant by-products in the water are also causing harmful effects by damaging or altering the urinary system. Patrick et al (2016) and Miao et al (2022) studied the health effects on pregnant women and found that DBPs can cause problems in the development of the fetus. High exposure to DBPs results in reduced genetic susceptibility to genes encoding them.

Samuel et al (2015) and Yan-Ling et al. (2019) found an increased concentration of THMs and HAAs in women with maternal exposure during pregnancy, however, so far, no studies have examined the effects of emerging DBPs in infants. Mohamed et al. (2020) and Kitis et al (2010) investigated water treatment effects on the natural characteristics of water. The researcher recorded that chlorination is the most common process used to disinfect water. A range of DBPs such as haloacetic acid, halogenated acetonitrile, and trihalomethane react with NOMs present in raw water which can cause adverse health effects on humans. It causes unwanted effects i.e., malignant growth of human organs (Li et al., 2021; Zhiwen et al., 2023).

6.5 Conclusion

There have been extensive investigations on DBP occurrence and found that there are some factors responsible for this in the urban water cycle. Therefore, it is recommended that future research regulate large-scale explorations, which may be helpful for better knowledge about the existence of DBPs in the non-rural water cycle. However, it is difficult to focus on the behavior of DBP as toxicological and epidemiological studies reveal its carcinogenic nature. The Science of Toxicology recommends that aromatic disinfectant by-products are required to establish regular monitoring and regulation. The predictive models are successfully evaluating the human health effects and ecological safety of DBPs. Various mathematical and predictive models are efficiently used for wastewater treatment in terms of water quality characteristics, physico-chemical components, and DBP removal as experimental variables. The present study concluded that scientific methods or predictive models for DBP removal should mainly focus on parameters related to disinfectant by-products. Other processes such as reclaimed wastewater management methods, disinfected hospital effluent, remedial treatment methods, and desalinated seawater are used to remove DBPs. These processes are considered to be modifications of the deodorized treatment to maintain pathogen removal against the larger volume of DBP evaluated. So, occurrence, abundance, and health risks must be evaluated. The evaporation process can potentially increase airborne or atmospheric DBP values, and the adsorption process leads to the settlement of DBPs as suspended particles, municipal sludge can participate in the addition of DBP to the land ecosystem through wastewater treatment. Consequently, in addition to water bodies, future research may also look at the presence and health hazards of DBPs in the atmosphere and lithosphere ecosystems.

6.6 Future Recommendation on DBPs Occurrence

In addressing the growing concern surrounding disinfection by-products (DBPs) in urban water cycles, future research should focus on several key aspects to effectively mitigate their eco-toxicological impact on human health and aquatic ecosystems. Firstly, there is a need for continued exploration of the formation mechanisms of DBPs and their occurrence in water treatment processes. Understanding the factors contributing to DBP formation, such as water source characteristics, disinfection methods, and treatment conditions, will aid in developing targeted mitigation strategies. Furthermore, the development of predictive models for assessing the impact of DBPs on an aquatic ecosystem, significance, removal, and mitigation of DBPs should be prioritized. These developed models can provide an easy way to integrate various parameters, such as environmental factors, treatment methods, and water quality data, to predict the distribution and concentration of DBPs in an aquatic body. By simulating different outcomes and scenarios, predictive models can guide the implementation of effective management strategies and help inform decision-making. In addition to aquatic ecosystems, future research should also explore the impact of DBPs on terrestrial and atmospheric environments. Water treatment techniques such as evaporation lead to an increase in the chances of airborne DBP values, while generated solid waste in the form of wastewater sludge from any effluent treatment plant increases the level in the terrestrial ecosystem. Understanding the pathways of DBP dispersion and their potential effects on soil and air quality is essential for comprehensive risk assessment and management. Besides, this is very important to provide regular and continuous monitoring for DBP concentrations in drinking water with regulatory standards for it. Stringent implementation, monitoring, and establishment of guidelines are very crucial for ensuring a good or pure water supply and safeguarding public health. In all parts, a comprehensive approach to DBP management requires interdisciplinary collaboration between safeguarding public health, policymakers, water utilities, public health agencies, and researchers. By summarizing the vast and complex challenges associated with DBPs from various sources or multiple environmental perspectives, we need to develop sustainable solutions to preserve the aquatic ecosystem along with biodiversity and protect human health and terrestrial ecosystems for future generations. This is very important to follow the guideline of standards of DBPs in drinking water.

Acknowledgements The authors are very much thankful to the School of Allied Science, Dev Bhoomi Uttarakhand University, Dehradun, and the School of Applied and Life Sciences, Uttaranchal University, Dehradun for their support in completing this work.

References

Abusallout, I., Rahman, S., & Hua, G. (2016). Effect of temperature and pH on dehalogenation of total organic chlorine, bromine and iodine in drinking water. *Chemosphere, 187*, 11–18.

Agus, E., & Sedlak, D. L. (2010). Formation and fate of chlorination by-products in reverse osmosis desalination systems. *Water Research, 44*(5), 1616–1626.

Alexandrou, L., Meehan, B. J., & Jones, O. A. H. (2018). Regulated and emerging disinfection by-products in recycled waters. *Science of the Total Environment*, 637–638.

Bond, T., Templeton, M. R., Kamal, N. H. M., Graham, N., & Kanda, R. (2015). Nitrogenous disinfection by products in English drinking water supply systems: Occurrence, bromine substitution and correlation analysis. *Water Research, 85*, 85–94.

Bougeard, C. M. M., Goslan, E. H., Jefferson, B., & Parsons, S. A. (2020). Comparison of the disinfection by-product formation potential of treated waters exposed to chlorine and monochloramine. *Water Research, 44*, 729–740.

Chuang, Y.-H., Parker, K. M., & Mitch, W. A. (2021). Development of predictive models for the degradation of halogenated disinfection by-products during the UV/H_2O_2 advanced oxidation process. *Environmental Science & Technology, 50*(20), 11209–11217.

Chen, Y., Wen, Y., Tang, Z., Li, L., Cai, Y., & Zhou, Q. (2014). Removal processes of disinfection by-products in subsurface-flow constructed wetlands treating secondary effluent. *Water Research, 51*, 163–171.

Chen, B., Lee, W., Westerhoff, P. K., Krasner, S. W., & Herckes, P. (2010). Solar photolysis kinetics of disinfection by-products. *Water Research, 44*(11), 3401–3409.

Chen, B., Jiang, J., Yang, X., Zhang, X., & Westerhoff, P. (2021). Roles and knowledge gaps of point-of-use technologies for mitigating health risks from disinfection by-products in tap water: A critical review. *Water Research, 200*, 117–265.

Chowdhury, S., Rodriguez, M. J., Sadiq, R., & Serodes, J. (2011). Modelling DBPs formation in drinking water in residential plumbing pipes and hot water tanks. *Water Research, 45*(1), 337–347.

Cui, H., Chen, B., Jiang, Y., Tao, Y., Zhu, X., & Cai, Z. (2021). Toxicity of 17 disinfection by-products to different trophic levels of aquatic organisms: Ecological risks and mechanisms. *Environmental Science & Technology, 55*(15), 10534–10541.

Diana, M., Felipe-Sotelo, M., & Bond, T. (2019). Disinfection by-products potentially responsible for the association between chlorinated drinking water and bladder cancer: A review. *Water Research, 162*, 492–504.

Ding, S., Chu, W., Krasner, S. W., Yu, Y., Fang, C., Xu, B., & Gao, N. (2018). The stability of chlorinated, brominated, and iodinated halo acetamides in drinking water. *Water Research, 142*, 490–500.

Ding, S., Deng, Y., Bond, T., Fang, C., Cao, Z., & Chu, W. (2019). Disinfection by-product formation during drinking water treatment and distribution: A review of unintended effects of engineering agents and materials. *Water Research, 160*, 313–329.

Du, Y., Lv, X. T., Wu, Q. Y., Zhang, D. Y., Zhou, Y. T., Peng, L., & Hu, H. Y. (2017) Formation and control of disinfection by products and toxicity during reclaimed water.

Fu, K. Z., Li, J. H., Vemula, S., & Moe, B. (2017). Effects of halo benzoquinone and halo acetic acid water disinfection by-products on human neural stem cells. *Journal of Environmental Sciences, 58*, 239–249.

Ged, E. C., Chadik, P. A., & Boyer, T. H. (2015). Predictive capability of chlorination disinfection by-products models. *Journal of Environmental Management, 149*, 253–262.

Gupta, H. (2018). Anthracene removal from water onto activated carbon derived from vehicular tyre. *Separation Science and Technology, 53*(4), 613–625.

Heng, B., Zhang, R., Wang, Z., Zhang, Y., Wang, Y., Song, Z., Liu, C., Sun, D., & Qi, F. (2021). Occurrence and risk assessment of volatile halogenated disinfection by-products in an urban river supplied by reclaimed waste-water. *Ecotoxicology and Environmental Safety, 211*, 111–127.

Huang, G., Jiang, P., & Li, X. F. (2017). Mass spectrometry identification of N-chlorinated dipeptides in drinking water. *Analytical Chemistry, 89*(7), 4204–4209.

Ike, I. A., Karanfil, T., Ray, S. K., & Hur, J. (2020). A comprehensive review of mathematical models developed for the estimation of organic disinfection by-products. *Chemosphere, 246*, 125–147.

Jasemizad, T., Sun, P., & Padhye, L. P. (2021). Aqueous N-nitrosamines: Precursors, occurrence, oxidation processes, and role of inorganic ions. *Critical Reviews in Environmental Science and Technology*, 1–47.

Jianrong, W., Bixiong, Y., Wuyi, W., Linsheng, Y., Jing, T., & Zhiyu, H. (2010). Spatial and temporal evaluations of disinfection by-products in drinking water distribution systems in Beijing, China. *Science of Total Environment, 408*(20), 4600–4606.

Jin, W., Zhou, J., Chen, B., Zhu, X., & Cui, C. (2012). Modeling volatilization and adsorption of disinfection byproducts in natural watersheds. *Journal of Environmental Monitoring : JEM, 14*(11), 2990–2999.

Keefe, S. H., Barber, L. B., Runkel, R. L., & Ryan, J. N. (2004). Fate of volatile organic compounds in constructed wastewater treatment wetlands. *Environmental Science & Technology, 38*(7), 2209–2216. https://doi.org/10.1021/es034661i

Khan, W. U., Ahmed, S., Dhoble, Y., & Madhav, S. (2023). A critical review of hazardous waste generation from textile industries and associated ecological impacts. *Journal of the Indian Chemical Society, 100*(1), 100829.

Kitis, M., Yigit, N. O., Harman, B. I., Muhammetoglu, H., Muhammetoglu, A., Karadirek, I. E., Demirel, I., Ozden, T., & Palanci, I. (2010). Occurrence of trihalomethanes in chlorinated groundwaters with very low natural organic matter and bromide concentrations. *Environmental Forensics, 11*, 264–274.

Krasner, S. W., Westerhoff, P., Chen, B., Rittmann, B. E., & Amy, G. (2010). Occurrence of disinfection by products in United States wastewater treatment plant effluents. *Environmental Science & Technology, 43*(21), 8320–8325.

Kumar, R., Gupta, B., Gupta, H., & Rani, M. (2014). Distribution of persistent organic pollutants in urban aquatic systems. *International Journal of Scientific Research in Environmental Sciences, 2*(7), 233–243.

Kumar, R., Rani, M., Gupta, H., Gupta, B., Park, D., & Jeon, B. H. (2017). Distribution of trace elements in flowing surface waters: Effect of seasons and anthropogenic practices in India. *International Journal of Environmental Analytical Chemistry, 97*(7), 637–656.

Kumari, M., & Gupta, S. K. (2018). Age dependent adjustment factor (ADAF) for the estimation of cancer risk through trihalomethanes (THMs) for different age group. An innovative approach. *Ecotoxicology and Environmental Safety, 148*, 960–968.

Lei, X., Lei, Y., Zhang, X., & Yang, X. (2021). Treating disinfection by products with UV or solar irradiation and in UV advanced oxidation processes: A review. *Journal of Hazardous Materials, 408*, 124–135.

Li, W. T., Cao, M. J., Young, T., Ruffino, B., Dodd, M., Li, A. M. & Kotshin, G. (2017). Application of UV absorbance and fluorescence indicators to assess the formation of biodegradable dissolved organic carbon and bromate during ozonation. *Water Research, 111*: 154–162.

Li, Z., Song, G., Bi, Y., Gao, W., He, A., Lu, Y., Wang, Y., & Jiang, G. (2021). Occurrence and distribution of disinfection by products in domestic wastewater effluent, tap water, and surface water during the SARS-COV-2 pandemic in. *Environmental Science & Technology, 55*(7), 4103–4114.

Li, J., Wang, W., Moe, B., Wang, H., & Li, X. F. (2015). Chemical and toxicological characterization of halo benzoquinones, an emerging class of disinfection by-products. *Chemical Research in Toxicology, 28*(3), 306–318.

Liu, J., Zhang, X., Li, Y., Li, W., Hang, C., & Sharma, V. K. (2019). Photo transformation of halo phenolic disinfection by-products in receiving seawater: Kinetics, products, and toxicity. *Water Research, 150*, 68–76.

Lv, X. T., Zhang, X., Du, Y., Wu, Q. Y., Lu, Y., & Hu, H. Y. (2017). Solar light irradiation significantly reduced cytotoxicity and disinfection by-products in chlorinated reclaimed water. *Water Research, 125*, 162–169.

Lu, J., Xiao, X., Xu, Z., Wang, C., Zhang, M., & Zhou, Y. (2021). The potential of virtual tourism in the recovery of tourism industry during the COVID-19 pandemic. *Current Issues in Tourism, 25*(3), 441–457. https://doi.org/10.1080/13683500.2021.1959526

Luan, X., Liu, X., Fang, C., Chu, W., & Xu, Z. (2020). Ecotoxicological effects of disinfected wastewater effluents: A short review of in vivo toxicity bioassays on aquatic organisms. *Environmental Science: Water Research & Technology, 6*(9), 2275–2286.

Madhav, S., Ahamad, A., Singh, A. K., Kushawaha, J., Chauhan, J. S., Sharma, S., & Singh, P. (2020). *Water pollutants: Sources and impact on the environment and human health. Sensors in water pollutants monitoring: Role of material* (pp. 43–62).

Mazhar, M. A., Madhav, S., Ahmed, S., Kumar, P., & Springer, C. (2024). Drinking water chlorination and disinfection by-products: Formation, history, and regulations. In *Drinking water disinfection by-products: Sources, fate and remediation* (pp. 21–34). Springer Nature Switzerland.

McCurry, D. L., Quay, A. N., & Mitch, W. A. (2016). Ozone promotes chloropicrin formation by oxidizing amines to nitro compounds. *Environmental Science and Technology, 50*(3), 1209–1217.

Meite, L., Fotsing, M., & Barbeau, B. (2015). Efficacy of ozone to reduce chlorinated disinfection by-products in Quebec (Canada) drinking water facilities. *Ozone Science Engineering, 37*: 1–12.

Mian, H. R., Hu, G., Hewage, K., Rodriguez, M. J., & Sadiq, R. (2018). Prioritization of unregulated disinfection by-products in drinking water distribution systems for human health risk mitigation: A critical review. *Water Research, 147*, 112–131.

Mohd, A., Nadeem, A. K., Sirajuddin, A., Afzal, H. K., Azhar, M., Nadeem, H., Rahisuddin, F. C., Mohd, Y., & Viola, V. (2020). Chlorination disinfection by-products in municipal drinking water—A review. *Journal of Cleaner Production, 273*, 123–159.

Mohamed, E.O., Sharaf, A.A., Talaat, D.M. and Nagui, D.A.. (2020). Remineralization effect of egg shell powder and novamine on initial caries-like lesions in young permanent teeth (in-vitro study). *Alexandria Dental Journal, 45*(2), 52–59.

Neale, P. A., Altenburger, R., Ait-Aissa, S., Brion, F., Busch, W., De Aragão Umbuzeiro, G., Denison, M. S., Du Pasquier, D., Hilscherova, K., Hollert, H., Morales, D. A., Novák, J., Schlichting, R., Seiler, T., Serra, H., Shao, Y., Tindall, A. J., Tollefsen, K. E., Williams, T. D., & Escher, B. I. (2017). Development of a bioanalytical test battery for water quality monitoring: Fingerprinting identified micropollutants and their contribution to effects in surface water. *Water Research, 123*, 734–750.

Pal, A., He, Y., Jekel, M., Reinhard, M., & Gin, K.Y.-H. (2014). Emerging contaminants of public health significance as water quality indicator compounds in the urban water cycle. *Environment International, 71*, 46–62.

Patrick, L., Yves, G., Moliere, N. M., Manuel, R., Celine, C., Robert, T., & Alexandre, B. (2016). Disinfection by-products exposure and intra-uterine growth restriction: Do genetic polymorphisms of *CYP2E1*or deletion of *GSTM1* or *GSTT1* modify the association? *Environment International, 92–93*, 220–231.

Peng, L., Wang, F., Zhang, D., Fang, C., van der Hoek, J. P., & Chu, W. (2021). Effect of oxidation ditch and anaerobic-aniconic processes on CX3R-type disinfection by-product formation during wastewater treatment. *Science of the Total Environment, 770*, 145–344.

Postigo, C., Emiliano, P., & Valero, F. (2020). High-throughput and reliable determination of 13 haloacetic acids and dalapon in water and evaluation of control strategies. *Environmental Science, Water Research & Technology, 6*(9), 2499–2509.

Postigo, C., Gil-Solsona, R., Herrera-Batista, M. F., Gago-Ferrero, P., Alygizakis, N., Ahrens, L., & Wiberg, K. (2021). A step forward in the detection of by-products of anthropogenic organic micropollutants in chlorinated water. *Trends in Environmental Analytical Chemistry, 32*, 148.

Richardson, S. D. (2011) Disinfection by-products: Formation and occurrence in drinking water. In J. O. Nriagu (Ed.), *Encyclopaedia of environmental health* (pp. 110–136).

Richardson S.D., Plewa, M.J., & Wagner, E.D. (2007). Occurrence, genotoxicity, and carcinogenicity of regulated and emerging disinfection by-products in drinking water: a review and roadmap for research. *Mutation Research 36*(1– 3):178–242.

Rong, X., Tian, O., Shunke, D., Chao, F., Zuxin, X., & Wenhai, C. (2022). Disinfection by-products as environmental contaminants of emerging concern: A review on their occurrence, fate and removal in the urban water cycle. *Critical Reviews in Environmental Science and Technology* (pp. 20–37).

Rostad, C. E., Martin, B. S., Barber, L. B., Leenheer, J. A., & Daniel, S. R. (2000). Effect of a constructed wetland on disinfection byproducts: Removal processes and production of precursors. *Environmental Science & Technology, 34*(13), 2703–2710. https://doi.org/10.1021/es9900407

Rostad, C. E., Martin, B. S., Barber, L. B., Leenheer, J. A., & Daniel, S. R. (2010). Effect of a constructed wetland on disinfection by-products: Removal processes and production of precursors. *Environmental Science & Technology, 34*(13), 2703–2710.

Samuel, I., P., Celine, C., Christelle, L., Raywat, D., Manuel, J., & Rodriguez, P. L. (2015). Women exposure during pregnancy to halo acetaldehydes and haloaceto nitriles in drinking water and risk of small-for-gestational-age neonate. *Environmental Research, 137*, 338–348.

Saxena, R., Gupta, H., & Dhiman, S. (2024). Availability and minimization of nitrogenous disinfectant by-products in drinking water. In S. Madhav, M. A. Mazhar, S. Ahmed, P. Kumar, & P. K. Mishra (Eds.), *Drinking water disinfection by-products*. Springer. https://doi.org/10.1007/978-3-031-49047-7_12

Seeger, E. M., Reiche, N., Kuschk, P., Borsdorf, H., & Kaestner, M. (2011). Performance evaluation using a three compartment mass balance for the removal of volatile organic compounds in pilot scale constructed wetlands. *Environmental Science & Technology, 45*(19), 8467–8474.

Shah, A. D., & Mitch, W. A. (2012). Halo nitroalkanes, halo nitriles, haloamides and N-nitrosamines: A critical review of nitrogenous disinfection by-product (N-DBP) formation pathways. *Environmental Science and Technology, 46*(1), 119–131.

Shiqing, Z., Shumin, Z., Yisheng, S., & Naiyun, G. (2015). Characteristics of C-, N-DBPs formation from algal organic matter: Role of molecular weight fractions and impacts of pre-ozonation. *Water Research, 72*, 381–390.

Soller, J. A., Eftim, S. E., & Nappier, S. P. (2019). Comparison of predicted microbiological human health risks associated with de facto, indirect, and direct potable water reuse. *Environmental Science & Technology, 53*(22), 13382–13389.

Szczuka, A., Parker, K. M., Harvey, C., Hayes, E., Vengosh, A., & Mitch, W. A. (2017). Regulated and unregulated halogenated disinfection by-product formation from chlorination of saline groundwater. *Water Research, 122*, 633–644.

Tang, H., Zhong, H., Pan, Y., Zhou, Q., Huo, Z., Chu, W., & Xu, B. (2021). A new group of heterocyclic nitrogenous disinfection by-products (DBPs) in drinking water: Role of extraction pH in unknown DBP exploration. *Environmental Science & Technology, 55*(10), 6764–6772.

Tak, S., & Vellanki, P. (2018). Natural organic matter as a precursor to disinfection by-products and its removal using conventional and advanced processes: State of the art review. *Journal Water Health, 16*, 681–703.

Thang, N.Q., Huy, B.T., & Van Tan, L. (2017). Lead and Arsenic Accumulation and Its Effects on Plasma Cortisol Levels in Oreochromis sp.. *Bulletin of Environmental Contamination and Toxicology, 99*, 187–193. https://doi.org/10.1007/s00128-017-2113-7

Uyak, V., Soylu, S., Topal, T., Karapinar, N., Ozdemir, K., Ozaydin, S., & Avsar, E. (2014). Spatial and seasonal variations of disinfection by-products (DBPs) in drinking water distribution systems of İstanbul city, Turkey. *Environmental Forensics, 15*, 190–205.

Van Huy, N., Murakami, M., Sakai, H., Oguma, K., Kosaka, K., Asami, M., & Takizawa, S. (2011). Occurrence and formation potential of N-nitroso dimethylamine in ground water and river water in Tokyo. *Water Research, 45*(11), 3369–3377.

Wang, X., Liu, Z., Wang, C., Ying, Z., Fan, W., & Yang, W. (2016). Occurrence and formation potential of nitrosamines in river water and groundwater along the Songhua River. *Journal of Environmental Sciences, 50*, 65– 71. https://doi.org/10.1016/j.jes.2016.05.021

Wang, L., Sun, Y., & Chen, B. (2018). Rejection of haloacetic acids in water by multi-stage reverse osmosis: Efficiency, mechanisms, and influencing factors. *Water Research, 144*, 383–392.

Wang, L., Zhang, X., Chen, S., Meng, F., Zhang, D., Liu, Y., Li, M., Liu, X., Huang, X., & Qu, J. (2021). Spatial variation of dissolved organic nitrogen in Wuhan surface waters: Correlation with the occurrence of disinfection by-products during the COVID-19 pandemic. *Water Research, 198*, 117–138.

Wagner, E. D., & Plewa, M. J. (2017). CHO cell cytotoxicity and genotoxicity analyses of disinfection by-products: An updated review. *Journal of Environmental Sciences, 58*, 64–76.

Watson, K., Shaw, G., Leusch, F. D. L., & Knight, N. L. (2012). Chlorine disinfection by-products in wastewater effluent: Bioassay-based assessment of toxicological impact. *Water Research, 46*(18), 6069–6083.

WHO. (2017). Potable reuse: Guidance for producing safe drinking water. Switzerland.

Wright, J. M., Evans, A., Kaufman, J. A., Rivera-Nunez, Z., & Narotsky, M. G. (2017). Disinfection by-product exposures and the risk of specific cardiac birth defects. *Environmental Health Perspectives, 125*(2), 269–277.

Xing-Fang, L., & William, A. M. (2018). Drinking water disinfection by-products (DBPs) and human health effects: Multidisciplinary challenges and opportunities. *Environmental Science and Technology, 52*(4), 1681–1689.

Xu, Z., Qu, Y., Wang, S., & Chu, W. (2021). Diagnosis of pipe illicit connections and damaged points in urban stormwater system using an inversed optimization model. *Journal of Cleaner Production, 292*, 126011.

Yang, Y., Shi, L., Fu, X., Ma, G., & Yang, Z. Li. Y. (2022). Metabolic and epigenetic dysfunctions underlie the arrest of in vitro fertilized human embryos in a senescent-like state. *PLoS Biol, 20*(6): 3001682. https://doi.org/10.1371/journal.pbio.3001682

Yan-Ling, D., Qiong, L., Xiao-Qiong, Y., Chong, L., Ting-Ting, L., Tian, S., Jia-Yue, Z., & Fei-Peng, C. (2019). Women attending an infertility clinic. *International Journal of Environmental Health, 24*, 113–231.

Yu, M., Min, Z., Pan-Pan, C., Yang-Cheng, Y., Lin-Jing, W., Hua-Hua, J., Qing-Yun, Y., Yu-Feng, L., Wen-Qing, L., & Qiang, Z. (2022). Associations between drinking water disinfection by-products and menstrual cycle characteristics: A cross-sectional study among.

Yi-Xin, W., Qiang, Z., Le, W., Yue-Hui, H., Zhi-Wei, L., Peng, W., Meng-Jie, H., Yi-Xin, W., Qiang, Z., Le, W., & Yue-Hui, H. (2014). Temporal variability in urinary levels of drinking water disinfection by-products dichloroacetic acid and trichloroacetic acid among men. *Environmental Research, 135*, 126–132.

Ying, Y., Xu, M., Ruya, C., Guiwei, L., Hui, T., & Baoyou, S. (2019). The occurrence and transformation behaviours of disinfection by-products in drinking water distribution systems in rural areas of eastern China. *Chemosphere, 228*, 101–109.

Yu, J., Hu, H., Agmon, A. and Svoboda, K. (2019). Recruitment of GABAergic interneurons in the barrel cortex during active tactile behavior. *Neuron, 104*(2), 412–427.

Yu, Y., & Reckhow, D. A. (2015). Kinetic analysis of halo acetonitrile stability in drinking waters. *Environmental Science & Technology, 49*(18), 11028–11036.

Zeng, T., Plewa, M. J., & Mitch, W. A. (2016). N-nitrosamines and halogenated disinfection by-products in U.S. full advanced treatment trains for potable reuse. *Water Research, 101*, 176–186.

Zeng, T., Glover, C. M., Marti, E., Woods, G., Karanfil, T., Mitch, W. A., & Dickenson, E. R. V. (2017). Relative importance of different water categories as sources of N-nitrosamine precursors. *Environmental Science and Technology, 50*(24), 13239–13248.

Zhang, Y., Chu, W., Yao, D., & Yin, D. (2018). Control of aliphatic halogenated DBP precursors with multiple drinking water treatment processes: Formation potential and integrated toxicity. *Journal of Environmental Sciences, 58*, 322–330.

Zhang, R., Wang, F., Fang, C., Luan, X., Yang, X., & Chu, W. (2021). Occurrence of CX3R-type disinfection by-products in drinking water treatment plants using DON-rich source water. *ACS ES&T Water, 1*(3), 553–561.

Zhiwen, C., Xiaobin, L., Ye, Y., Liu, H., Zixiang, H., Yingying, D. F. H. Y., Xubo, S., Tongxan, X., Yuefeng, X., & Wendong, W. (2023). Analysis of rainwater storage and use recommendations: From the perspective of DBPs generation and their risks. *Journal of Hazardous Materials*, 130–233.

Zhou, Y., Tigane, T., Li, X., Truu, M., Truu, J., & Mander, U. (2013). Hexachlorobenzene dichlorination in constructed wetland mesocosms. *Water Research, 47*(1), 102–110.

Chapter 7
Surfactants and Flame Retardants in Water Resources

Shreya Kotnala, Richa Saxena, Arunima Nayak, and Brij Bhushan

Abstract As industrial and domestic activities continue to grow, the release of surfactants and flame retardants into water bodies poses a significant threat to the quality of freshwater resources. The chapter provides a comprehensive exploration of the impact, presence, and management of surfactants and flame retardants in aquatic environments. Various sources and pathways via which these anthropogenic compounds enter into water sources are discussed. The chapter deals with the physicochemical properties of surfactants and flame retardants, their behavior in aquatic ecosystems, and their ecotoxicological effects on the environment and aquatic life. The chapter also explores several analytical techniques employed for the analysis of both surfactants and flame retardants in water systems. Several mitigation techniques including physical and chemical methods for the removal of surfactants and flame retardants have been discussed. The present chapter also deals with the challenges, future perspectives, and several analytical as well as mitigation strategies employed for the eradication of surfactants and flame retardants. The chapter provides a detailed overview of sources, properties, and negative impact of surfactants and flame retardants that could be useful for researchers and academicians to develop analytical methods for the removal of these contaminants and hence foster the sustainability of water ecosystems.

Keywords Surfactants · Flame retardants · Adsorption · Ecotoxicological effects · Sustainability

---

S. Kotnala
Department of Chemistry, School of Basic and Applied Sciences, Shri Guru Ram Rai University, Dehradun, India

R. Saxena (✉)
Department of Physics, School of Basic and Applied Sciences, Shri Guru Ram Rai University, Dehradun, India
e-mail: saxena.richa23@gmail.com

A. Nayak · B. Bhushan
Department of Chemistry, Graphic Era University, Dehradun 248002, India

H. Gupta et al. (eds.), *Emerging Contaminants in Water and Wastewater*, Environmental Science and Engineering, https://doi.org/10.1007/978-3-031-82579-8_7

7.1 Introduction

Due to ever increasing population and rapid industrialization scarcity of water is becoming a major issue. It has been estimated that the freshwater resources will decrease from 7370 $m^3year^{-1}capita^{-1}$ to 5717 $m^3year^{-1}capita^{-1}$ in the year 2025 due to increase in water consumption by 70% (Jena et al., 2023). The release of harmful contaminants such as dyes, pesticides, surfactants, flame retardants, cosmetics, etc. has resulted in the contamination of freshwater resources (Khan et al., 2023; Kumar et al., 2014, 2017; Madhav et al., 2024; Singh et al., 2020, 2022). All these contaminants are soluble in nature and they are persistent in the environment therefore they pose detrimental effects on humans as well as the environment (Kumar et al., 2023; Saxena et al., 2024; Siyal et al., 2020). Surfactants are called amphiphilic compounds or surface-active agents, and they are usually known for their capability to decrease the surface tension between a pair of phases, often between water and some other substance (Bolan et al., 2023; Nunes & Teixeira, 2022). Surfactants are used in various fields such as agrochemicals, cleaning agents, pharmaceuticals, personal care products, detergents, industrial products, etc. Surfactants are classified into four categories: cationic, anionic, non-ionic amphoteric, and amphoteric surfactants. Based on their hydrophilic and hydrophobic component, they may exhibit different properties and applications. Cationic surfactants are positively charged, anionic surfactants are negatively charged, non-ionic are neutral, and amphoteric surfactants carry both positive and negative charges (Mitru et al., 2020). On the other hand, flame retardants are a variety of chemicals, which belong to different classes, that are designed to prevent or delay ignition, combustion, and/or spread of fire in various materials like textiles, plastics, foams, electronic devices including construction materials (Innes & Innes, 2011).

Surfactants and flame retardants are omnipresent in the environment due to their widespread use. They are found in surface water, groundwater, and drinking water sources (Jena et al., 2023). There are several pathways via which these chemicals enter into aquatic environments such as urban runoff, effluents from domestic water, discharge from industries, and agricultural run-off (Bautista-Toledo et al., 2014; Gogoi et al., 2018). These contaminants pose a potent threat not only to the environment but also to human beings. As these chemicals are frequently released into water bodies they may bioaccumulate and can also lead to eutrophication as they enhance solubility (Jena et al., 2023). Due to their toxic effect, they must be eliminated from water resources by adopting mitigation strategies (Badmus et al., 2021). Surfactants such as linear alkyl benzene sulfonate and benzalkonium chloride are known to affect the ability of aquatic organisms to reproduce and hence have a negative impact on their development.

Flame retardants consist of diverse groups of chemicals such as bromine, chlorine, and phosphorus and they are regarded as endocrine disruptors (Feiteiro et al., 2021). These have been utilized commercially since the 1970s and are used to reduce the flammability of materials (van der Veen & de Boer, 2012; Zota et al., 2018). They tend to bioaccumulate and have a negative impact on human health as the human population has been exposed to FRs for decades as they are present in consumer products, food, and indoor dust (Feiteiro et al., 2021; Zota et al., 2018). Brominated flame retardants (BFRs) are one of the major classes of FRs that are widely used in industrial chemicals and account for about 21% of total FRs produced. These are identified as toxic chemicals and are persistent in nature. They have the tendency to bioaccumulate and are found at a considerable level in both wildlife and human beings (Feiteiro et al., 2021; Waaijers & Parsons, 2016). Polybrominated diphenyl ethers (PBDE), hexabromocyclododecane (HBCD), and tetrabromobisphenol A (TBBA) are widely used in industries and Several studies have reported that their level is increasing in humans, the environment and wildlife owing to their ability to persist, bioaccumulate and biomagnify leading to several negative impacts on the health of wildlife and humans (Kim & Oh, 2014).

Surfactants and FRs are persistent substances that are highly toxic in nature. Various technologies can be adopted to reduce the level of these contaminants in water bodies in order to lessen their harmful impact on aquatic ecosystems. The current chapter provides a basic overview of several sources and pathways via which these chemicals are introduced into water bodies. Various physicochemical properties are elaborated in order to explain the characteristics of surfactants and FRs. The chapter provides a detailed discussion of the ecotoxicological effects of these chemicals on aquatic organisms, plants, and humans. The properties, behavior, and nature of surfactants and FRs must be kept in mind while developing various methods for their analysis and detection. Several analytical techniques for the identification and detection of these chemicals has been discussed. The present chapter also delves into several mitigation strategies including physical and chemical methods adopted for the eradication of these contaminants from water bodies. This ensures the development of policies and practices that could be implemented to safeguard the environment and pave the way towards sustainability.

7.2 Sources and Pathways

There are several sources and pathways via which surfactants and flame retardants can enter water resources and pose a significant threat to human health and aquatic ecosystems. Surfactants mainly enter water bodies by the outflow of wastewater from homes and businesses. They are most frequently found in detergents, soaps, and industrial cleaners (Bautista-Toledo et al., 2014; Jena et al., 2023; Li et al., 2018; Moura et al., 2019). The most common sources of environmental pollution are agrochemicals those are pesticides, biocides, and herbicides, it also includes various products of industry like wetting agents, personal care items, paper, emulsifiers, fabric, detergents, and treatments of carpets, which are important sources of origin for the pollution created to the environment because of surfactants. Additionally, household activities like washing clothes, disinfection, and fumigation are also the reason for presence of the surfactants in the living world (Badmus et al., 2021). When the use of these products is done for cleaning purposes, the surfactants they contain

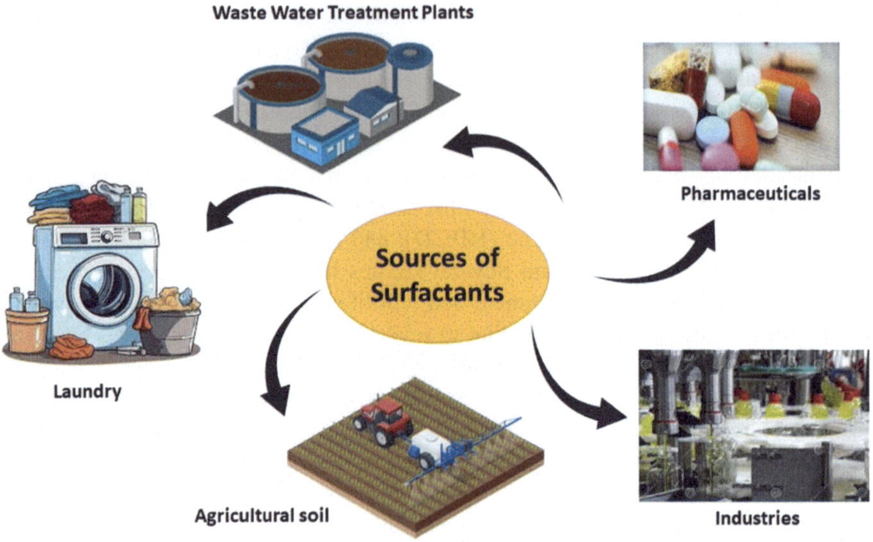

Fig. 7.1 Various sources of surfactants

are rinsed off and washed down drains, ultimately reaching sewage treatment plants or being directly discharged into water bodies. Additionally, agricultural runoff from fields treated with pesticides and herbicides containing surfactants can carry these compounds into surface water sources during rainfall or irrigation events. Several sources of surfactants are depicted in Fig. 7.1.

Organophosphorus FRs (OPFRs) are released into the environment via abrasion, dissolution, leaching, and volatilization (Tian et al., 2023; Yao et al., 2021) and hence can be found in air, sediments, dust, and water bodies (Cao et al., 2019; Fromme et al., 2014). The OPFRs can be transformed into OP diesters which are harmful to the environment and exist in urine, surface water, and effluents from wastewater treatment plants (Ding et al., 2015; Fu et al., 2020; Tian et al., 2023; Xu et al., 2019). Aryl phosphate FRs are used in plastics, furniture, electronic products and foaming products and degradation of these products may result in their release into water bodies and sediments resulting in toxic effects to the environment and human health (Jin et al., 2024). Halogenated FRs are widely used in products of commercial use. Two major classes of halogenated FRs Polybrominated diphenyl ethers (PBDE) and Hexabromocyclododecane (HBDD) are widely used with a global production of 1,900,000 and 6,000,000 tonnes respectively (Abbasi et al., 2019; Li & Wania, 2018; Ma et al., 2023). Since these are extensively used, they are usually found in considerable amounts in the environment and humans (Ma et al., 2023). The brominated FRs are found in water bodies and they could enter these waterbodies via several pathways such as discharge from the industries, effluent from wastewater treatment plants, and leaching from landfills (Kerric et al., 2023). The presence of various types of FRs in the environment raises an alarming concern as they

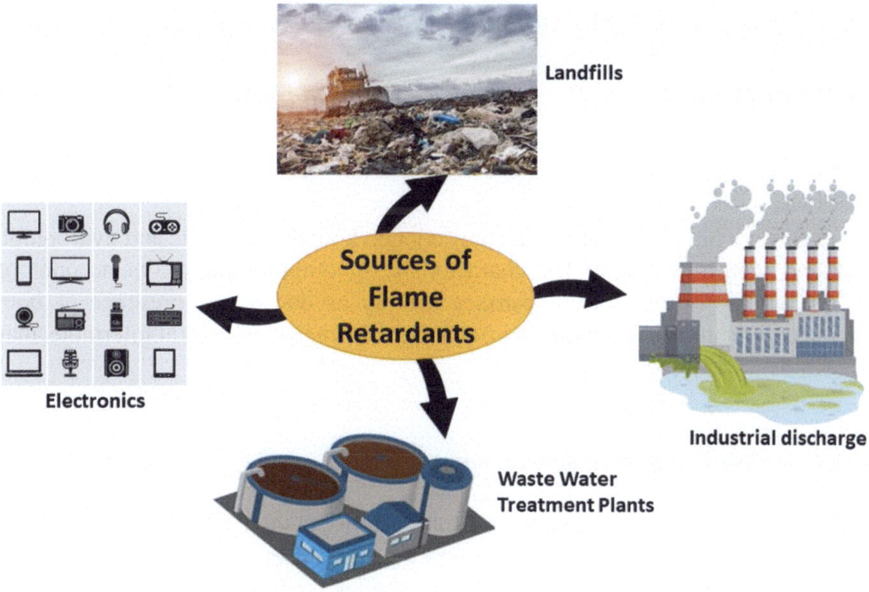

Fig. 7.2 Various sources of Flame retardants

are associated with endocrine disruption, developmental issues, and neurotoxicity (Zapata-Corella et al., 2023). Various sources via which these contaminants are released into the environment is depicted in Fig. 7.2.

The presence of both surfactants and FRs in the environment is a concern and their eradication is a must in order to lessen their impact on aquatic life and humans. Therefore several mitigation strategies must be developed or adopted in order to maintain water quality and to rectify their harmful impact. Several methods such as the development of eco-friendly methodologies, use of less harmful chemicals, innovative treatment methodologies, and strict regulations for the usage of these chemicals must be embraced so as to safeguard the environment thereby promoting sustainability.

7.3 Physicochemical Properties

7.3.1 Physicochemical Properties of Surfactants

7.3.1.1 Amphiphilicity

Surfactants are bipolar in nature i.e., they contain both hydrophobic and hydrophilic parts and therefore exhibit the characteristic of amphiphilicity. On account of this characteristic, they exhibit surface activity as due to the presence of both water-repelling and water-loving portions they can engage in several intermolecular interactions (Jena et al., 2023).

7.3.1.2 Critical Micelle Concentration (CMC)

At minimal concentration, surfactants display a CMC, at this concentration they form micelles spontaneously within the solution. As the concentration of the surfactants exceeds a certain value i.e. threshold value micelles are formed, and because of this surface tension is reduced. In various industries like mineral processing, personal care, food formulation, drug delivery, and surfactant remediation, the CMC plays a vital role. Surfactants are commonly used for lowering interfacial tension, stabilization of suspension, emulsification, enhancing the stability of foam, behave as delivery vehicles. They are used at greater concentrations than CMC as these are most pronounced when a higher concentration of micelle is present (Schramm et al., 2003; Tripathy et al., 2018).

7.3.1.3 Krafft and Cloud Points

The temperature at which the surfactant's capability of forming the solubility of micelle increases is known as the Krafft point. At the temperature below this Krafft point, the surfactant's low solubility prevents micelle formation, so the concentration of surfactant monomers is determined solely by solubility. Micelles are formed in a considerable amount at a temperature higher than Kraft point thereby enhancing the solubility. In the case of non-ionic surfactants, the solubility decreases with the increase in temperature. Also, such surfactants do not possess Krafft's point. These surfactants may lose their surface-active properties above the transition temperature known as the cloud point, They may lose their surface-active qualities above the cloud point as a phase rich in surfactants and inflated micelles separates, evidently increasing the turbidity of the dispersion (Schramm et al., 2003).

7.3.1.4 Change in Wetting Behavior

Surfactants adsorb onto solid surfaces, reducing interfacial tension and altering the wetting behavior of water or oil on the surface. When the surfactant's hydrophobic tail groups face away from the surface as well as along the surface, there is a decrease in water-wetting and there is an increase in oil-wetting. Conversely, if the polar head group faces away from the surface, water-wetting increases. Other than these inherent properties of molecules of surfactant, the specific structural features can affect how closely packed the surfactant molecules are, influencing surface wetting. There is a shift in wettability due to surfactant adsorption to water-wetting by the oil wetting as demonstrated in various studies (Schramm et al., 2003).

7.3.1.5 Interfacial Tension Reduction

Because surfactants have surface-active characteristics, they preferentially gather at liquid–liquid and liquid–air interfaces, for example, where two immiscible phases meet (Siyal et al., 2020). Adsorbing at these interfaces, surfactants lessen the force that minimizes the surface area between the two phases. This force is known as the interfacial tension. Interfacial tension is lowered, which makes it easier for colloidal systems to emulsify, disperse, and stabilize (Tripathy et al., 2018).

7.3.1.6 Dispersibility

Dispersibility is the ability of surfactants to prevent the particles from clumping and settling in the solution. It is a crucial characteristic of surfactants which makes it useful in several applications such as in detergents, cosmetics, and pharmaceuticals (Tripathy et al., 2018).

7.3.2 Physicochemical Properties of Flame Retardants

7.3.2.1 Structure

On the basis of chemical structure flame retardants are divided into several classes such as brominated flame retardants (BFRs), organophosphate flame retardants (OPFRs), halogen-free flame retardants, and others (Birnbaum & Staskal, 2004).

7.3.2.2 Solubility

Generally, the flame retardants are insoluble in water. It has been observed that with the increase in bromination the solubility decreases in case of polybrominated biphenyls. They are usually lipophilic substances (Guerra et al., 2010). Polybrominated diphenyl ethers are insoluble in water.

7.3.2.3 Octanol–Water Partition Coefficient (K_{OW})

It is an important characteristic which is used to assess the environmental fate and behaviour of chemicals. It is inversely related to solubility. FRs having low value of K_{OW} are highly insoluble in water and they have the tendency to bioaccumulate (Salih Al-Omran, 2019).

7.3.2.4 Thermal Stability

Flame retardants must be thermally stable to withstand high temperatures encountered during combustion. This property ensures that they remain effective in suppressing or retarding the ignition and spread of flames.

7.3.2.5 Toxicity

The toxicity of flame retardants is an important consideration, especially in applications where human exposure is possible. Ideally, flame retardants should exhibit low toxicity to humans and the environment while still providing effective fire protection. The Physiological properties of Surfactants and Flame Retardants are shown in Table 7.1.

7.4 Ecotoxicological Effects

Surfactants and flame retardants are ubiquitous in various consumer products, industrial processes, and household items. They serve essential functions such as enhancing cleaning efficacy, improving material properties, and reducing fire hazards. However, their widespread use raises concerns about their potential ecotoxicological effects on aquatic ecosystems, particularly when they enter water resources through various pathways. Ecotoxicological effects of surfactants and Flame Retardants are presented in Table 7.2.

Table 7.1 Physiological properties of surfactants and flame retardants

Physiological properties	Surfactants	Flame retardants
Amphiphilicity	Amphiphilic molecules containing both hydrophilic and hydrophobic regions, enabling adsorption at interfaces between immiscible phases	Diverse classes such as brominated, organophosphate, and halogen-free, exhibiting various chemical structures influencing properties
Critical micelle concentration (CMC)	Exhibits CMC, the minimum concentration for micelle formation, maximizing effects above this threshold	N/A
Krafft and cloud points	Krafft point influences micelle formation temperature; cloud point affects surfactant's surface-active properties	Influenced by chemical structure and thermal stability, important for effective fire suppression
Change in wetting behavior	Adsorb onto solid surfaces, altering wetting behavior depending on orientation of hydrophobic and hydrophilic groups	N/A
Interfacial tension reduction	Reduce interfacial tension, facilitating processes like emulsification and dispersion	N/A
Dispersibility	Prevent clumping and settling of particles in solution, crucial for applications in detergents, cosmetics, and pharmaceuticals	N/A
Solubility	Varies in solubility, often soluble in water or organic solvents	Demonstrates lipophilic properties, often resistant to dissolution in water
Octanol–Water Partition coefficient (Kow)	N/A	Denotes potential for the bioaccumulation, lower values preferred for lower environmental impact
Thermal stability	N/A	Essential for effective fire protection, must withstand high temperatures during combustion
Toxicity	Can vary widely in toxicity, some may be harmful if ingested or inhaled	Toxicity varies, ideally exhibiting low toxicity to humans and the environment while providing effective fire protection

Table 7.2 Ecotoxicological effects of surfactants and flame retardants

Ecotoxicological effects	Surfactants	Flame retardants
Toxicity to aquatic organisms	• Disruption of chemoreceptors in fish species at low concentrations (0.5 ppm) (Bardach et al., 1965) • Chronic and sublethal toxicities at concentrations above 0.1 mg/L (Rebello et al., 2014) • Disruption of cell membranes leading to cellular damage and dysfunction • Adverse effects on orans of fish such as gills, liver, kidney, spleen, and intestines (Mustapha & Bawa-Allah, 2020)	• Induced oxidative stress in fish species (Carassius auratus) at higher doses and prolonged exposure (Feng et al., 2013) • Toxic effects on water fleas (Daphnia magna) with higher concentrations causing greater mortality (Hardy et al., 2012) • DNA damage in liver cells of fathead minnows after exposure to specific flame retardant mixtures (Bearr et al., 2010)
Effect on aquatic plants	• Disruption of surface tension and infiltration of cell membranes causing structural damage (Jovcic et al., 2010) • Inhibition of respiration and photosynthesis, impeding essential physiological processes • Disruption of nutrient uptake leading to imbalances in plant health (Rebello et al., 2014)	• Lipophilic nature leading to accumulation in fatty tissues of aquatic organisms • Biomagnification across trophic levels, resulting in elevated concentrations in apex predators (Yao et al., 2021) • Bioaccumulation in plants, animals, and humans, inducing cytotoxicity and increasing cancer risk (Yao et al., 2021)
Impact on microorganisms	• Inhibition of microbial activity leading to imbalances in nutrient cycling and degradation of organic matter (Lima et al., 2011) • Detrimental effects on bacteria and algae growth, motility, and photosynthetic ability Aizdaicher & Markina, 2006) • Adverse effects on duckweeds, reducing cell density, motility, and chlorophyll content (Azizullah et al., 2012)	• Endocrine-disrupting effects on organisms, including antagonistic effects on androgen receptors and disruption of thyroid function (Rosenmai et al., 2021) • Bioaccumulation and biomagnification in aquatic ecosystems, with potential cytotoxic and carcinogenic effects in exposed individuals (Yao et al., 2021)

7.4.1 *Ecotoxicological Effects of Surfactants*

7.4.1.1 Toxicity to Aquatic Organisms

The surfactants produce harmful effects on aquatic life and this has been studied since the late 1960s when studies revealed that exposure to detergent levels as low as 0.5 ppm disrupted chemoreceptors in fish species such as *Ictalurus natalis* (Bardach et al., 1965). Aquatic animals experience incurable and sublethal toxicities at surfactant concentrations above 0.1 mg/L, with more pronounced adverse effects at higher concentrations (Rebello et al., 2014). Surfactants can disrupt the membranes of the cells of aquatic organisms, which can cause cellular damage and dysfunction. Surfactant exposure affects fish organs like gills, liver, kidney, spleen, and intestines. High concentrations disrupt gill epithelia, causing asphyxiation or failure of the osmoregulatory, while sublethal levels induce hyperplasia, oxidative stress, and damage of the mucus layer, and susceptibility to infections increases (Mustapha & Bawa-Allah, 2020). Overall, surfactant contamination poses significant risks to aquatic ecosystems, the new and effective advancements in the analysis techniques are required so that their harmful effect on the aquatic life must be minimized.

7.4.1.2 Effect on Aquatic Plants

Surfactants disrupt the surface tension of water, infiltrating the cell membranes of aquatic plants and causing structural damage that compromises their essential physiological processes. Surfactants may interfere with the process of respiration thereby inhibiting the uptake of oxygen by the plants resulting in suffocation and stunted growth. Furthermore, they interfere with photosynthesis by causing structural disruptions to chloroplasts, which reduces the plant's capacity to generate energy and essential chemicals (Jovcic et al., 2010). Also, they may interfere with the intake of nutrients, causing imbalances in various plant functions that may worsen plant health.

7.4.1.3 Impact on Microorganisms

They detrimentally affect bacteria like phosphate solubilizing *Acinetobacter junii*, autotrophic ammonia-oxidizing *Nitrosomonas* and *Nitrosospira* strains (Brandt et al., 2001), and bioluminescent *Vibrio fischeri* (Lima et al., 2011). These effects arise from various reactions occurring at the surface of cells (such as membrane depolarization, reduced nutrient absorption, and altered oxygen consumption). Surfactants can delay the secretion of harmful metabolic products thereby leading to their accumulation and finally resulting in the death of the organism (Rebello et al., 2014). Surfactants have a significant negative impact on algae especially affecting their growth, motility, and ability to photosynthesize. It was reported that the growth and the motility of several algal species such as *Plagioselmis prolonga* and *Thalassiosira pseudonana* were affected by the higher concentration of surfactants (Aizdaicher & Markina, 2006; Aizdaicher & Reunova, 2002). The harmful effects of surfactants were also observed for duckweeds. When concentrations of detergents such as Ariel exceed 1 mg/L, *Euglena gracilis* exhibits significant reductions in motility, density of cells, swimming velocity, and chlorophyll content (Azizullah et al., 2012). The aforementioned results emphasize the multifarious and deleterious consequences of surfactants on microbial populations, underscoring the necessity for the development of several mitigation strategies to eradicate surfactants from water resources thereby protecting and safeguarding the microorganisms.

7.4.2 Ecotoxicological Effects of Flame Retardants

7.4.2.1 Toxicity to Aquatic Organism

Several researchers have reported the toxicity of brominated flame retardants in aquatic organisms (Xiong et al., 2019). In one study, *Carassius auratus* (goldfish) exposed to DBDPE (Decabromodiphenyl Ethane) at different levels (10, 50, and 100 mg/kg) for extended periods of time experienced oxidative stress. This shows that enhanced dosage and prolonged exposure time can result in oxidative stress in *Carassius auratus* (Feng et al., 2013). Additionally, the evaluation of toxicity of DBDPE for *Daphnia magna* (water flea) produced a Furthermore, 48-h EC50 value of 19 μg/L. From this study, it was evident that a higher concentration of DBDPE and less exposure time can result in increased mortality in water fleas (Hardy et al., 2012). Tetrabromobisphenol A (TBBB) and tris(1,3-dichloro-2-propyl) phosphate (TBPH) were evaluated for their genotoxic effects in fathead minnows. The liver cells exhibited evidence of DNA damage when the oral dosage of 1 mg/fish/day was administered in Firemaster 550 and Firemaster BZ-54 having both TBB and TBPH. Additionally, Firemaster 550, Firemaster BZ-54, and TBPH were shown to be highly lethal under US EPA requirements after acute toxicity experiments on the freshwater crustacean Daphnia magna revealed their high toxicity (Bearr et al., 2010).

7.4.2.2 Endocrine Disruption

Numerous OPFRs have been shown to significantly affect an organism's endocrine system. According to (Rosenmai et al., 2021), one study emphasizes the possibility of endocrine disruption caused by these compounds. Specifically, it notes that aryl compounds have antagonistic effects on the androgen receptor (AR), while chlorinated OPFRs have little influence on transthyretin binding. Tri(2-chloroethyl) phosphate, or TMPP, has been demonstrated to interact with estrogen receptors in MCF-7 breast cancer cells to mediate endocrine effects and stimulate cell proliferation (Böckers et al., 2020).

7.4.2.3 Bioaccumulation and Biomagnification

FRs have the tendency to bioaccumulate and biomagnify thereby posing risk to higher organisms. Brominated FRs are omnipresent in omnipresent in environment and they have the tendency to accumulate in living systems and may undergo biomagnification in food chain (Sun et al., 2015; Tao et al., 2019). Biomagnification is an important criterion for assessing the ecological effect of these harmful chemicals (Cristale et al., 2013). Few studies have been carried out on the assessment of biomagnification potential of brominated FRs. Biomagnification of DBDPE has been observed in predatory fishes found in lake Winnipeg in Canada and in food web (mangrove) in China (Sun et al., 2015). BTBPE also demonstrated biomagnification potential in prey fishes such as walleye or white fish and emerald shiner or zooplankton in Winnipeg Lake (Mo et al., 2012). Also, biomagnification was observed in predatory fish, *Channa argus* by Tao et al. (2019). Hexabromocyclododecane (HBCD) and Tetrabromobisphenol A (TBBP-A) can bioaccumulate in plants, animals and humans and can cause adverse effects. They may induce cytotoxicity and may increase the risk of cancer development in exposed individuals (Yao et al. 2021).

7.5 Analytical Techniques

7.5.1 Analytical Techniques for Surfactants

7.5.1.1 Titrimetry

Because of the investigation of non-ionic surfactants as well as for the ion-ionic surfactants analysis, the most commonly used technique is titrimetry. Sample pretreatment is not required, this is the advantage of this technique (Beneito-Cambra et al., 2013). It works on the principle of off-line analysis of the analytes of the target. The advantage of titrimetry is that it is simple to operate and there is no need for any advancement in the analytical instruments. This uses the specific analyte and colored matrix.

7.5.1.2 Ion Selectivity Potentiometry

Based on ionophores non-ionic surfactants and determination of cationic surfactants takes place by utilizing ion selective electrodes. Kulapin & Arinushkina (2000) determined various types of surfactants, using solid contact potentiometric sensors. pairs of ions of Cetylpyridinium, dodecylsulfate, cetylpyridinium, and tetraphenylborate were the active components of the membrane. These electrodes were employed for the determination of cationic surfactants in the rinsing of hair and in shampoos the anionic and non-ionic surfactants (Bartolomé et al., 2005).

7.5.1.3 Chromatography Technique

In an environmental sample for the assessment of surfactants, chromatographic techniques is the most advanced method. With in a shortest possible time, this technique is specific, sensitive and possess an excellent capability to identify, separate each surfactant in environmental matrix. Liquid chromatography and gas chromatography are the two types of chromatographic techniques.

7.5.1.4 Gas Chromatography

Because of large number of disadvantages of traditional analytical techniques, gas chromatography is more beneficials for analysing low molecular mass metabolites and separating low complex matrices. The metabolites of compound derivatives are needed for the GC technique (Petrović & Barceló, 2000). In recent time some common detectors, tandem mass spectroscopy, chemical ionization, single quadrupole electron impact, flame ionization detector are used (Clara et al., 2007). Gas chromatography—mass spectroscopy is commonly applied for the separation and investigation of linear alkylbenzene sulfonate in various samples. GC analysis is the most appropriate technique for analysing various variety of surfactants as it applies the hybrid of the capillary column and flame ionization detector but for the investigation of cationic surfactants the use of gas chromatography is not been available in detail in literature.

7.5.1.5 Liquid Chromatography (LC)

Liquid chromatography is applied for the investigation of low molecular weight and non -volatile surfactants like biodegraded and bio surfactants. There is no need of derivatization in case of liquid chromatography as it is needed in gas chromatography for the investigation of environmental samples. For surfactants, reverse phase column is usually used in liquid chromatography (Bassarab et al., 2011; Lara-Martín et al., 2006; León et al., 2000; Merino et al., 2004; Rico-Rico et al., 2009). The typical detectors used in analysis of surfactants are Mass spectrometry (MS), triple quadrupole Ms(MS-MS), ion-trap MS(MSn) and quadrupole ToF MS(Q-ToF-MS) (Lunar et al., 2004). The most advanced type of liquid chromatography in liquid mobile phase is the high-Performance liquid chromatography (HPLC) that make use of high pressure. To analyse the analytes of trace quantities within a short period of time, the most advanced technique is ultra power LC (UPLC) that is coupled with Q-ToF-MS.

7.5.1.6 Fourier Transform Infrared Spectrometry (FTIR)

The eminent approach used for the extraction and concentration of non-ionic surfactants is Fourier transform infrared spectrometry (Carolei & Gutz, 2005; Gartshore et al., 2000) the salting out methodology the non-ionic surfactants can be used (Andrew, 1993). For a specific class of surfactants this process is easy, fast ad sensitive (Beneito-Cambra et al., 2013). For the determination of absolute value of non-ionic surfactants this FTIR method is not appropriate. Development of soap and FTIR efficiency method interfere with each other.

7.5.1.7 Attenuated Total Reflectance Fourier Transform Infrared Spectroscopy (ATR-FTIR)

For the determination of multicomponent complex mixtures one of the important and popular technique is Attenuated total reflectance Fourier transform infrared spectroscopy (ATR-FTIR) (Kargosha et al., 2008). In infrared region, this technique uses the principle of absorbance of the spectrum. This technique is used in mixture for determination of different surfactants at the same time (Carolei And Gutz, 2005). For short term on-stream analysis this technique is authentic and steady. To analyse a trace compound, lack of suitability and instrumental cost is the major disadvantage of this technique. In this technique, in the range of absorption water has a substantial interference when used as a solvent. During analysis use of paper substrate is recent development in this technique.

7.5.1.8 Spectrophotometric Analysis

Spectrophotometric analysis is an easy and quick method for analysis of surfactants. The highly popular spectroscopic technique used for the analysis is the UV–Vis. In ultraviolet visible region, the use of electronic transmission band is made in UV–Vis spectrophotometer. This technique mainly focuses on development of surfactants with reagent of pair targeted ion reagent and their removal into admissible hydrophobic solvent is the main focus of this technique (Kurrey et al., 2019). This method was applied for the evaluation of the concentration of linear alkylbenzene sulfonate and is dependent on the ionic pair surfactant methyl blue formation (Hu et al., 2019). In the investigation of surfactants by making use of UV–Vis during the spectrometry experiment, different bands are shown by different compounds. A lot of toxic wastes generates during this technique. Chromophores must be found in the samples is the major drawback in this technique.

7.5.1.9 High Performance Liquid Chromatography (HPLC)

For the analysis of surfactants, High-performance liquid chromatography is the most popular chromatographic technique that is used for the surfactant analysis in prohibited environmental matrix. This technique is widely used in the investigation of analytes having low and high molecular weights. It is used most commonly applied for the analysis of low and high molecular weight analytes. In this analysis derivatization is not required. It is also useful in biodegraded products, different surfactants groups and non-volatile analytes. Most of the disadvantages of previous technique is get control of by HPLC. Before the analysis, the main issue involves in HPLC techniques that is applied for the separation and extraction of analytes. For the investigation of the target analytes, systems of different types of solid liquid partitioning is used by HPLC analytes. The detection limit of HPLC is low and 0.0002 mg/L low only can be detected (Liu et al., 2006). For the investigation and identification of various surfactants HPLC detectors like MS, UV, FLD, MS–MS, conductivity, fluorescence, refracting index are used.

7.5.1.10 Potentiometric Analysis

All the variety of ionic as well as non-ionic surfactants components must be analysed completely by the use of potentiometric titration. After adding the different titrants volume, the switch of the electromotive force cell is responsible for the function of this technique. At the end of titration, the detection is ion-selective. For the analysis of individual components in the mixture of surfactants this technique is not appropriate. The drawbacks of this technique are stability of signal and reproducibility of data. In the analysis of non-ionic cationic is done by ion-selective potentiometric. The investigation and analysis method are based on ionophores by the application of ion selective electrodes as observed by Sak-Bosnar et al. (2015). Potentiometric analysis is used for the investigation of anionic surfactants and also sodium lauryl ether sulphate (SLES) (Masadome et al., 1999). Thereby, this technique is based on ion selective electrodes and is applied to different variety of environmental matrices. The advantages of potentiometric analysis are higher sensitivity, non-disastrous to analytes authenticity, less expensive, less monitoring time in comparison to chromatographic analytical techniques and spectrometric analytical techniques. The modification of nanomaterials and carbon nanotubes are the recent modification in the method of ion-selective potentiometric analysis.

7.5.1.11 Supercritical Fluid Chromatography (SFC)

For the detachment and identification of surfactants, the utilisation of supercritical fluid chromatography (SFC) is a current advancement for the investigation of surfactants (Ma et al., 2019). A fluid is assumed to be supercritical fluid (SF), if no other definite liquid or gas is at critical point. The best benefits of both the liquid phases are combined because supercritical fluids are having the same density with liquid as well as having same viscosity with gas phase, therefore the best advantages of both of the fluid phases are combined. Thereby strong points of both HPLC and GC are shared by SFC. SFC is comparatively more sensitive and efficient whereas HPLC has better selectivity. Inherent flammability and toxicity linked with liquid chromatography can be avoided by SFC. As compared to liquid chromatography it functions at higher flow rates than liquid chromatography and less cheap and more environment friendly. For mobile phase CO_2 is used because of its good solvent properties, reasonable cost, non-toxicity, UV inertness, low critical temperature and easy to attain pressure. Along with SFC large range of detectors can be used. Fourier transform infrared spectroscopy, flame ionization detector, ultra violet visible detection, corona charged aerosol detection are the standard detectors can be used (Shah et al., 2015). The employment of superficial fluid chromatography is a noticeable advancement for the investigation and analysis of pentaerythritol fatty acid esters. On contrary, a high quality and purity of chemicals are required for SFC, that is not easy to make available because of high cost in the small laboratories.

7.5.1.12 Biological Technique

Biological technique makes use of algae, fungi and bacteria to break compounds having the large molecular weights into compounds of lower molecular weights (Garcia-Rodríguez et al., 2014). To analyse the surfactants examining their health risk and toxicological impacts, some biological techniques are used. The electrochemical and conventional analytical techniques takes too much time and need sophisticated and technical of the well-known instruments. But simple, quick and reliable technique are required for the investigation of animals, food samples and other environment samples. For the analysis of surfactants the common biological techniques like biosensors, molecular and immunochemistry. A modern immunochemistry technique, enzyme linked immunosorbent assay (ELISA) is cost effective, highly sensitive, needs fewer technical skills, short time analysis and give reliable analytical result. The most relevant technique is the biological technique in which gyrase chain reaction or polymerase method is employed mostly for the analysis of alkylphenol, ethoxylated alkylphenol found in the animals. DNA sequences amplification is done by a process of thermocycling of PCR. The molecular techniques that is polymerase chain reaction technique (PCR) is very fast, trustable, highly sensitive and for the detection of gene sequence of interest is accurate. This method entirely depends upon effects of surfactants on the expression of the gene for the prime neuroendocrine system. For the analysis of the gene expression of cytochrome P450N, PCR has been successfully used because of the outcome of nonylphenol on that of the aromatase transcript of Atlantic salmon.

7.5.1.13 Volammetric Technique

For the quantitative analysis of environmental samples, electroanalytical procedures of volammetric techniques are used. In surface water for the accurate assessment of non-ionic surfactants, sewage etc. (Nesměrák & Němcová, 2006; Szymanski et al., 1996) this technique in comparison to chromatographic, spectrophotometric and electrophoretic these techniques are less significant. In environmental samples voltammetry provides a best specific analysis for the surfactants. The technique depends on changes in the transport charges of the analytes and ionic conductivity. Therefore, this technique is used for the measurement of conduction charge transport of electrolyte interface. Voltametric techniques is employed for the quantisation of the CMC of surfactants due to micelles potential to adsorbed electroactive species. During preconcentration, because of the adsorption of the substances by the HDME surface therefore the measurement was made possible.

7.6 Mitigation Strategies

7.6.1 Mitigation Strategies of Surfactants

The existance of surfactants in aquatic life as well as in the terrestrial environments creates a very harmful effect on the life of human being, microorganisms, plants and animals. As the large number of use of surfactants, their adverse effect on human health and authoritative regulatory policies for putting restricted limit demands on their restriction in the living world. This is subject and topic of research for the investigators. This is because lot of mitigation strategies are suggested for the cleaning of these type of surfactants. In earlier year for the removal of surfactants bioremediation was used. But now physical, chemical, biological and membrane techniques are used. The suitable mitigation technique is affected by various elements like effluent, time of operation, treatment efficiency, influent characteristics and the use of power. These are physical and chemical methods used for removal of surfactants.

7.6.1.1 Physical Methods

Surfactants can be removed effectively by using the adsorption which is the is the most common physical technique. Many materials are used as adsorbents for example clay, carbon, nanomaterials, zeolites and resins. Because of their high cost they are not widely used. The activated carbon is used as the best adsorbent material (Valizadeh et al., 2016). Because of high cost of activated carbon, several research have been carried out to create cheap, safe and highly efficient adsorbent in place of activated carbon. Green adsorbent developed from waste water is recent advancement for surfactant removal from waste water. Many varied materials are used as waste adsorbents (Hossain et al., 2020). Because of mostly using these materials this increases waste management and also support economic growth. Different carbon nanotubes are used for comparison of rectification of three surfactants that are hexadecyltrimethylammonium bromide (CTAB), dodecyl dimethyl benzyl ammonium chloride (DDBAC), tetradecyl dimethyl benzyl ammonium chloride (TDBAC), pure functionalized multiwalled carbon nanotubes and OH/COOH -modified CNTs (Gao et al., 2020). It was noted that efficiency of CNTs increases with large surface area and porosity. To gain the maximum efficiency CNTs is required in optimum quantity. Excluding cationic surfactants, non-ionic and anionic surfactants have been found to be removed by multiwalled carbon nanotubes (Bai et al., 2010; Yang et al., 2010). The ability of carbon nanomaterial to remove cationic surfactants depends on the physiochemical properties of nanomaterials and the composition of cationic surfactants. Surfactants can be minimised from waste water by another important technique called foam fractionation by injection of air into the fractionation column that contain contaminated water with surfactants. On the basis of difference of mass concentration, the surfactants are separated. Foam rich surfactants layer is obtained at the top and aqueous rich surfactant layer is formed at the bottom of the column. This process depends on the nanomaterials that are utilised by foam stabilization and this progress is used for removal of surfactants from wastewater. The property of foam is influenced by formability and foam stability. The rise of viscosity of activated SNP affects the formability. This reduces the efficiency of surfactants removal. Foam fractionation is a low-cost technology for minimising and recovering surfactants from wastewater (Burghoff, 2012). Many research has been observed the liquid height, effect of time, foam height, air flow rate, feed flow rate, for the minimization of anionic surfactants (Srinet et al., 2017). More findings are required for foam fractionation and its upscaling.

7.6.1.2 Chemical Technique

The alternate for removal of surfactants from waste system is chemical technique. Many valuable products like nitrogen, carbon di oxide and water are created by using chemical technique. Chemical techniques are of different kinds and forms. Other effective method is coagulation. Because of larger volumes and more substantial diversity formed by sludge carrying aluminium, aluminium based coagulants are the best. Air flocculation and Fenton process is combined with other methods. For removing the pollutants coagulation flocculation is very effective method.

According to various studies as compared with other removal process Fenton process is less efficient. Coagulation process efficiency is also affected by the type of the coagulant. If the coagulants are not properly discharged, that creates another waste material. There care must be taken in choosing the coagulant so that waste surfactants can be removed from water without polluting influent. For anionic surfactants, an alternative remedial technique is electrochemical advanced oxidation technique. From waste water, for the degradation of recalcitrant surfactants the technique of electrochemistry is used. The three electrochemical processes (EO, EF and EOH_2O_2) for removing anionic surfactants are compared by (Ganiyu et al., 2018). Anionic surfactants are completely removed from wastewater by these processes. The increase in the applied current increases the removal efficiency. There is an adverse consumption of energy because of increase of current density. Anaerobic/oxic processes are also used for the minimization of surfactants (Gao et al., 2017). This technique consists of six compartments that are divided into four oxic conditions and two anoxic conditions. For the removal of surfactants, a critical role is played by anoxic treatment. The efficiency of removal is affected by change in the climate in the considerable region or country. In aqueous solutions, to generate various species irradiation-based beam of electron is used. For the degradation of aqueous pollutants these generated species are used. Minimum use of chemicals, easy operation mode, free from sludge creation and byproducts and friendly with environment are various advantages for remediation of surfactants by sonoreactor. Advanced oxidation treatment is an efficient process because of its low cost and maintenance. Figure 7.3 shows various mitigation strategies for removal of surfactants including physical method, chemical method, acoustical method, biological method and hybrid method.

7.7 Conclusion and Future Perspectives

Surfactants and flame retardants are the contaminants in water resources that have become more and more noticeable because of their widespread distribution in the environment. Wastewater treatment plants, laundry, agricultural soil, industries, and pharmaceuticals are the various sources of surfactants while sources of flame retardants in water resources include landfills, electronics, industrial discharge, wastewater treatment plants, etc. A lot of research has to be done on these emerging contaminants to know their harmful impact on the environment and human health. Low-cost, highly effective, and eco-friendly adsorbents in the form of pellets must be used so that they can be easily utilized in the adsorption column for the removal of contaminants. Future research must be done to develop more effective techniques for the detection and removal of these contaminants. Less data is present on the removal of flame retardants. Research needs to be done on their removal techniques. Various analytical, detection, and mitigation strategies must be developed. Extraction, purification, and pre-concentration can be used for better analysis of surfactants. The chromatography technique is the best method for the detection and analysis of surfactants. Physical, chemical, biological, acoustical, and hybrid methods are various mitigation techniques. Advanced oxidation is the low-cost and effective method for treatment but the presence of degradation intermediate is very harmful in this method. Therefore, achievable techniques must be used to treat surfactants and flame retardants.

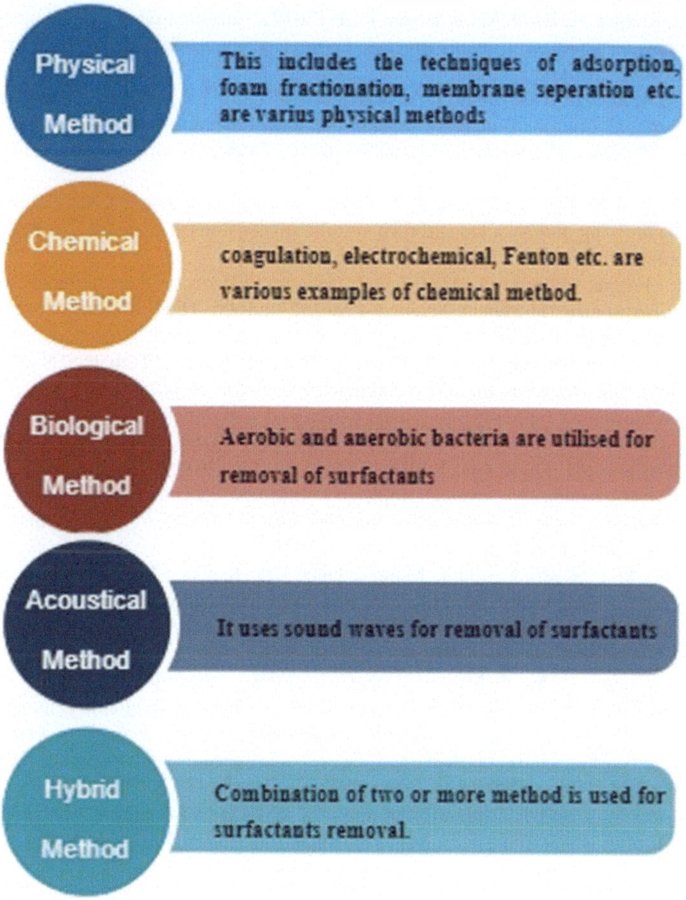

Fig. 7.3 Various mitigation strategies

References

Abbasi, G., Li, L., & Breivik, K. (2019). Global historical stocks and emissions of PBDEs. *Environmental Science & Technology, 53*(11), 6330–6340. https://doi.org/10.1021/acs.est.8b07032

Aizdaicher, N. A., & Markina, Zh. V. (2006). Toxic effects of detergents on the alga Plagioselmis prolonga (Cryptophyta). *Russian Journal of Marine Biology, 32*(1), 45–49. https://doi.org/10.1134/S1063074006010068

Aizdaicher, N. A., & Reunova, Yu. A. (2002). Effects of detergents on in vitro growth of diatom alga Thalassiosira pseudonana. *Russian Journal of Marine Biology, 28*(5), 324–328. https://doi.org/10.1023/A:1020907501713

Andrew, B. E. (1993). Determination of non-ionic surfactants in waste water by direct extraction with Fourier transform infrared spectroscopic detection. *The Analyst, 118*(2), 153. https://doi.org/10.1039/an9931800153

Azizullah, A., Richter, P., Jamil, M., & Häder, D.-P. (2012). Chronic toxicity of a laundry detergent to the freshwater flagellate Euglena gracilis. *Ecotoxicology, 21*(7), 1957–1964. https://doi.org/10.1007/s10646-012-0930-3

Badmus, S. O., Amusa, H. K., Oyehan, T. A., & Saleh, T. A. (2021). Environmental risks and toxicity of surfactants: Overview of analysis, assessment, and remediation techniques. *Environmental Science and Pollution Research, 28*(44), 62085–62104. https://doi.org/10.1007/s11356-021-16483-w

Bai, Y., Lin, D., Wu, F., Wang, Z., & Xing, B. (2010). Adsorption of Triton X-series surfactants and its role in stabilizing multi-walled carbon nanotube suspensions. *Chemosphere, 79*(4), 362–367. https://doi.org/10.1016/j.chemosphere.2010.02.023

Bardach, J. E., Fujiya, M., & Holl, A. (1965). Detergents: Effects on the Chemical Senses of the Fish *Ictalurus natalis* (le Sueur). *Science, 148*(3677), 1605–1607. https://doi.org/10.1126/science.148.3677.1605

Bartolomé, L., Cortazar, E., Raposo, J. C., Usobiaga, A., Zuloaga, O., Etxebarria, N., & Fernández, L. A. (2005). Simultaneous microwave-assisted extraction of polycyclic aromatic hydrocarbons, polychlorinated biphenyls, phthalate esters and nonylphenols in sediments. *Journal of Chromatography A, 1068*(2), 229–236. https://doi.org/10.1016/j.chroma.2005.02.003

Bassarab, P., Williams, D., Dean, J. R., Ludkin, E., & Perry, J. J. (2011). Determination of quaternary ammonium compounds in seawater samples by solid-phase extraction and liquid chromatography–mass spectrometry. *Journal of Chromatography A, 1218*(5), 673–677. https://doi.org/10.1016/j.chroma.2010.11.088

Bautista-Toledo, M. I., Rivera-Utrilla, J., Méndez-Díaz, J. D., Sánchez-Polo, M., & Carrasco-Marín, F. (2014). Removal of the surfactant sodium dodecylbenzenesulfonate from water by processes based on adsorption/bioadsorption and biodegradation. *Journal of Colloid and Interface Science, 418*, 113–119. https://doi.org/10.1016/j.jcis.2013.12.001

Bearr, J. S., Stapleton, H. M., & Mitchelmore, C. L. (2010). Accumulation and DNA damage in fathead minnows (*Pimephales promelas*) exposed to 2 brominated flame-retardant mixtures, Firemaster® 550 and Firemaster® BZ-54. *Environmental Toxicology and Chemistry, 29*(3), 722–729. https://doi.org/10.1002/etc.94

Beneito-Cambra, M., Herrero-Martínez, J. M., & Ramis-Ramos, G. (2013). Analytical methods for the characterization and determination of nonionic surfactants in cosmetics and environmental matrices. *Analytical Methods, 5*(2), 341–354. https://doi.org/10.1039/C2AY25847A

Birnbaum, L. S., & Staskal, D. F. (2004). Brominated flame retardants: Cause for concern? *Environmental Health Perspectives, 112*(1), 9–17. https://doi.org/10.1289/ehp.6559

Böckers, M., Paul, N. W., & Efferth, T. (2020). Organophosphate ester tri-o-cresyl phosphate interacts with estrogen receptor α in MCF-7 breast cancer cells promoting cancer growth. *Toxicology and Applied Pharmacology, 395*, 114977. https://doi.org/10.1016/j.taap.2020.114977

Bolan, S., et al. (2023). Surfactant-enhanced mobilization of persistent organic pollutants: Potential for soil and sediment remediation and unintended consequences. *Journal of Hazardous Materials, 443*, 130189. https://doi.org/10.1016/j.jhazmat.2022.130189

Brandt, K.K., Hesselso/e, M., Roslev, P., Henriksen, K. and So/rensen, J. (2001). Toxic Effects of Linear Alkylbenzene Sulfonate on Metabolic Activity, Growth Rate, and Microcolony Formation of *Nitrosomonas* and *Nitrosospira* Strains. *Applied and Environmental Microbiology* 67(6), pp. 2489–2498. https://doi.org/10.1128/AEM.67.6.2489-2498.2001

Burghoff, B. (2012). Foam fractionation applications. *Journal of Biotechnology, 161*(2), 126–137. https://doi.org/10.1016/j.jbiotec.2012.03.008

Cao, D., et al. (2019). Presence and human exposure assessment of organophosphate flame retardants (OPEs) in indoor dust and air in Beijing, China. *Ecotoxicology and Environmental Safety, 169*, 383–391. https://doi.org/10.1016/j.ecoenv.2018.11.038

Carolei, L., & Gutz, I. (2005). Simultaneous determination of three surfactants and water in shampoo and liquid soap by ATR-FTIR. *Talanta, 66*(1), 118–124. https://doi.org/10.1016/j.talanta.2004.10.005

Clara, M., Scharf, S., Scheffknecht, C., & Gans, O. (2007). Occurrence of selected surfactants in untreated and treated sewage. *Water Research, 41*(19), 4339–4348. https://doi.org/10.1016/j.watres.2007.06.027

Cristale, J., Katsoyiannis, A., Sweetman, A. J., Jones, K. C., & Lacorte, S. (2013). Occurrence and risk assessment of organophosphorus and brominated flame retardants in the River Aire (UK). *Environmental Pollution, 179*, 194–200. https://doi.org/10.1016/j.envpol.2013.04.001

Ding, J., Shen, X., Liu, W., Covaci, A., & Yang, F. (2015). Occurrence and risk assessment of organophosphate esters in drinking water from Eastern China. *Science of the Total Environment, 538*, 959–965. https://doi.org/10.1016/j.scitotenv.2015.08.101

Feiteiro, J., Mariana, M., & Cairrão, E. (2021). Health toxicity effects of brominated flame retardants: From environmental to human exposure. *Environmental Pollution, 285*, 117475. https://doi.org/10.1016/j.envpol.2021.117475

Feng, M., Li, Y., Qu, R., Wang, L., & Wang, Z. (2013). Oxidative stress biomarkers in freshwater fish *Carassius auratus* exposed to decabromodiphenyl ether and ethane, or their mixture. *Ecotoxicology, 22*(7), 1101–1110. https://doi.org/10.1007/s10646-013-1097-2

Fromme, H., et al. (2014). Organophosphate flame retardants and plasticizers in the air and dust in German daycare centers and human biomonitoring in visiting children (LUPE 3). *Environment International, 71*, 158–163. https://doi.org/10.1016/j.envint.2014.06.016

Fu, L., et al. (2020). Tracing the occurrence of organophosphate ester along the river flow path and textile wastewater treatment processes by using dissolved organic matters as an indicator. *Science of the Total Environment, 722*, 137895. https://doi.org/10.1016/j.scitotenv.2020.137895

Ganiyu, S. O., Vieira dos Santos, E., de Araújo, T., Costa, E. C., & Martínez-Huitle, C. A. (2018). Electrochemical advanced oxidation processes (EAOPs) as alternative treatment techniques for carwash wastewater reclamation. *Chemosphere, 211*, 998–1006. https://doi.org/10.1016/j.chemosphere.2018.08.044

Gao, D., Li, Z., Guan, J., & Liang, H. (2017). Seasonal variations in the concentration and removal of nonylphenol ethoxylates from the wastewater of a sewage treatment plant. *Journal of Environmental Sciences, 54*, 217–223. https://doi.org/10.1016/j.jes.2016.02.005

Gao, Q., et al. (2020). Chemical composition-dependent removal of cationic surfactants by carbon nanotubes. *Science of the Total Environment, 716*, 137017. https://doi.org/10.1016/j.scitotenv.2020.137017

Garcia-Rodríguez, A., Matamoros, V., Fontàs, C., & Salvadó, V. (2014). The ability of biologically based wastewater treatment systems to remove emerging organic contaminants—A review. *Environmental Science and Pollution Research, 21*(20), 11708–11728. https://doi.org/10.1007/s11356-013-2448-5

Gartshore, J., Lim, Y. C., & Cooper, D. G. (2000). Quantitative analysis of biosurfactants using Fourier Transform Infrared (FT-IR) spectroscopy. *Biotechnology Letters, 22*(2), 169–172. https://doi.org/10.1023/A:1005670031432

Gogoi, A., Mazumder, P., Tyagi, V. K., Tushara Chaminda, G. G., An, A. K., & Kumar, M. (2018). Occurrence and fate of emerging contaminants in water environment: A review. *Groundwater for Sustainable Development, 6*, 169–180. https://doi.org/10.1016/j.gsd.2017.12.009

Guerra, P., Alaee, M., Eljarrat, E., & Barceló, D. (2010). *Introduction to brominated flame retardants: commercially products, applications, and physicochemical properties* (pp. 1–17). https://doi.org/10.1007/698_2010_93.

Hardy, M. L., Krueger, H. O., Blankinship, A. S., Thomas, S., Kendall, T. Z., & Desjardins, D. (2012). Studies and evaluation of the potential toxicity of decabromodiphenyl ethane to five aquatic and sediment organisms. *Ecotoxicology and Environmental Safety, 75*, 73–79. https://doi.org/10.1016/j.ecoenv.2011.08.005

Hossain, N., Bhuiyan, M. A., Pramanik, B. K., Nizamuddin, S., & Griffin, G. (2020). Waste materials for wastewater treatment and waste adsorbents for biofuel and cement supplement applications: A critical review. *Journal of Cleaner Production, 255*, 120261. https://doi.org/10.1016/j.jclepro.2020.120261

Hu, N., Li, Y., Yang, C., Wu, Z., & Liu, W. (2019). In-situ activated nanoparticle as an efficient and recyclable foam stabilizer for enhancing foam separation of LAS. *Journal of Hazardous Materials, 379*, 120843. https://doi.org/10.1016/j.jhazmat.2019.120843

Innes, A., & Innes, J. (2011). Flame retardants. In: *Applied plastics engineering handbook* (pp. 469–485). Elsevier. https://doi.org/10.1016/B978-1-4377-3514-7.10027-3.

Jena, G., Dutta, K., & Daverey, A. (2023). Surfactants in water and wastewater (greywater): Environmental toxicity and treatment options. *Chemosphere, 341*, 140082. https://doi.org/10.1016/j.chemosphere.2023.140082

Jin, Y., et al. (2024). Ecological and human health risk of aryl-phosphate flame retardants (APFRs): Sources, distribution, and toxicity. *Critical Reviews in Environmental Science and Technology, 54*(15), 1117–1137. https://doi.org/10.1080/10643389.2023.2293425

Jovcic, B., Venturi, V., Davison, J., Topisirovic, L., & Kojic, M. (2010). Regulation of the sdsA alkyl sulfatase of *Pseudomonas* sp. ATCC19151 and its involvement in degradation of anionic surfactants. *Journal of Applied Microbiology, 109*(3), 1076–1083. https://doi.org/10.1111/j.1365-2672.2010.04738.x.

Kargosha, K., Ahmadi, S. H., Mansourian, M., & Azad, J. (2008). Simultaneous determination of one nonionic and two anionic surfactants using Fourier transform infrared spectrometry and multivariate analysis. *Talanta, 75*(2), 589–593. https://doi.org/10.1016/j.talanta.2007.11.065

Kerric, A., Mazerolle, M. J., Sorais, M., Giroux, J.-F., & Verreault, J. (2023). Impact of landfill characteristics on the atmospheric exposure to halogenated flame retardants in gulls. *Chemosphere, 343*, 140207. https://doi.org/10.1016/j.chemosphere.2023.140207

Khan, W. U., Ahmed, S., Dhoble, Y., & Madhav, S. (2023). A critical review of hazardous waste generation from textile industries and associated ecological impacts. *Journal of the Indian Chemical Society, 100*(1), 100829.

Kim, U.-J., & Oh, J.-E. (2014). Tetrabromobisphenol A and hexabromocyclododecane flame retardants in infant–mother paired serum samples, and their relationships with thyroid hormones and environmental factors. *Environmental Pollution, 184*, 193–200. https://doi.org/10.1016/j.envpol.2013.08.034

Kulapin, A. I., & Arinushkina, T. V. (2000). Solid-contact potentiometric sensors based on ion pairs of cetylpyridinium with dodecylsulfate and tetraphenylboratc for various surfactants. *Journal of Analytical Chemistry, 55*(11), 1096–1101. https://doi.org/10.1007/BF02757340

Kumar, R., Gupta, B., Gupta, H., & Rani, M. (2014). Distribution of persistent organic pollutants in urban aquatic systems. *International Journal of Scientific Research in Environmental Sciences, 2*(7), 233–243.

Kumar, R., Rani, M., Gupta, H., Gupta, B., Park, D., & Jeon, B. H. (2017). Distribution of trace elements in flowing surface waters: Effect of seasons and anthropogenic practices in India. *International Journal of Environmental Analytical Chemistry, 97*(7), 637–656.

Kumar, N., Kumar, B., Gupta, H., & Gupta, A. (2023). Development and evaluation of cellulose/graphene-oxide based composite for removing phenol from aqueous solutions. *Polymers, 15*(3), 572. https://doi.org/10.3390/polym15030572

Kurrey, R., Mahilang, M., Deb, M. K., & Shrivas, K. (2019). Analytical approach on surface active agents in the environment and challenges. *Trends in Environmental Analytical Chemistry, 21*, e00061. https://doi.org/10.1016/j.teac.2019.e00061

Lara-Martín, P. A., Gómez-Parra, A., & González-Mazo, E. (2006). Development of a method for the simultaneous analysis of anionic and non-ionic surfactants and their carboxylated metabolites in environmental samples by mixed-mode liquid chromatography–mass spectrometry. *Journal of Chromatography A, 1137*(2), 188–197. https://doi.org/10.1016/j.chroma.2006.10.009

León, V. M., González-Mazo, E., & Gómez-Parra, A. (2000). Handling of marine and estuarine samples for the determination of linear alkylbenzene sulfonates and sulfophenylcarboxylic acids. *Journal of Chromatography A, 889*(1–2), 211–219. https://doi.org/10.1016/S0021-9673(00)00569-0

Li, L., & Wania, F. (2018). Elucidating the variability in the hexabromocyclododecane diastereomer profile in the global environment. *Environmental Science & Technology, 52*(18), 10532–10542. https://doi.org/10.1021/acs.est.8b03443

Li, X., Doherty, A. C., Brownawell, B., & Lara-Martin, P. A. (2018). Distribution and diagenetic fate of synthetic surfactants and their metabolites in sewage-impacted estuarine sediments. *Environmental Pollution, 242*, 209–218. https://doi.org/10.1016/j.envpol.2018.06.064

Lima, T. M. S., Procópio, L. C., Brandão, F. D., Leão, B. A., Tótola, M. R., & Borges, A. C. (2011). Evaluation of bacterial surfactant toxicity towards petroleum degrading microorganisms. *Bioresource Technology, 102*(3), 2957–2964. https://doi.org/10.1016/j.biortech.2010.09.109

Liu, X., Pohl, C. A., & Weiss, J. (2006). New polar-embedded stationary phase for surfactant analysis. *Journal of Chromatography A, 1118*(1), 29–34. https://doi.org/10.1016/j.chroma.2006.03.080

Lunar, L., Rubio, S., & Pérez-Bendito, D. (2004). Differentiation and quantification of linear alkyl benzenesulfonate isomers by liquid chromatography-ion-trap mass spectrometry. *Journal of Chromatography A, 1031*(1–2), 17–25. https://doi.org/10.1016/j.chroma.2003.08.071

Ma, Q., Zhang, Y., Zhai, J., Chen, X., Du, Z., Li, W., & Bai, H. (2019). Characterization and analysis of non-ionic surfactants by supercritical fluid chromatography combined with ion mobility spectrometry-mass spectrometry. *Analytical and Bioanalytical Chemistry, 411*(13), 2759–2765. https://doi.org/10.1007/s00216-019-01777-3

Ma, Y., et al. (2023). Socio-economic factors impact US dietary exposure to halogenated flame retardants. *Environmental Science & Technology Letters, 10*(6), 478–484. https://doi.org/10.1021/acs.estlett.3c00224

Madhav, S., Mishra, R., Kumari, A., Srivastav, A. L., Ahamad, A., Singh, P., Ahmed, S., Mishra, P. K., & Sillanpää, M. (2024). A review on sources identification of heavy metals in soil and remediation measures by phytoremediation-induced methods. *International Journal of Environmental Science and Technology, 21*(1), 1099–1120.

Masadome, T., Imato, T., & Asano, Y. (1999). End-point detection of the potentiometric titration of anionic polyelectrolytes using an anionic surfactant-selective plasticized poly (vinyl chloride) membrane electrode and an anionic surfactant as a marker ion. *Fresenius' Journal of Analytical Chemistry, 363*(3), 241–245. https://doi.org/10.1007/s002160051181

Merino, F., Rubio, S., & Pérez-Bendito, D. (2004). Evaluation and optimization of an on-line admicelle-based extraction-liquid chromatography approach for the analysis of ionic organic compounds. *Analytical Chemistry, 76*(14), 3878–3886. https://doi.org/10.1021/ac049736v

Mitru, D., et al. (2020). Impact of various surfactant classes on the microorganism community used for WWTP biodegradation treatment. *Romanian Journal of Ecology & Environmental Chemistry, 2*(2), 210–220. https://doi.org/10.21698/rjeec.2020.226

Mo, L., Wu, J., Luo, X., Zou, F., & Mai, B. (2012). Bioaccumulation of polybrominated diphenyl ethers, decabromodiphenyl ethane, and 1,2-bis(2,4,6-tribromophenoxy) ethane flame retardants in kingfishers (*Alcedo atthis*) from an electronic waste–recycling site in South China. *Environmental Toxicology and Chemistry, 31*(9), 2153–2158. https://doi.org/10.1002/etc.1929

Moura, A. G. L., Centurion, V. B., Okada, D. Y., Motteran, F., Delforno, T. P., Oliveira, V. M., & Varesche, M. B. A. (2019). Laundry wastewater and domestic sewage pilot-scale anaerobic treatment: Microbial community resilience regarding sulfide production. *Journal of Environmental Management, 251*, 109495. https://doi.org/10.1016/j.jenvman.2019.109495

Mustapha, D. S., & Bawa-Allah, K. A. (2020). Differential toxicities of anionic and nonionic surfactants in fish. *Environmental Science and Pollution Research, 27*(14), 16754–16762. https://doi.org/10.1007/s11356-020-08212-6

Nesměrák, K., & Němcová, I. (2006). Determination of critical micelle concentration by electrochemical means. *Analytical Letters, 39*(6), 1023–1040. https://doi.org/10.1080/00032710600620302

Nunes, R. F., & Teixeira, A. C. S. C. (2022). An overview on surfactants as pollutants of concern: Occurrence, impacts and persulfate-based remediation technologies. *Chemosphere, 300*, 134507. https://doi.org/10.1016/J.CHEMOSPHERE.2022.134507

Petrović, M., & Barceló, D. (2000). Determination of anionic and nonionic surfactants, their degradation products, and endocrine-disrupting compounds in sewage sludge by liquid chromatography/mass spectrometry. *Analytical Chemistry, 72*(19), 4560–4567. https://doi.org/10.1021/ac0003060

Rebello, S., Asok, A. K., Mundayoor, S., & Jisha, M. S. (2014). Surfactants: Toxicity, remediation and green surfactants. *Environmental Chemistry Letters, 12*(2), 275–287. https://doi.org/10.1007/s10311-014-0466-2

Rico-Rico, Á., Droge, S. T. J., Widmer, D., & Hermens, J. L. M. (2009). Freely dissolved concentrations of anionic surfactants in seawater solutions: Optimization of the non-depletive solid-phase microextraction method and application to linear alkylbenzene sulfonates. *Journal of Chromatography A, 1216*(15), 2996–3002. https://doi.org/10.1016/j.chroma.2009.01.090

Rosenmai, A. K., et al. (2021). Organophosphate ester flame retardants have antiandrogenic potential and affect other endocrine related endpoints in vitro and in silico. *Chemosphere, 263*, 127703. https://doi.org/10.1016/j.chemosphere.2020.127703

Sak-Bosnar, M., Madunić-Čačić, D., Grabarić, Z., & Grabarić, B. (2015). *Potentiometric determination of anionic and nonionic surfactants in surface waters and wastewaters*, pp. 157–176. https://doi.org/10.1007/978-3-662-44034-6_7.

Salih Al-Omran, L. (2019). Physiochemical properties and environmental levels of legacy and novel brominated flame retardants. In *Flame retardants*. IntechOpen. https://doi.org/10.5772/intechopen.79823.

Saxena, R., Gupta, H., & Dhiman, S. (2024). Availability and minimization of nitrogenous disinfectant by-products in drinking water. In S. Madhav, M. A. Mazhar, S. Ahmed, P. Kumar & P. K. Mishra (Eds.), *Drinking water disinfection by-products*. Springer, Cham. https://doi.org/10.1007/978-3-031-49047-7_12

Schramm, L. L., Stasiuk, E. N., & Marangoni, D.G. (2003). *2 Surfactants and their applications*. https://doi.org/10.1039/B208499F.

Shah, I., Adnan, R., Wan Ngah, W. S., & Mohamed, N. (2015). Iron impregnated activated carbon as an efficient adsorbent for the removal of methylene blue: Regeneration and kinetics studies. *PLoS ONE, 10*(4), e0122603. https://doi.org/10.1371/journal.pone.0122603

Singh, S., Gupta, H., Dhiman, S., & Sahu, N. K. (2022). Decontamination of cationic dye brilliant green from the aqueous media. *Applied Water Science, 12*(4), 61. https://doi.org/10.1007/s13201-022-01596-5

Singh, S., Kumar, A., & Gupta, H. (2020). Activated banana peel carbon: A potential adsorbent for Rhodamine B decontamination from aqueous system. *Applied Water Science, 10*, 185. https://doi.org/10.1007/s13201-020-01274-4

Siyal, A. A., Shamsuddin, M. R., Low, A., & Rabat, N. E. (2020). A review on recent developments in the adsorption of surfactants from wastewater. *Journal of Environmental Management, 254*, 109797. https://doi.org/10.1016/j.jenvman.2019.109797

Srinet, S. S., Basak, A., Ghosh, P., & Chatterjee, J. (2017). Separation of anionic surfactant in paste form from its aqueous solutions using foam fractionation. *Journal of Environmental Chemical Engineering, 5*(2), 1586–1598. https://doi.org/10.1016/j.jece.2017.02.008

Sun, Y.-X., Zhang, Z.-W., Xu, X.-R., Hu, Y.-X., Luo, X.-J., Cai, M.-G., & Mai, B.-X. (2015). Bioaccumulation and biomagnification of halogenated organic pollutants in mangrove biota from the Pearl River Estuary, South China. *Marine Pollution Bulletin, 99*(1–2), 150–156. https://doi.org/10.1016/j.marpolbul.2015.07.041

Szymanski, A., Jaroszynski, T., Jeszka, P., & Lukaszewski, Z. (1996). Tensammetric studies of biodegradation of Marlipal 1618/25 in the river water die-away test. *Water Research, 30*(10), 2465–2471. https://doi.org/10.1016/0043-1354(96)00137-6

Tao, L., et al. (2019). Biomagnification of PBDEs and alternative brominated flame retardants in a predatory fish: Using fatty acid signature as a primer. *Environment International, 127*, 226–232. https://doi.org/10.1016/j.envint.2019.03.036

Tian, Y. X., Chen, H. Y., Ma, J., Liu, Q. Y., Qu, Y. J., & Zhao, W. H. (2023). A critical review on sources and environmental behavior of organophosphorus flame retardants in the soil: Current knowledge and future perspectives. *Journal of Hazardous Materials, 452*, 131161. https://doi.org/10.1016/j.jhazmat.2023.131161

Tripathy, D. B., Mishra, A., Clark, J., & Farmer, T. (2018). Synthesis, chemistry, physicochemical properties and industrial applications of amino acid surfactants: A review. *Comptes Rendus Chimie, 21*(2), 112–130. https://doi.org/10.1016/j.crci.2017.11.005

Valizadeh, S., Younesi, H., & Bahramifar, N. (2016). Highly mesoporous K_2CO_3 and KOH/activated carbon for SDBS removal from water samples: Batch and fixed-bed column adsorption process. *Environmental Nanotechnology, Monitoring & Management, 6*, 1–13. https://doi.org/10.1016/j.enmm.2016.06.005

van der Veen, I., & de Boer, J. (2012). Phosphorus flame retardants: Properties, production, environmental occurrence, toxicity and analysis. *Chemosphere, 88*(10), 1119–1153. https://doi.org/10.1016/j.chemosphere.2012.03.067

Waaijers, S. L., & Parsons, J. R. (2016). Biodegradation of brominated and organophosphorus flame retardants. *Current Opinion in Biotechnology, 38*, 14–23. https://doi.org/10.1016/j.copbio.2015.12.005

Xiong, P., Yan, X., Zhu, Q., Qu, G., Shi, J., Liao, C., & Jiang, G. (2019). A review of environmental occurrence, fate, and toxicity of novel brominated flame retardants. *Environmental Science & Technology, 53*(23), 13551–13569. https://doi.org/10.1021/acs.est.9b03159

Xu, F., et al. (2019). Human exposure pathways to organophosphate flame retardants: Associations between human biomonitoring and external exposure. *Environment International, 127*, 462–472. https://doi.org/10.1016/j.envint.2019.03.053

Yang, K., Jing, Q., Wu, W., Zhu, L., & Xing, B. (2010). Adsorption and conformation of a cationic surfactant on single-walled carbon nanotubes and their influence on naphthalene sorption. *Environmental Science & Technology, 44*(2), 681–687. https://doi.org/10.1021/es902173v

Yao, C., Yang, H., & Li, Y. (2021). A review on organophosphate flame retardants in the environment: Occurrence, accumulation, metabolism and toxicity. *Science of the Total Environment, 795*, 148837. https://doi.org/10.1016/j.scitotenv.2021.148837

Zapata-Corella, P., Ren, Z.-H., Liu, Y.-E., Rigol, A., Lacorte, S., & Luo, X.-J. (2023). Presence of novel and legacy flame retardants and other pollutants in an e-waste site in China and associated risks. *Environmental Research, 216*, 114768. https://doi.org/10.1016/j.envres.2022.114768

Zota, A. R., et al. (2018). Polybrominated diphenyl ethers (PBDEs) and hydroxylated PBDE metabolites (OH-PBDEs) in maternal and fetal tissues, and associations with fetal cytochrome P450 gene expression. *Environment International, 112*, 269–278. https://doi.org/10.1016/j.envint.2017.12.030

Chapter 8
Dye and Dye-Containing Hazardous Waste in Water Resource

Jasvinder Kaur, Sanjana Tewari, Amanpreet Kaur, and Rajdeep Malik

Abstract The use of synthetic colours in the textile sector is a significant contributor to water pollution. As a result of their inability to establish a strong link with the cloth, textile dyes are released into the water bodies in the form of effluent. Consequently, there are major adverse effects on human and environmental health due to the continuous outflow of untreated wastewater from several textile enterprises. The colours used in textiles have the ability to be poisonous to aquatic species, which can subsequently make their way into the food chain. This chapter will cover the consequences that dyes used in textiles have not only on human health but also on the health of aquatic flora and natural water bodies. In addition to increasing the biochemical and chemical oxygen demand, preventing photosynthesis, inhibiting plant growth, providing recalcitrance and bioaccumulation, and possibly promoting toxicity, mutagenicity, and carcinogenicity, textile dyes significantly diminish the aesthetic values of bodies of water. As a result, wastewater that contains dyes has to be adequately treated by employing environmentally friendly methods in order to prevent adverse impacts on the surroundings, mankind's wellness, and the natural resources of water. The most recent methods that are routinely used to remove colour from textile effluent are compared and contrasted in this research. The chapter also centres on the benefits and downsides of the different techniques that are currently in use.

Keywords Industrial wastewater · Dye removal · Hazardous · Toxicity · Environmental safety

---

J. Kaur (✉) · S. Tewari · A. Kaur
Department of Chemistry, School of Sciences, IFTM University, Moradabad, Uttar Pradesh 244102, India
e-mail: jasvinderkaur2911@gmail.com

J. Kaur · R. Malik
Department of Chemistry, Gurukula Kangri (Deemed to be University), Haridwar, Uttarakhand 249404, India

H. Gupta et al. (eds.), *Emerging Contaminants in Water and Wastewater*, Environmental Science and Engineering, https://doi.org/10.1007/978-3-031-82579-8_8

8.1 Introduction

The appeal of any texture depends on its hue. Since 3500 BCE, regular colours made from herbs, plants, flowers, a few creepy crawlies, and fish have been found in real records. In 1856, W. H. Perkins was the first scientist to introduce manufactured hues. The colour of the pattern is made up of unique hues. In addition, they showed defenceless shade with no resistance to sunlight or cleaning (Alothman et al., 2024). Modern sectors include business paper, plastics, food processing, cowhide tanning, luxury care items, cosmetics, printing, and shade generation using engineered dyes. In industries, only a small amount of the pigment applied to fabric stays on it; the rest gets thrown into the water as harmful waste (Ali et al., 2019). Thus, removing colour from wastewater is considered more important than removing soluble, colourless organic molecules. Many articles address this topic, including the crystal violet adsorption onto a residue of agricultural waste, an emerging dye abatement method, the biodegradation of synthetic dyes, the methylene blue adsorption on inexpensive adsorbents, and the wastewater removal of synthetic dyes. There are other ways to tackle environmental problems (Allaoui et al., 2016; Gupta & Gupta, 2018; Khan et al., 2023; Madhav et al., 2020; Singh et al., 2018).

It is necessary to handle dye waste as one of the environmental challenges that must be addressed (Singh et al., 2020, 2022). CO_2 emissions from fossil fuels are a major factor in the phenomenon known as global warming. The effects of global warming are not limited to a reduction in agricultural production; it also bring about additional biological and societal problems. The $ZnIn_2S_4$-based Z-scheme heterojunction and its applications in water splitting, CO_2 reduction, and pollutant degradation were the subject of a review that was published (Almroth et al., 2021). There is a widespread belief that renewable energy resources are an excellent alternative. By this term, we are referring to energy sources that are not sustainable, such as coal, which is the most extensively used source of energy. As the population of the globe continues to increase, there will be a greater demand for energy to power our homes, businesses, and communities. In spite of the fact that the development and dissemination of renewable energy sources is absolutely necessary in order to maintain a sustainable level of energy and protect our planet from the phenomenon of climate change, renewable resources are now unable to satisfy the requirements of the international community. The use of solar photovoltaic (PV) as a cost-free alternative to diesel generators is a strategy that is both productive and efficient in terms of fuel consumption. The literature was reviewed, and it was discovered that bioenergy has the potential to make a considerable contribution to the grid balancing that is required in a system that is powered solely by renewable energy (Al-Tohamy et al., 2022). While there are a number of variables that contribute to air pollution, the transportation industry is a significant contributor. Critically important are either the reduction of emissions from the transportation industry or the use of alternative energy sources as inputs. The textile, light, chemical, and other sectors in China are all significantly impacted by the dye business. The National Bureau of Statistics has recently disclosed that China has risen to the top of the world rankings in terms of fuel production and trade volume. It is worth noting that around 70–80% of China's fuel output is comprised of azo dyes. Based on the report published by the Technical Policy on Waste-Water Pollution Prevention and Control in the Textile Dyeing and Printing and enterprises, the textile industry may release up to 900 million metric tonnes of wastewater a year (Basak et al., 2019).This indicates that the textile industry ranks sixth among all industrial discharges. The complicated, high concentration, high colour, and numerous compounds that are difficult to biodegrade in these industries make it frequently impossible to accomplish the required impact by employing a single-step approach (Basak et al., 2019). This is because of the large number of substances that are difficult to biodegrade. In addition, the waste water contains a large variety of organic chemicals that are hazardous to living organisms and have three characteristic properties: carcinogenic, teratogenic, and mutagenic activity (Gupta, 2018; Gupta & Gupta, 2016). As an illustration, the physicochemical treatments have many drawbacks, such as high processing expenses, restricted processing capabilities, and challenging working circumstances. On the other hand, traditional biological treatments have problems with low processing efficiency and are even useless. Consequently, in order to meet the ever-increasingly stringent discharge standards and to achieve the objective of comprehensive treatment, it is absolutely necessary to devise a technology for the treatment of this kind of waste water that is both innovative and efficient. A large amount of environmental, theoretical, and practical application value may be derived from the use of CWO technology for the treatment of dye-containing water impede, particularly water containing azo dye waste (Behera et al., 2021).

Many nations have effectively implemented legislation to assist electric vehicles (EVs), which are regarded as a technical alternative that has the potential to be successful. Regarding electric vehicles, it is essential to take into account the batteries, charging techniques, and enhanced power output. There were a great number of articles that predicted the application of grapheme material and investigated these difficulties in great detail. As a result of variances in battery cells, lithium-ion batteries, which are used in electric vehicles, are unable to undergo a complete charging process (Coria-Oriundo et al., 2021). The methods of charging Li-ion batteries were the subject of an exhaustive inquiry that Zhang and his colleagues carried out. Additionally, they provide an alternate equalisation control mechanism that assists the battery group as a whole in achieving State of Charge (SOC) equalisation. This is accomplished by generating a series battery group for the purpose of charging. Magnesium-based batteries are also one of the electrochemical energy storage chemistries that are effectively developing (Abu-Nada et al., 2021). This is due to the fact that metallic magnesium has a large theoretical volumetric capacity.

Textile effluent has been shown to include a variety of poisonous dyes, heavy metallic elements (including cadmium, chromium, arsenic, and lead) required to make textile dye colour pigments, and aromatic compounds. In order to create textile dye colour pigments, heavy metals, including arsenic, cadmium, chromium, and mercury must be present throughout the synthesis process. These dangerous materials are transported across huge distances with wastewater. After that, they linger in the soil and water for a long time, which is bad for aquatic life and reduces the fertility

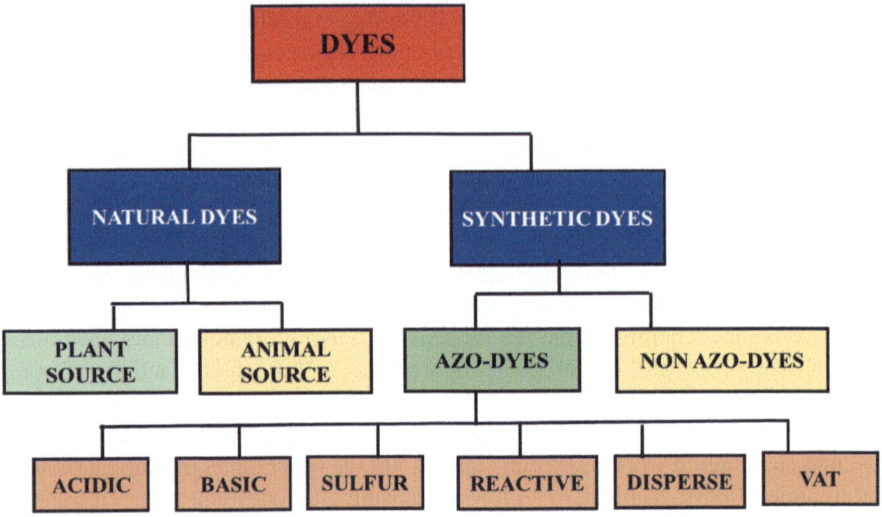

Fig. 8.1 Classification of dyes

of the soil and photosynthetic activity in aquatic flora, which results in the development of anoxic circumstances (Akpomie et al., 2020). Dyeing textiles diminish the aesthetic values of bodies of water by interfering with photosynthesis, stunting plant growth, entering the food chain, causing recalcitrance and bioaccumulation, and possibly increasing toxicity, mutagenicity, and carcinogenicity. They also raise the biochemical and chemical oxygen demand. Fabric manufacturing requires a lot of water, which results in effluent that has high concentrations of organics, metals, salts, dissolved solids, and refractory colours. Due to the subsequent contaminants and the poor removal of organic matter following discoloration, standard treatment procedures are often useless when treating synthetic dyes since they are very soluble in water and extremely lasting (Siwińska-Stefańska et al., 2018). Therefore, sophisticated approaches must be used. This chapter will provide in-depth information on the harmful consequences that wastewater from textiles containing dyes has on natural ecosystems and living beings. It will also go over a number of state-of-the-art treatment methods that are being used to better manage textile wastewater in an effort to protect the environment (Fig. 8.1).

8.2 The Impact and Toxicity of Textile Dyes

Textile colours, the most frequent organic waste, can be found in untreated textile sector discharges. The majority of textile dyes (about 60%) are azo dyes, which are based on structure made up by couple of azo groups. The discharge of fifteen to fifty percent of unbonded azo dyes into wastewater is a consequence of textile dying methods that are ineffective. The failure of certain textile factories to clean their effluent before releasing it into water bodies creates serious ecological hazards and harmful effects on aquatic life. In order to lower the amount of excess azo dyes released into the surroundings environment, other textile mills cleanse their effluent. The state of soil and crop viability rates were negatively impacted, when agriculturalists in developing countries used wastewater especially waste from factories to water their crops (Ali et al., 2022). The outflow of azo dyes into waterways reduces the amount of light reaching aquatic plants like algae. Additionally, when organisms like fish ingest dyes, their bodies may create toxic intermediates, which puts the fish as well as their attackers at risk (Elgarahy et al., 2021). Some industrial effluents include azo dyes, which can be harmful to people and other animals if they consume them or come into touch with them directly (Manickam et al., 2021). Intestinal microbes are responsible for the transformation of azo dyes into amino acids that are poisonous and may cause injury to a wide variety of organs and tissues throughout the body (Feng et al., 2012). Furthermore, it has been shown that human skin-grown bacteria may break down azo dyes and generate amines that might cause cancer. It is critical to employ appropriate and efficient management strategies to maintain clean ecosystems and increase sustainability, given an abundance of textile firms and huge amount of waste of water that contains colours. When deciding on the most effective treatment method, it is important to consider the existence of other organic and inorganic components, the toxicity of those elements, and the relevant environmental discharge levels (Dotto et al., 2019).

8.3 Impacts of Textile Dyes on Human Health

Tounsadi et al. (2020) state that textile dyes are extremely dangerous and may even result in cancer. It has been shown that these dyes are responsible for a broad variety of ailments in humans as well as in wildlife. A variety of diseases, including as dermatitis and problems of the central nervous system, have been associated with the use of textile dyes. It is possible that these problems are brought on by the replacement of enzyme cofactors, which leads to the enzymatic activity themselves becoming inactive (Wu et al., 2021). When textile dyes are ingested or breathed, they have the potential to trigger inflammation to the human body, particularly if they are come in to contact to fine particles (Dutta et al., 2022). When workers are exposed to reactive dyes, they put themselves at risk for allergic responses involving, nasal congestion, conjunctive irritation, skin eruptions, and industrial asthma (Elgarahy et al., 2021). Immunoglobulin antibodies are produced when albumin from human serum and the chemically reactive dye mix to form a conjugate. This conjugation serves as an antigen, and it is responsible for the generation of antibodies that bind to histamine (Suryavathi et al., 2005), found that some of the potentially toxic substances that workers in the textile industry are exposed to have the ability to interfere with the process of spermatogenesis and stimulate ovulation. There is a significant need for azole dyes in the textile, paper, and leather sectors. These colours are manufactured from benzidine and its derivatives. Consequently, its toxicity has been focus of a significant amount of study, and it has been associated with the development of prostate cancer in human beings. Because intestinal bacteria in animals metabolise azo colours into their respective parent amines, which the stomach readily are absorbed, it is known that azo dyes may be detected in both vertebrate and animal urine. This is due to the fact that the microbes in the animal's digestive tract metabolise azo pigments. Reactive Blue 2, Disperse Red 1, and Reactive Green 19 are examples of textile dyes that have long-term genotoxic consequences on the wellness of humans. Based on their findings, Leme et al. (2015) found that Reactive Green 19 exhibited a dose-associated genotoxicity, but Reactive Blue 2 and Disperse Red 1 showed no genotoxicity at all. As stated by Hossain et al. (2018), the environmental presence of dyes is a result of their widespread use. In addition, freshwater plants, such as fish and algal species, may biologically collect dyes in their food chains (Hossain et al., 2018). Based on research, (Korpi et al., 2009), the quantity of harmful organic compounds that are present in individuals is one thousand times more than the amount that they had when these chemicals were first found in water. During the process of treatment of water, trihalomethanes are produced when chlorine combine with molecule of organic material. These trihalomethanes are outcome of chlorination that are utilised to eradicate harmful microorganisms. (Tounsadi et al., 2020) have conducted investigations to expose these compounds for their tendency to weaken humans' immune systems and increase the likelihood of developing colorectal, colon, and bladder cancer in the last few years. A wide variety of dyes, particularly azo dyes, have the potential to trigger mutations in some organisms. From this perspective, it was shown that intestinal microbes were accountable for the enzymatic conversion of Sudan I, a lipophilic pigment containing azo groups that is used in the food and fabric sectors, into volatile amines that possess carcinogenic. The possibility that the colouring agent and its precursors might cause cancer was raised in a research by (Piątkowska et al., 2018). In human lymphocytes and hepatoma (HepG2) cells, it has been shown that Disperse Red 1 dye is mutagenic (Will et al., 2016). This is due to the fact that it has the potential to raise the rate of micronuclei, a sign of chromosomal mutagenic action. Disperse Orange 1 dye may also cause damage to DNA by base-pair substitution and frameshift genetic changes, both of which change the reading frame. These are two other mechanisms via which the dye might interrupt with DNA. The dye known as Disperse Orange 1 possesses cytotoxic qualities that make it possible for it to trigger apoptosis in HepG2 cells upon contact reported by (Ferraz et al., 2011).

8.4 Impacts of Textile Dyes on Terrestrial and Aquatic Ecosystems

The scarcity of pure and premium potable water remains an prominent challenge for humanity, in spite the fact that water makes up more than 71% of the surface of earth's. Estimates show that houses use 11% of the water, businesses use 19%, and agriculture uses 70%. The need for safe potable water is growing at frightening pace. Human pollution has greatly affected natural environments on both land and water. The clothing sector is responsible for a significant percentage of the degradation of water quality, leading to permanent environmental contamination. Consumption of fish and other aquatic species with unnatural colors are root of cramping, a high body temeprature, and stress in humans. Wastewater containing dyes and textile pigments has a highly colored pH, a varied number of suspended particles, total organic carbon, chemical oxygen demand, and biological oxygen demand. The suspended materials obstruct gas exchange in fish, affecting their gills and potentially slowing their growth or causing death. The long-term exposure to textile effluents has been found to have harmful effects on fish. It decreases the amount of feed that fish consume, which leads to a decrease in their development rate and the amount of protein, carbohydrates, and fat they contain. Research suggests that reactive azo dyes found in industrial wastewater may trigger DNA impact in adult fish. These dyes cause an upregulation of erythrocytic micronuclei creation in adult fish and the development of gill micronuclei in juvenile fish. Fish are susceptible to various illnesses due to hypoxia, which negatively affects their immune systems and physiological reactions. This ultimately affects the health of people who consume polluted seafood. Additionally, in stagnant aquatic habitats, organic matter breakdown by bacteria can lead to under-oxygenation, while the agglomeration of organic materials in watercourses can cause bad taste, bacterial growth, unpleasant smells, and abnormal colours. In accordance with (Franca et al., 2020), the minimum concentration of dye mass required for water coloration is 5 g/L. Aquatic plants may have trouble carrying out photosynthesis if colourants block the water's light pathway. Synthetic organic dyes are difficult to degrade using conventional biological techniques. The persistence of dyes is linked to their chemical reaction capability, with unsaturated molecules being less enduring than single bond substances. The endurance of aromatic compounds is increased by the number of substituents, whereas the persistence of dyes that include alkyl groups is improved by the presence of halogen substituents at higher concentrations. Textile dye phytotoxicity is evaluated using aquatic macrophytes as natural ecological markers. Various metrics, such as dry weight, total frond area, number of fronds, and chlorophyll content, are used to measure the strong development of plants. Textile dyes have been shown to impact the development rates and growth of aquatic macrophytes by altering every parameter of the macrophyte. Either stress by oxidation or a modification to the plant's light-system caused by a reduction in electron transport in the chloroplasts. Lemna minor, a vital link in the food chain, has been studied to investigate the impact of Red dye Congo and Gentian Violet dyes on its production of biogenic amines. These dyes have been found to prevent L. minor from growing, producing biomass, and producing chlorophyll. Furthermore, Gentian Violet was found to be more deadly than Red dye Congo as it obstructed the biosynthesis formation for bio generated amines. Based on the data, it seems that the quantities of biogenic amines and the activity of decarboxylase are sensitive and early indicators of the phytotoxic effects that these textile dyes have on Lactobacillus minor. Contaminated wastewater with higher solids concentration distrupt the osmotic balance of seeds and reduces the amount of dissolved oxygen that seedling need while they are growing. Furthermore, plants lose chlorophyll as the concentration of dissolved substances rises (Daneshgar et al., 2024). Lower chlorophyll levels affect light dependent activity, hinders the growth of plant, and lessen the build-up material that is dry in plants. Extremly high amounts of abscisic acid may degrade chlorophyll and prevent plants from growing by inhibiting the production of new chlorophyll. Additionally, plants may accumulate more proline due to textile dye effluents, indicating that the dyes are stressing the plants. The DNA impact of industrial waste include azo dyes on Allium cepa seedlings include micronucleus and chromosomal abnormalities.

8.5 Diverse Approaches and Technologies Are Employed to Remediate Wastewater Containing Textile Dyes

Due to humanity action and limited resources, obtaining safe potable water is most significant issues that people all over the globe are facing in the present day. By the year 2030, around 47% of the world's population would grapple with a scarcity of drinkable water. There is a rising drive to eradicate bacteria and organic pollutants such as textile dyes in order to reach the objective of having clean and pure water to consume. This is being done in order to achieve the achievement of the goal. There are approximately 10,000 distinct artificial colours and pigments that are used in the fabric and paper manufacturing sectors. These dyes and pigments have a negative impact on both human well-being and their surroundings. Furthermore, groundwater resources and the environment are exposed to textile pigments flowing into the water over long periods of time. Therefore, it is essential to develop sustainable methods to improve colour removal technology (Fig. 8.2).

There is currently no worldwide agreement on the release of wastewater containing dyes and the associated textile effluents, especially azo dyes. The prohibitions on the material and chemical factors of treated wastewater often occurs concurrently with the establishment of restrictions on azo dyes. Wastewater can be treated using chemical, physical, biological, or combined techniques, depending on its characteristics.

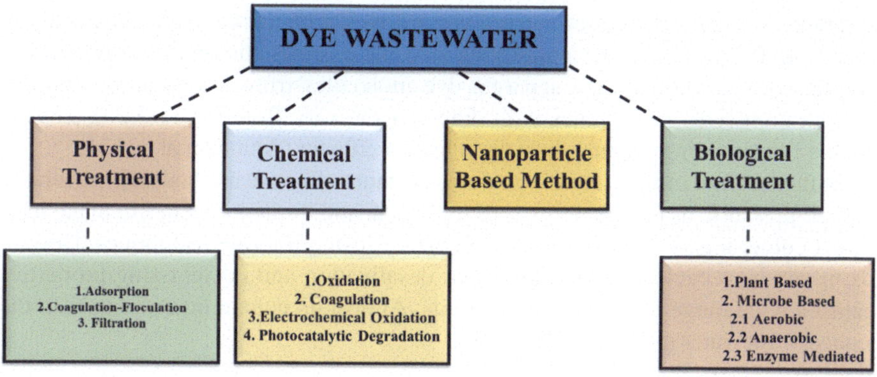

Fig. 8.2 Treatment strategies of dye-containing wastewater

8.5.1 Physical Methods for Treating Wastewater Contaminated by Textile Dyes

The removal efficiency of physical techniques for treating wastewater containing dyes ranges from 85 to 99%. These techniques include adsorption, ion exchange, and membrane filtering, and all rely by movement of mass working technique. Physical strategies have several benefits: they are easy to operate, cost-effective, require fewer chemicals, and are not affected by the presence of harmful substances (He et al., 2021). However, there are also disadvantages to these methods, such as the production of sludge and toxic byproducts, as well as their limited applicability, which makes them less desirable. Additionally, their use in textile wastewater treatment can be hindered by factors like COD, BOD, high temperatures, pH, colour, and heavy metals.

8.5.1.1 Filtration Technology

When it comes to the purification of wastewater, filtration technology is among the most significant physical treatment technologies on the market. Reverse osmosis, ultra filtering, nanofiltration, and micro filtering are some of the methods that are included in this category. This technology has shown promising results in removing color from wastewater. One cannot exaggerate the significance of the function that individual membranes play in the process of water purification. While microfiltration method is not good for treating contaminants water due to its enormous particle size, ultrafiltration and nanofiltration can effectively elimante a wide range of colors (Islamet al., 2021).

The mentioned approach is effective but requires periodic membrane replacement. Microfiltration, often known as MF, is a method of filtration that involves the use of a conventional membrane in order to eliminate colours and suspended particles from wastewater with ranging from 0.1 to 10 μm. Nanofiltration (NF) is another advanced technique used to purify wastewater containing dyes, with membrane width generally range from 0.5 to 0.2 nm. By employing size and electrostatic repulsion principles, nanofiltration technology can separate dye molecules from wastewater solutions. Ultrafiltration (UF) membranes with diameter the vary from 0.1 to 0.001 μm are suitable for removing organic dyes from textile effluent (Ahmadet al., 2024).

For the removal of salts, Reverse osmosis is more effective methods. However, the elevated pressure, usage of energy, cell expenses, and limited system life make hard to treat colourants or remove organic impurities using filtering techniques. Despite this, it can be recycled and has excellent desalination and decolorizing properties against a wide range of dye waste materials. Almost no contaminants remain in the cleaned wastewater (Jamee et al., 2019).

8.5.1.2 Adsorption

A process known as the adsorption, is the process by which molecule that are have been dissolved are able to form bonds with outer most layer of adsorbant material applying both chemical and physical forces. Carbon activated adsorption enables wastewater treatment by removing colorants found in wastewater through ion exchange or adsorption alone. This method is recommended for removing various dyes due to its effectiveness, high adsorption capacity, and ease of use. However, its high cost has limited its use, leading researchers to seek more affordable alternatives to activated carbon (Ahamad et al., 2024).

In order to remove dye from wastewater, a number of different types of adsorbents are often used. These include zeolites, silica gel, silica gel, and alumina. The industrial sector is one of the most common applications for activated carbon material. Using the Biopolymer/ZSM–5 zeolite adsorbent, wastewater that included pigments such as Basic Fuchsin (15 mg per litre; pH 9.0),Crystal Violet (15 mg per litre; pH 7.5), and Methylene Blue (15 mg per litre; pH 8.0) was treated. The results of this treatment resulted in decolorization % of 81.2%, 75.3%, and 86.6%, correspondingly. In addition, ZnO@Ze composite particles were able to obtain a decolorization rate of up to 90% for Congo Red at dye concentrations varying from 25 to 500 mg per litre and an adsorbent dose ranging from 0.025 to one grammes per litre (Khalaj et al., 2018).

8.5.1.3 Ion Exchange

In recent years, the ion exchange approach has been more popular for the treatment of industrial fabric waste and waste materials. This is mostly owing to the several advantages that this technology offers, including lower cost, ease of use, versatility, and outstanding performance. Robust bindings are formed between the solutes and the resins in an enclosed bed reactor during the ion exchange process. These bonds make it easier to achieve successful separation since they are stronger. Within the context of the ion exchange process that is used in dye removal, the relationship between the ions of molecules of dye and the functional structures of resins is of primary importance. For instance, in order to extract Acid Orange 10 from effluent at a dye concentration of Ten − 2 M, an anion-exchange resin was used, which resulted in a removal rate of 97%. In addition, a resin cation exchanger was able to remove 91.7% of Disperse Violet 28 when the dye concentration ranged from 10 to 500 mg per litre respectively (Chandanshive et al., 2020).

8.5.2 *Chemical Methods for Treating Wastewater Contaminated by Textile Dyes*

Wastewater treatment involves various chemical processes widely used to enhance water quality and purity, including oxidation, Coagulation-f, electro-oxidation, and photocataytic degradation.

8.5.2.1 Oxidation

Oxidation is a commonly used technique for chemical decolorization due to its simplicity. Oxidation methods can involve the use of living organisms, ozone, H_2O_2, NaClO, and even acids. This manner breaks down dyes into smaller molecules through oxidation, resulting in decolorization (Wang et al., 2024).

Numerous methods have been studied for treating wastewater containing dyes. One of these methods, known as advanced oxidation, involves the formaation of hydroxyl radicals (OH·) in situ. These radicals are potent oxidizing agents used to purify runoff with dyes. Advanced oxidative procedure include photo-catalyst, Fenton, Photo-Fenton, ozone formation, and oxidative electrochemical processes. In challenging conditions, these techniques can effectively remove color without generating sludge. In spite of this, they have negative aspects such as higher costs, pH dependency, and the production of unwanted byproducts. Two well known advanced oxidation methods that are used to for break down organic contaminants, including molecule of dye, are the Fenton and photo-Fenton processes, which rely on a mixture of soluble iron (II) salts and Fenton's reagent (H_2O_2).The Fenton technique involves several steps. First, OH· is created, typically through the photo-Fenton process. It is also possible for this radical to be produced via the ozonation process, which is used to purify wastewater that contains colours. The technique consists of three main steps: start, propagation, and termination (Kishor et al., 2021). A new method for treating dye-containing wastewater, called electrochemical technology, has recently been developed. In this method, OH· radicals are generated at the anode surface via photo electrocatalysis. Additionally, anodic oxidation can result in the production of OH· radicals through sonoelectrolysis or electro-Fenton.

8.5.2.2 Coagulation–Flocculation Method

Reactive and disperse pigments are commonly utilised in the fabric industry to dye cotton and polyester, but these chemicals have emerged as significant ecological concern. Because of low water solubility, nonionic dispersion dyes can be effectively removed using the coagulation-flocculation procedure. Coagulants, include metal salts and polymers, aid in the simple separation of wastewater with high concentrations of pollutants and toxins before it is disposed of. Meanwhile, flocculants are polymers that encourage the aggregation of flocks. Natural coagulants such as Azadirachta and Tamarindus indica are effective in eliminating colour from textile wastewater effluents. The coagulation-flocculation process utilizes well-known chemical coagulants such as ferric sulphate, ferric chloride, and others to maintain the purity of the removal system. Additionally, magnesium carbonate and hydrated lime are chemical coagulants that help absorb colors and dye products. This process involves adding a coagulant to a solution and vigorously mixing it, causing precipitates to form and trap organic pollutants and impurities, resulting in clear, treated water.

8.5.2.3 Electrochemical Oxidation

There are three approaches have been used in recent time for the purpose eliminating organic contaminants, such as fabric pigments, from water waste. These three techniques are electro-Fenton, anodic oxidation, and electrocoagulation. Neither the production of sludge nor the inclusion of chemicals is necessary for the use of electrochemical treatment methods. However, such methods have few downsides, including the fact that they are less successful than other treatment technologies and that they need a significant amount of money to be spent on electricity (Li et al., 2022), drawbacks, such as being less effective than other treatment technologies and incurring expensive electrical expenses (Li et al., 2022). The electro-Fenton method, considered one of the most widely used advanced oxidation techniques, combines two stages of oxidation and coagulation to eliminate organic contaminants. This process stands out due to the lack of hazardous byproducts, environmental preservation, and the minimal use of chemicals. The electrocoagulation method utilizes an electrocoagulation process to treat runoff include dyes and two metal based electrodes for direct power delivery and the production of in-situ coagulant tiny particles. In contrast, strong oxidants like O_3 and H_2O_2 are generated electrochemically by the indirect oxidative procedure. This approach effectively removes pigments and other organic pollutants but has certain disadvantages, such as expensive maintenance and poor stability.

8.5.2.4 Photocatalytic Degradation

One effective method for breaking down dye compounds is photocatalytic degradation, which involves the use of UV/H_2O_2 and UV/TiO_2 to mineralize the compounds.

This method is advantageous as it does not require additional disposal procedures. When dealing with Methylene Blue dye, photocatalysis has shown very efficient method for its breakdown. The presence of an electron scavenger showed to improve the photocatalytic activity. Various parameters like pH, H_2O_2 concentration and starting dye concentration, effected deterioration. Both homogeneous and heterogeneous photocatalytic degradation can occur in this process (Liu et al., 2021; Lobiuc et al., 2018).

8.5.3 Biological Methods for Treating Wastewater Contaminated by Textile Dyes

Numerous physicochemical techniques have been recently used, but they are costly and generate a substantial amount of waste after cleaning process. The least expensive option is biological therapy when contrast to other physical and chemical procedure. Microbes include yeast, algae, fungus, and bacteria may absorb and degrade variety of pigments (Madan et al., 2019). Therefore, industrial pollutant cleanup techniques often involve fungus decolorization, microbiological decolorization, adsorption by microbial biomass (either alive or dead), and bioremediation systems. Biological therapy techniques are classified into two treatment groups depending the amount of oxygen required: aerobic and anaerobic. Aerobic biological therapy is the most often utilised approach because of its great effectiveness (Zarei et al., 2018).

Microorganisms are often used in bioremediation to remove pollutants from the environment. When exposed to hazardous wastes, microbes can develop new, resistant strains that transform dangerous substances into less harmful forms. In this microbial system, stubborn compounds break down as a result of biotransformation enzymes.

The efficiency of microbiological discoloration is determined on the chosen microorganisms' activity and adaptability. As a result, researchers have investigated a broad range of taxa for dye removal and mineralization. Identifying and eliminating dominant species is an intriguing biological aspect of wastewater treatment. The Microbes including actinomycetes, Fungus yeasts, and bacteria may decolorize a variety of colours (Manickam et al., 2021).

8.5.3.1 Enzyme-Mediated Degradation of Wastewater Containing Dyes

The expensive price of enzymes make them an unpopular option are for treating water waste. On the other hand, commercial enzymes are unique because they are affordable, reliable, efficient, and readily available in liquid availability. Textile effluent containing azo colours may be effectively treated by enzymes such azo reductases and laccases, which break down the dyes into simpler compounds and then remove them from the wastewater by means of flocculation. Both the chemical and biotech sectors make extensive use of enzyme. Nevertheless, denaturation-induced biocatalyst inactivation is a major obstacle to enzymatic degradation (Marin et al., 2019).

The oxidoreductase enzyme class has been extensively studied in relation to its activity during the degradation of azo dyes. Numerous studies have been conducted recently on this subject. Among the most accessible enzymes that can withstand a wide range of commercial dyes are peroxidase enzymes. These enzymes also have greater temperature tolerance and a larger pH range of operation. Entrapping enzymes on an appropriate matrix can enhance the durability and biodegradability of dye treatments. Nevertheless, the substrate, supply, and class of the enzyme all affect the relevant enzyme activity (Mishra et al., 2019). The process of immobilisation makes an enzyme reusable several times, and studies have shown that immobilised enzymes are more successful in decolorizing dyes than soluble enzymes. Furthermore, achieving a high decolorization response requires the optimization of the dye decolorization process, which is influenced by temperature, pH level, salt content, co-substrate, and donor electron (Oyeniran et al., 2021). During the process of dye removal, certain enzymes such as manganese peroxidase, azoreductase, and other related enzymes can generate chemical intermediates like amino acids and phenols, which are more toxic than the primary pigments molecules. In this scenario, toxic intermediates can be transformed into insoluble end products by the action of two or more oxidoreductase enzymes.

8.5.3.2 Degradation of Dyes by Plants (Phytoremediation)

Dye waste pollutants from various sources can be treated using phytoremediation. This novel technique shows promise in effectively and economically removing organic pollutants and heavy metals from soils and groundwater. Phytoremediation involves using plants and microorganisms connected to plant root systems to eliminate pollutants in the form of organic and inorganic wastes, thus protecting the ecosystem. One of the major advantages of phytoremediation is that it is an autotrophic process with low nutritional requirement. Additionally, due to its pleasing looks and sustainability for the environment, phytoremediation is simple to maintain and well accepted by the general public. Plants' genetic adaptability enables them to tolerate toxins from polluted environments.

Phaseolus mungo, *Sorghum vulgare*, and *Brassica juncea* are three plant species with different agricultural implications. Their capacity to eliminate azo dyes from waste fabrics has been investigated. The result showed the effectiveness of above plants are, 79%, 57%, and 53%, respectively. Additionally, it has been demonstrated that the plant *Blumeama lcommi* can break down colors in textiles. *Tagetes patula* L. hairy root cultures can effectively decolorize Reactive Red 198, and the responsible enzymatic mechanism has been found. Huge-scale phytoremediation requires extensive treatment areas, the evaporation of aromatic organic compounds, the bioavailable proportion of contaminants. Table 8.1 showed a contrast of different treatment technologies of dye containing wastewaters (Adesanmi et al., 2022).

Table 8.1 Comparison of different treatment technologies of dye containing wastewaters

Techniques	Advantages	Disadvantages
Physical treatment		
Utilising activated carbon for adsorption	Highly capable to remove a broad range of dyes	Price and overpriced
Ion exchange method	Adsorbent loss is prevented by regeneration	Efficient for small number of dyes
Electro-coagulation	Affordable and doable	Large scale sludge production
Membrane filtration	Reusable, water recovering and suitable for ant type of dye	Costly to setup
Reverse osmosis	Generating pure water that is effective for a wide range of dyes and helpful for recycling water	Expensive and enormous amount of pressure
Irradiation	At the laboratory scale efficient and optimised	High oxygen requirements
Biological treatment		
Microbial biomass-based adsorption	A high affinity for microbial biomass enables certain dyes to bind with it	Efficient for a small quantity of dyes
Enzyme degradation	It is highly efficient nontoxic, and reusable	Unreliable enzyme production
Aerobic-anaerobic (conventional) method	Cheap and effective for decolorizing a wide variety of dye	Yields sludge
Microbial cultures	Decolorization takes 24–30 h	Not suitable for all dyes
Chemical treatment		
Fenton reaction	Eliminates toxins, effective for eliminating dyes	Production for iron sludge
Photochemical	Wih no waste	Production of multiple byproducts
Oxidation	Simple system responds quickly	Expensive
Electrochemical destruction	No presence of sludge	Generate harmful material

8.5.3.3 Fungal-Mediated Breakdown of Wastewater Containing Dyes

Many different fungal strains or consortia have a significant impact on the breakdown and mineralization of fabric colours. The main benefit of emplyoing fungus for treating wastewater containing dyes is the ability to expedite fungal metabolism to achieve ideal environmental conditions (Shindhal et al., 2021).. Enzymes, including intracellular and extracellular ones like diarylpropane oxygenase, laccase, and manganese peroxidase, may enhance intracellular activity and assist in treating polluted water containing dyes. White-rot fungi have been found to play a role in breaking down resistant organic pollutants, such as azo dyes, since the early 1990s due to unspecific lignin-modifying enzymes. Phanerochaetechrysosporium can break down various contaminants found in wastewater from the paper, pulp, and textile industries (Taher et al., 2024).

8.5.3.4 Degradation of Wastewater Containing Dyes with Algae Assistance

The pollutants in question contain polycyclic aromatic hydrocarbons. Aspergillus niger demonstrated the capability to biodegrade Red Congo within 144 h of cultivation, achieving a remarkable removal rate of 97%). It was found that 27% of the dye can be removed through biosorption with just one gm of newly formed material, while the process of biodegradations may eliminate 70% of the colour due to the collaborative activity of lignin peroxidase, manganese peroxidase, and potentially deaminase. Studies have shown that the fungus laccase is capable of removing 70–88% of dye molecules. Additionally, diarylpropane oxygenase, laccase, and manganese have been used together to effectively break down color molecules (Pandi et al., 2019).

8.5.3.5 Degradation of Wastewater Containing Dyes with Yeast Assistance

When it comes to breaking down textile colors, yeasts are superior to filamentousfungi and bacteria in several ways. Yeasts might replace other methods for treating wastewater containing dyes because they grow quickly and can withstand unfavourable circumstances in the environment, including low pH (Ruscasso et al., 2021). However, only a few papers emphasize how yeasts break down and remove colors. Wastewater containing dyes has been shown to be broken down by several types of yeasts, such as, Saccharomyces, Debaryomyces and Candida among ascomycetous yeast species, and Rhodotorula and Trichosporon among basidiomycetous yeast genera. Many yeasts, such as *Scheffersomyces spartinae*, *Pichia occidentalis*, and *Sterigmatomyces halophilus*, have shown the capacity to withstand adverse circumstances, such as elevated concentrations of salts in fabric polluted water, and to segmenting fabric colours, including azo dyes. Reactive Brilliant Red K-2BP was eliminated by twelve strains of yeasts, involved, *Saccharomyces uvarum*, *S. cerevisiae*, and *Torulopsis candida* through biosorption. Another test was conducted on *S. cerevisiae* for the elimination of Ramazole Blue. Using biosorption, the bacteria could reduce the COD and color absorbance of actual textile effluent by 100% and 61.82%, respectively, without the need to use expensive substances such as activated carbon. As a biosorbent for the fabric dyes Reactive Red 141 and Reactive Blue 19, Ruscasso et al. (2021) explored the use of Antarctic yeast *Debaryomyces hansenii* F39A biomass (Parmar et al., 2022). The findings showed that no hazardous or toxic substances were released during the first 15 min of the process, despite a significant reduction in dye concentration. In conclusion, compared to more costly materials like activated carbon, using organic material from certain yeasts as a biosorbent is a practical, economical, and effective option.

8.6 Future Perspectives

Nearly 800 million individuals throughout the globe do not have access to safe potable water that meets household standards. The presence of water is essential for the survival of all forms of life. The issue of extensive pollution of natural water resources by inorganic and organic pollutants has recently emerged in a number of nations. The textile industry's wastewater includes high concentrations of many harmful xenobiotics, endangering both ecosystems and human health. Harmful dye-containing wastewater is a major problem since textile mills throughout the globe discharge millions of gallons of severely polluted wastewater every day. Conversely, there is not any single workable solution to the problem of treating wastewater that contains colours, making the task very complex. There is documentation of a number of well-established and novel approaches of treating effluent that contains pigments. Dye removal and degradation from dirty water may be accomplished using both chemical and physical approaches.

However, these treatments often come with high operating costs and unwanted side effects. In contrast, the microbial technique for remediating dye-containing wastewater is widely accepted as more ecologically friendly and economically viable than physical and chemical alternatives. While biological techniques require a longer duration to be effective, they are generally considered to be less efficient.

8.7 Conclusion

Worldwide, environmental pollution is becoming a major issue due to the release of hazardous and destructive substances. The textile industry is currently facing two significant issues: dye degradation and decolorization. The purpose of this study was to review published research in order to understand different methods and approaches for eliminating contaminants or decolorizing wastewater from the textile sector, particularly dyestuffs. However, these methods cannot completely eliminate the colour from effluent. So a cost-effective, environmentally friendly, and efficient method for decolorizing wastewater must be developed. However, challenges such as selecting strains for maximum enzyme production and successfully cultivating them on a large scale need to be addressed. Additionally, a variety of colors were successfully removed using dye removal methods based on silver nanoparticles. Nonetheless, there are significant obstacles that the nanoparticle-based strategy must quickly overcome in the dye wastewater treatment process. Despite these limitations, the study suggests that the combination of innovative nanotechnologies and enzyme degrading methods may offer a next generation solution to the problem of poisonous discharge from the clothing sector. The simultaneous use of nanotechnology and traditional enzyme degrading methods will be necessary for environmental protection.

References

Abu-Nada, A., Abdala, A., & McKay, G. (2021). Removal of phenols and dyes from aqueous solutions using graphene and graphene composite adsorption: A review. *Journal of Environmental Chemical Engineering, 9*(5), 105858.

Adesanmi, B. M., Hung, Y. T., Paul, H., & Huhnke, C. (2022). Comparison of dye wastewater treatment methods: A review. *GSC Advanced Research and Reviews, 10*(2), 126.

Ahmad, A., Khawar, M. R., Ahmad, I., Javed, M. H., Ahmad, A., Rauf, A., Younas, U., Nazir, A., Choi, D., & Karami, A. M. (2024). Green synthesis of ZnO nanocubes from Ceropegiaomissa H. Huber extract for photocatalytic degradation of bisphenol an under visible light to mitigate water pollution. *Environmental Research, 249*, 118093.

Akpomie, K. G., & Conradie, J. (2020). Advances in application of cotton-based adsorbents for heavy metals trapping, surface modifications and future perspectives. *Ecotoxicology and Environmental Safety, 201*, 110825.

Ali, S. S., Al-Tohamy, R., & Sun, J. (2022). Performance of Meyerozymacaribbica as a novel manganese peroxidase-producing yeast inhabiting wood-feeding termite gut symbionts for azo dye decolorization and detoxification. *Science of the Total Environment, 806*, 150665.

Ali, S. S., Al-Tohamy, R., Sun, J., Wu, J., & Huizi, L. (2019). Screening and construction of a novel microbial consortium SSA-6 enriched from the gut symbionts of wood-feeding termite, Coptotermesformosanus and its biomass-based biorefineries. *Fuel, 236*, 1128–1145.

Allaoui, M., Berradi, M., & Elharfi, A. (2016). Synthesis of a new asymmetric semi permeable membrane and based alloy of two polymers PVC and PSU. Application in the treatment of a colored solution. *Moroccan Journal of Chemistry, 4*(1).

Almroth, B. C., Cartine, J., Jönander, C., Karlsson, M., Langlois, J., Lindström, M., Lundin, J., Melander, N., Pesqueda, A., Rahmqvist, I., & Renaux, J. (2021). Assessing the effects of textile leachates in fish using multiple testing methods: From gene expression to behavior. *Ecotoxicology and Environmental Safety, 207*, 111523.

Alothman, A. A., Ahmad, N., Albaqami, M. D., Alothman, Z. A., Alqahtani, K. N., Khan, M. R., & Vasanthakumar, V. (2024). Bismuth-doped lanthanum ferrite perovskites: A promising approach for enhancing photo-Fenton degradation of organic dye pollutants. *Ceramics International*.

Al-Tohamy, R., Ali, S. S., Li, F., Okasha, K. M., Mahmoud, Y. A. G., Elsamahy, T., Jiao, H., Fu, Y., & Sun, J. (2022). A critical review on the treatment of dye-containing wastewater: Ecotoxicological and health concerns of textile dyes and possible remediation approaches for environmental safety. *Ecotoxicology and Environmental Safety, 231*, 113160.

Basak, B., Jeon, B. H., Kurade, M. B., Saratale, G. D., Bhunia, B., Chatterjee, P. K., & Dey, A. (2019). Biodegradation of high concentration phenol using sugarcane bagasse immobilized Candida tropicalis PHB5 in a packed-bed column reactor. *Ecotoxicology and Environmental Safety, 180*, 317–325.

Behera, M., Nayak, J., Banerjee, S., Chakrabortty, S., & Tripathy, S. K. (2021). A review on the treatment of textile industry waste effluents towards the development of efficient mitigation strategy: An integrated system design approach. *Journal of Environmental Chemical Engineering, 9*(4), 105277.

Chandanshive, V., Kadam, S., Rane, N., Jeon, B. H., Jadhav, J., & Govindwar, S. (2020). In situ textile wastewater treatment in high rate transpiration system furrows planted with aquatic macrophytes and floating phytobeds. *Chemosphere, 252*, 126513.

Coria-Oriundo, L. L., Battaglini, F., & Wirth, S. A. (2021). Efficient decolorization of recalcitrant dyes at neutral/alkaline pH by a new bacterial laccase-mediator system. *Ecotoxicology and Environmental Safety, 217*, 112237.

Daneshgar, H., Sojdeh, S., Salehi, G., Edrisi, M., Bagherzadeh, M., & Rabiee, N. (2024). Comparative study of synthesis methods and pH-dependent adsorption of methylene blue dye on UiO-66 and NH2-UiO-66. *Chemosphere*, 141543.

Dotto, J., Fagundes-Klen, M. R., Veit, M. T., Palácio, S. M., & Bergamasco, R. (2019). Performance of different coagulants in the coagulation/flocculation process of textile wastewater. *Journal of Cleaner Production, 208*, 656–665.

Dutta, S., & Bhattacharjee, J. (2022). A comparative study between physicochemical and biological methods for effective removal of textile dye from wastewater. In *Development in wastewater treatment research and processes* (pp. 1–21). Elsevier.

Elgarahy, A. M., Elwakeel, K. Z., Mohammad, S. H., & Elshoubaky, G. A. (2021). A critical review of biosorption of dyes, heavy metals and metalloids from wastewater as an efficient and green process. *Cleaner Engineering and Technology, 4*, 100209.

Feng, J., Cerniglia, C. E., & Chen, H. (2012). Toxicological significance of azo dye metabolism by human intestinal microbiota. *Frontiers in Bioscience (Elite Edition), 4*, 568.

Ferraz, E. R., Grando, M. D., & Oliveira, D. P. (2011). The azo dye disperse orange 1 induces DNA damage and cytotoxic effects but does not cause ecotoxic effects in *Daphnia similis* and *Vibrio fischeri. Journal of Hazardous Materials, 192*(2), 628–633.

Franca, F.R.D.G., Vieira, A., Carvalho, G., Oehmen, A., Pinheiro, H. M., Crespo, M. T. B., & Lourenco, N. D. (2020). Oerskoviapaurometabola can efficiently decolorize azo dye acid red 14 and remove its recalcitrant metabolite. *Ecotoxicology and Environmental Safety, 191*, 110007.

Gupta, H. (2018). Anthracene removal from water onto activated carbon derived from vehicular tyre. *Separation Science and Technology, 53*(4), 613–625.

Gupta, H., & Gupta, B. (2016). Adsorption of polycyclic aromatic hydrocarbons on banana peel activated carbon. *Desalination and Water Treatment, 57*(20), 9498–9509.

Gupta, H., & Gupta, B. (2018). Vehicular tire as potential adsorbent for the removal of polycyclic aromatic hydrocarbons. *Polycyclic Aromatic Compounds, 38*(4), 354–368.

He, H., Xin, X., Qiu, W., Li, D., Liu, Z., & Ma, J. (2021). Waste sludge disintegration, methanogenesis and final disposal via various pretreatments: Comparison of performance and effectiveness. *Environmental Science and Ecotechnology, 8*, 100132.

Hossain, M. S., Das, S. C., Islam, J. M., Al Mamun, M. A., & Khan, M. A. (2018). Reuse of textile mill ETP sludge in environmental friendly bricks–effect of gamma radiation. *Radiation Physics and Chemistry, 151*, 77–83.

Islam, A., Teo, S. H., Taufiq-Yap, Y. H., Ng, C. H., Vo, D. V. N., Ibrahim, M. L., Hasan, M. M., Khan, M. A. R., Nur, A. S., & Awual, M. R. (2021). Step towards the sustainable toxic dyes removal and recycling from aqueous solution—A comprehensive review. *Resources, Conservation and Recycling, 175*, 105849.

Jamee, R., & Siddique, R. (2019). Biodegradation of synthetic dyes of textile effluent by microorganisms: an environmentally and economically sustainable approach. *European Journal of Microbiology and Immunology, 9*(4), 114–118.

Khalaj, M., Kamali, M., Khodaparast, Z., & Jahanshahi, A. (2018). Copper-based nanomaterials for environmental decontamination—An overview on technical and toxicological aspects. *Ecotoxicology and Environmental Safety, 148*, 813–824.

Khan, W. U., Ahmed, S., Dhoble, Y., & Madhav, S. (2023). A critical review of hazardous waste generation from textile industries and associated ecological impacts. *Journal of the Indian Chemical Society, 100*(1), 100829. https://doi.org/10.1016/j.jics.2022.100829

Kishor, R., Purchase, D., Saratale, G. D., Saratale, R. G., Ferreira, L. F. R., Bilal, M., Chandra, R., & Bharagava, R. N. (2021). Ecotoxicological and health concerns of persistent coloring pollutants of textile industry wastewater and treatment approaches for environmental safety. *Journal of Environmental Chemical Engineering, 9*(2), 105012.

Korpi, A., Järnberg, J., & Pasanen, A. L. (2009). Microbial volatile organic compounds. *Critical Reviews in Toxicology, 39*(2), 139–193.

Leme, D. M., Primo, F. L., Gobo, G. G., & da Costa, C. R. V., Tedesco, A. C., & de Oliveira, D. P. (2015). Genotoxicity assessment of reactive and disperse textile dyes using human dermal equivalent (3D cell culture system). *Journal of Toxicology and Environmental Health, Part A, 78*(7), 466–480.

Li, F., Katz, L., & Hu, Z. (2019). Adsorption of major nitrogen-containing components in microalgal bio-oil by activated carbon: equilibrium, kinetics, and ideal adsorbed solution theory (IAST) model. *ACS Sustainable Chemistry & Engineering, 7*(19), 16529–16538.

Li, F., Zhao, K., Ng, T. S. A., Dai, Y., & Wang, C. H. (2022). Sustainable production of bio-oil and carbonaceous materials from biowaste co-pyrolysis. *Chemical Engineering Journal, 427*, 131821.

Liu, J., Chen, J., Zuo, K., Li, H., Peng, F., Ran, Q., Wang, R., Jiang, Z., & Song, H. (2021). Chemically induced oxidative stress improved bacterial laccase-mediated degradation and detoxification of the synthetic dyes. *Ecotoxicology and Environmental Safety, 226*, 112823.

Lobiuc, A., Olaru, S., Hancu, E. I., Costica, N., Fortuna, M. E., Zamfirache, M. M., & Constantinescu, G. (2018). Toxicity and removal of direct red 28 diazo dye in living polymeric systems. *Revista de Chimie, 69*(7), 1628–1635.

Madan, S., Shaw, R., Tiwari, S., & Tiwari, S. K. (2019). Adsorption dynamics of Congo red dye removal using ZnO functionalized high silica zeolitic particles. *Applied Surface Science, 487*, 907–917.

Madhav, S., Ahamad, A., Singh, A. K., Kushawaha, J., Chauhan, J. S., Sharma, S., & Singh, P. (2020). *Water pollutants: Sources and impact on the environment and human health. Sensors in water pollutants monitoring: Role of material* (pp. 43–62).

Manickam, P., & Vijay, D. (2021). Chemical hazards in textiles. In *Chemical management in textiles and fashion* (pp. 19–52). Woodhead Publishing.

Marin, N. M., Pascu, L. F., Demba, A., Nita-Lazar, M., Badea, I. A., & Aboul-Enein, H. Y. (2019). Removal of the acid orange 10 by ion exchange and microbiological methods. *International Journal of Environmental Science and Technology, 16*, 6357–6366.

Mishra, S., & Maiti, A. (2019). Applicability of enzymes produced from different biotic species for biodegradation of textile dyes. *Clean Technologies and Environmental Policy, 21*, 763–781.

Oyeniran, D. O., Sogbanmu, T. O., & Adesalu, T. A. (2021). Antibiotics, algal evaluations and subacute effects of abattoir wastewater on liver function enzymes, genetic and haematologic biomarkers in the freshwater fish, *Clarias gariepinus. Ecotoxicology and Environmental Safety, 212*, 111982.

Pandi, A., Kuppuswami, G. M., Ramudu, K. N., & Palanivel, S. (2019). A sustainable approach for degradation of leather dyes by a new fungal laccase. *Journal of Cleaner Production, 211*, 590–597.

Parmar, S., Daki, S., Bhattacharya, S., & Shrivastav, A. (2022). Microorganism: An ecofriendly tool for waste management and environmental safety. In *Development in wastewater treatment research and processes* (pp. 175–193). Elsevier.

Piątkowska, M., Jedziniak, P., Olejnik, M., Żmudzki, J., & Posyniak, A. (2018). Absence of evidence or evidence of absence? A transfer and depletion study of Sudan I in eggs. *Food Chemistry, 239*, 598–602.

Ruscasso, F., Bezus, B., Garmendia, G., Vero, S., Curutchet, G., Cavello, I., & Cavalitto, S. (2021). *Debaryomyces hansenii* F39A as biosorbent for textile dye removal. *Revista Argentina De Microbiología, 53*(3), 257–265.

Shindhal, T., Rakholiya, P., Varjani, S., Pandey, A., Ngo, H. H., Guo, W., Ng, H. Y., & Taherzadeh, M. J. (2021). A critical review on advances in the practices and perspectives for the treatment of dye industry wastewater. *Bioengineered, 12*(1), 70–87.

Singh, S., Gupta, H., Dhiman, S., & Sahu, N. K. (2022). Decontamination of cationic dye brilliant green from the aqueous media. *Applied Water Science, 12*(4), 61. https://doi.org/10.1007/s13201-022-01596-5

Singh, S., Kumar, A., & Gupta, H. (2020). Activated banana peel carbon: A potential adsorbent for Rhodamine B decontamination from aqueous system. *Applied Water Science, 10*, 185. https://doi.org/10.1007/s13201-020-01274-4

Singh, S., Parveen, N., & Gupta, H. (2018). Adsorptive decontamination of rhodamine-B from water using banana peel powder: A biosorbent. *Environmental Technology and Innovation, 12*, 189–195. https://doi.org/10.1016/j.eti.2018.09.001

Siwińska-Stefańska, K., Kubiak, A., Piasecki, A., Goscianska, J., Nowaczyk, G., Jurga, S., & Jesionowski, T. (2018). TiO_2–ZnO binary oxide systems: Comprehensive characterization and tests of photocatalytic activity. *Materials, 11*(5), 841.

Suryavathi, V., Sharma, S., Sharma, S., Saxena, P., Pandey, S., Grover, R., Kumar, S., & Sharma, K. P. (2005). Acute toxicity of textile dye wastewaters (untreated and treated) of Sanganer on male reproductive systems of albino rats and mice. *Reproductive Toxicology, 19*(4), 547–556.

Taher, T., Yu, Z., Melati, E. K. A., Munandar, A., Aflaha, R., Triyana, K., Wibowo, Y. G., Khairurrijal, K., Lesbani, A., & Rianjanu, A. (2024). Enabling dual-functionality material for effective anionic and cationic dye removal by using Nb2O5/MgAl-LDH nanocomposites. *Journal of Hazardous Materials Letters*, 100103.

Tounsadi, H., Metarfi, Y., Taleb, M., El Rhazi, K., & Rais, Z. (2020). Impact of chemical substances used in textile industry on the employee's health: Epidemiological study. *Ecotoxicology and Environmental Safety, 197*, 110594.

Wang, Y., Cao, L., Lu, Y., Liao, J., Lu, Y., Su, C., & Gao, S. (2024). Impact analysis of hydraulic residence time and dissolved oxygen on performance efficiency and microbial community in N, N-dimethylformamide wastewater treated by an AnSBR-ASBR. *Environmental Pollution, 344*, 123326.

Will, Y., McDuffie, J. E., Olaharski, A. J., & Jeffy, B. D. (Eds.). (2016). *Drug discovery toxicology: From target assessment to translational biomarkers*. Wiley.

Wu, L., Xu, Y., Lv, X., Chang, X., Ma, X., Tian, X., Shi, X., Li, X., & Kong, X. (2021). Impacts of an azo food dye tartrazine uptake on intestinal barrier, oxidative stress, inflammatory response and intestinal microbiome in crucian carp (*Carassius auratus*). *Ecotoxicology and Environmental Safety, 223*, 112551.

Zarei Mahmudabadi, T., Ebrahimi, A. A., Eslami, H., Mokhtari, M., Salmani, M. H., Ghaneian, M. T., Mohamadzadeh, M., & Pakdaman, M. (2018). Optimization and economic evaluation of modified coagulation–flocculation process for enhanced treatment of ceramic-tile industry wastewater. *AMB Express, 8*, 1–12.

Chapter 9
Plastics and Microplastics: Emerging Pollutants in the Water Systems and Potential Remedies

Paulami Sahu and Priyanka Varma

Abstract Overconsumption and mismanagement of plastic, particularly single-use plastic, is a growing menace across the globe and it has become an emerging contaminant for terrestrial as well as aquatic ecosystems challenging their sustainability. Owing to the characteristics of lightweight and high durability, plastic materials have the potential to be transported across long distances from their source without any degradation and finally deposited in lentic or lotic ecosystems. Out of the global production of 300 million metric tons/annum of plastic waste, approximately 0.8–30 million metric tonnes, finds its final destination in the aquatic environment annually. The predominant source of plastic pollution is terrestrial, with its transportation occurring primarily through the rivers/canals and ultimately leading to the contamination of near-shore sediments as well as the marine environment. In the end, plastic products find their way into the food chain, creating a grave threat to the well-being of aquatic life forms and their consumers. The primary contributors to this extensive accumulation of plastic waste include a deficiency in technical expertise, inadequate infrastructure development, and, most importantly, a lack of user awareness. The present chapter provides an overview of plastic and microplastics; their sources; pathways to the aquatic systems and their detrimental impact on human health and other terrestrial and aquatic biota, along with the comprehensive mitigation strategies employed to address this issue of global interest. This chapter also delves into some treatments available to mitigate plastic pollution including physical, thermal, chemical and biological, as well as advanced techniques like membrane-based technologies and advanced oxidation processes. Implementation of plastic waste management practices like the Circular Economy approach, 4R's principles, adoption of Extended Producer Responsibility (EPR) etc. which can indirectly reduce the plastic load of the aquatic ecosystems, and possible environmental-friendly plastic pollution-reducing strategies are also discussed.

P. Sahu (✉) · P. Varma
School of Environment and Sustainable Development, Central University of Gujarat, Gandhinagar, Gujarat 382030, India
e-mail: paulamisahu@gmail.com

H. Gupta et al. (eds.), *Emerging Contaminants in Water and Wastewater*, Environmental Science and Engineering, https://doi.org/10.1007/978-3-031-82579-8_9

Keywords Plastic and microplastic waste · Surface water · 4R's strategies · Mitigation

9.1 Introduction

Plastics are affordable, user-friendly, and exhibit numerous desirable properties such as high-impact strength, lightweight, flexibility, rigidity, vibrant color options, malleability, and resistance to heat, corrosion, UV light, and chemicals (Galafassi et al., 2019). It is widely used in diverse industries such as packaging, electrical appliances, furniture, residential applications, agriculture, and more. Due to excessive use of plastic, plastic becomes an emerging contaminant for land as well as water. Plastic pollution knows no boundaries, affecting not only megacities, cities, and towns but also reaching into slums, rural communities, and remote settlements (Blettler et al., 2019; Mihai et al., 2018). The issue of plastic and microplastic pollution poses a substantial environmental challenge that has a widespread impact across the globe. These plastic materials have the potential to be transported across long distances from their initial location, so exacerbating the worldwide issue of marine plastic pollution. The physical forces of waves and currents, along with environmental factors like solar radiation and temperature, play a role in breaking down plastics into smaller sizes in aquatic systems. The process of plastic waste deteriorating into extremely small particles is referred to as microplastics. The minuscule particles, with their harmful characteristics, pose a danger to the natural environment and have the ability to contaminate the food chain. According to a study, coastal countries contribute 4.8–12.7 million metric tons of untreated plastic waste to the oceans each year (Jambeck et al., 2015). Microplastics have been identified in various environmental components, including soil, water, and air, prompting concerns regarding potential health consequences. In the present study, attempts are made to suggest alternative solutions to mitigate the plastic from the aquatic environment.

9.1.1 Plastic and Its Types

The amount of plastic production and deposition has increased significantly since 1945, and in recent years, it has become more prevalent to find articles reporting the impact of microplastic contamination on living organisms. Celluloid, the initial plastic material, was created by John Wesley Hyatt during the 1860s. After conducting numerous trials, he determined that combining cellulose nitrate and camphor under heat resulted in a resilient, adaptable, water-resistant plastic that could be molded into different forms. In 1907, chemist Leo Hendrik Baekeland developed Bakelite, the first synthetic plastic, a thermosetting plastic that is durable, heat resistant, chemically resistant, and electrically non-conductive (Punyauppa-path 2020). The words "plastic" and "plasticus," which imply "formable or moldable in a cast," are derived from Greek and Latin, respectively. Plastics are synthetic or semi-synthetic organic polymers that are produced from natural gas, coal and petroleum as a byproduct. Plastic waste is often toxic, non-biodegradable, and takes more than 500 years to decompose. The production of plastic is also associated with the utilization of fossil feedstock and petrochemical processes. Approximately 99% of the overall feedstock for plastic manufacturing derives from fossil fuels, non-renewable resources that contribute to nearly 8–9% of global gas and oil consumption (Nielsen et al., 2020; Walker et al., 2021). Based on their thermal characteristics, they tend to be divided into two groups: Thermoplastics (polymers that can be re-melted) and Thermosetting (Polymers that exhibit a permanent solid state upon completion of the hardening process). Plastic debris that is smaller than 5 mm in size is known as microplastic. It is categorized into primary microplastics and secondary microplastics, as explained by (Conley et al., 2019).

9.1.2 *Sources of Plastic and Microplastics in Aquatic Environments*

Primary and secondary microplastics are derived from various sources.

9.1.2.1 Primary Sources of Microplastics in Aquatic Environments

Microplastic pollution is primarily caused by industrial abrasives, medical products like personal care products, and improper disposal. Aquatic-based sources of microplastic pollution include tidal flow (reversible flow from estuaries and sea), wastewater treatment plants, atmospheric deposition, marine activities like fishing, aquaculture industry, shipping (accidental spillage of plastic pellets), and river flow (flow to another fresh system) (Thushari & Senevirathna, 2020). In addition, microplastic pollution can be caused by the inadequate disposal of microplastic products from industries such as electronics, automotive, textiles, and paints, ultimately contaminating water bodies and river catchment areas (Kay et al., 2018; Wang et al., 2019).

9.1.2.2 Secondary Sources of Microplastics in Aquatic Environments

The formation of secondary microplastics occurs when discarded plastics, fishing gear, textiles, and hygiene products degrade. This degradation can occur through physical, chemical, and biological processes in the environment. These processes can also include abrasion in landfills and recycling facilities, the release of fibers from synthetic textiles, and litter generated by ships. The formation of secondary microplastics adds to the pollution and contamination of the environment (Duis & Coors, 2016; da Costa et al., 2019).

9.2 Pathway and Distribution of Plastic and Microplastic Waste in Water

Globally, around 300 million metric tons of plastic waste is being generated annually (Tulashie et al., 2020). According to model estimates, a staggering amount of plastic waste, ranging from 0.8 to 30 million metric tonnes, finds its way into the aquatic environment every year worldwide (Borrelle et al., 2020; Lau et al., 2020; Meijer et al., 2021). All of the plastic found in freshwater compartments originates on land, which contributes approximately 80 percent to marine plastic pollution (Bauer-Civiello et al., 2019; Li et al., 2016). This includes plastic that degrades in landfills and seeps into the soil and water table. Additionally, plastic pollution can also come from ocean-based sources, such as shipping and transportation activities, which can lead to plastic litter entering waterways. Plastic pollution originates mostly on land and finds its way into groundwater (Varma et al., 2022) or is carried by rivers, eventually reaching coastal sediments and infiltrating the marine ecosystem (Lebreton & Andrady, 2019). Another pathway for small pieces or particles of plastic is through freshwater systems, as plastic can travel from land to freshwater and then into marine water. Waste plastics in the aquatic environment release chemicals, additives, and other components, notably PAHs (polycyclic aromatic hydrocarbons). Ultimately, plastic products enter the food chain and pose serious health risks, as depicted in Fig. 9.1. When ingested by humans, fish, and other animals, plastic persists in the digestive system and can cause gastrointestinal blockages (Gall & Thompson, 2015). Aquatic organisms can also become entangled or die as a result of plastic pollution (Hofmeyr et al., 2006; Lithner et al., 2009). Moreover, blockages in sewage systems, particularly in developing countries, can lead to the spread of disease vectors and the production of offensive odors (Adane & Muleta, 2011; Ellis et al., 2005). The pollution of aquatic bodies by plastic is directly linked to several Sustainable Development Goals (SDGs), namely SDG6 "Clean Water and Sanitation" and SDG14 "Life Below Water," which specifically addresses the issue of microplastic pollution in the environment (mostly marine), primarily in aquatic ecosystems. Achieving Sustainable Development Goal 12 (SDG12), which emphasizes the" Responsible production and consumption" target, is crucial for the plastic industry, urban areas, and rural communities (Nations, 2015).

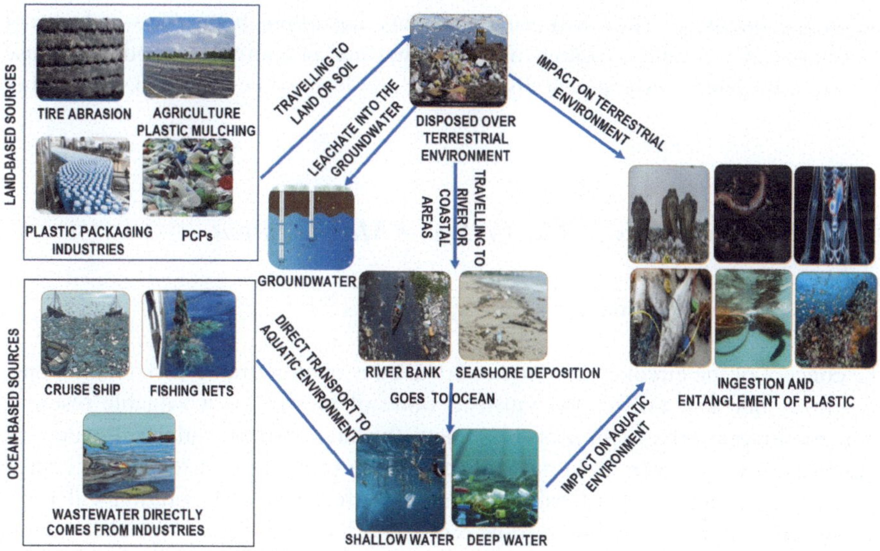

Fig. 9.1 *Source*, pathways and impact of plastic waste on the environment

9.3 Plastic Pollution Mitigation Strategies

Technologies that address these problems focus on either (1) collecting and removing existing plastic pollution or (2) directly stopping plastic from entering waterways.

9.3.1 Technology and Innovation

In current waste management technologies, there are two primary scientifically based strategies, including recycling and a degradation-based approach (Brems et al., 2013). These include physical methods including UV and photo-oxidant (Masry et al., 2021; Pospísil et al., 2006) and Recycling (Ragaert et al., 2017; Vollmer et al., 2020); Thermal treatment techniques covering Incineration (Netzer et al., 2021) Pyrolysis (Fivga & Dimitriou, 2018; Miranda et al., 2013; Pandey et al., 2020; Verma et al., 2016; Wong et al., 2015) Gasification (Nikolaidis et al., 2017) and Hydrocracking (Iqbal et al., 2023; Kasar et al., 2020) and Waste to Energy (Mohanraj et al., 2017), Chemical methods such as Chemical oxidation treatment (Ariza-Tarazona et al., 2019) Coagulation treatment (Larue et al., 2003) and Electrocoagulation (Liu et al., 2023; Sun et al., 2019); Biological methods such as Bio-methanation (Shah et al., 2008), Bioremediation (Bernard, 2021; Iranzo et al., 2001) and Advanced treatment methodology such as Membrane-based technology (Poerio et al., 2019; Ziajahromi et al., 2017), Advanced oxidation process (Liu et al., 2019; Shen et al., 2022) details in Table 9.1. Most often, chemical and physical remediation techniques are expensive and energy-intensive. They also usually produce hazardous byproducts and are not environmentally friendly. Biological approaches are an appropriate option because they are affordable, environmentally friendly, and have some drawbacks than other methods.

9.3.2 *Best Practices in Plastic Waste Management*

9.3.2.1 Circular Economy

The concept of the circular economy refers to an economic framework that not only recognizes but also utilizes the intrinsic value of plastics as a valuable resource while minimizing adverse impacts on society, the environment, and the climate. The objectives of the circular economy encompass the promotion of renewable energy utilization as a substitute for fossil fuels, as well as fostering collaboration with other specialized economies such as the blue economy, sharing economy, social economy, and green economy (D'Amato et al., 2017). The concept of the circular economy challenges the traditional linear model of "extract-make-use-throw away" by advocating for a restorative and regenerative approach known as "extract-make-use-recover and restore" (Foundation, 2017). The "New Plastics Economy" program has garnered the collaboration of more than 1000 organizations globally, with the shared objective of attaining a circular economy. The vision of a new plastics economy is to establish a framework in which plastic is never disposed of as waste or pollution. To achieve this vision and establish a circular economy for plastic, three specific actions need to be taken (Macarthur and Heading, 2019) which is mentioned in Fig. 9.2.

9.3.2.2 Strategies of 4R's Rules

For the management of any contaminant, including plastics, public procurement of Reduce, Reuse, Recycle and Recover is a common practice which is shown in Fig. 9.3.

The current process of plastics recycling may result in recycled products containing unsafe levels of contaminants for specific applications (Groh et al., 2019), such as food packaging. Manage plastic waste by developing innovative plastic substitutes, emphasizing the 4R's waste management tactics to reduce plastic pollution. Plastic waste separation technologies are used in recycling plants to remove contaminants and obtain high-purity polymers. These technologies include gravity-based sorting, electrostatic separation, magnetic density separation, and flotation processes which are detailed in Table 9.2.

Table 9.1 Treatment technologies for plastic waste management

Types of methods	Process	Definition	Advantages	Disadvantages	References
Physical methods	Photo-oxidation	UV radiation, known as photooxidation, can accelerate the oxidation process by producing free radicals in the presence of oxygen	Solar UV radiation photooxidation is generally regarded as the most significant environmental degradation mechanism	Some disadvantages such that Photo-oxidation is a surface-only process, compared to thermal oxidation, which takes place at the plastic's surface and inner layers	Masry et al. (2021) and Pospísil et al. (2006)
Physical methods	Recycling	Recycling is an effective waste management technique that aims to convert waste materials into raw materials that can be used in the production of a wide range of useful products	Recycling techniques support resource preservation, energy reduction, pollution control, sustainability, and economic growth	Recycling has environmental drawbacks, like chemical discharge, global warming, and health risks due to volatile gas release	Ragaert et al. (2017) and Vollmer et al. (2020)
Thermal methods	Incineration	Incineration refers to the process of fully burning waste in the presence of oxygen, resulting in the release of water molecules and carbon dioxide into the atmosphere	There are various advantages including reduced amount of waste, generation of heat and electricity for various activities, and reduction of environmental pollution	Incineration is costly and emits toxic gases like dioxins and furans, posing risks to people and the environment	Netzer et al. (2021)

(continued)

Table 9.1 (continued)

Types of methods	Process	Definition	Advantages	Disadvantages	References
Thermal methods	Pyrolysis	Pyrolysis refers to the thermal decomposition of plastic waste in an oxygen-free environment, which yields liquid oil and other value-added products and occurs at temperatures ranging between (300–900 °C)	The process of pyrolysis yields a variety of significant products, including liquid oil, gas, etc. However, because hydrocarbons are a component of polymers like PE, PP, and PET, pyrolysis is effective for treating them	Plastic pyrolysis is a complex, multi-step process with drawbacks like high viscosity and large output polymer molecules, influenced by feedstocks and multi-step reactions	Fivga and Dimitriou (2018), Wong et al. (2015), Miranda et al. (2013), Verma et al. (2016) and Pandey et al. (2020)
Thermal methods	Gasification	Gasification is the process in which carbonated organic or fossil fuels undergo high-temperature conversion (800–1100 °C) into common gases like carbon monoxide, hydrogen, and carbon dioxide	The process of gasification yields a variety of products solid, liquid and gas	The drawbacks associated with gasification products, like pyrolysis products, are significantly impacted by factors such as temperature, waste input, and the overall process framework. The production of tar hinders the gasification of plastic waste, leading to lower gas yields and productivity	Nikolaidis et al. (2017)

(continued)

Table 9.1 (continued)

Types of methods	Process	Definition	Advantages	Disadvantages	References
Thermal methods	Hydrocracking	Hydrocracking, also known as hydrogenation, is a process that converts long-chain hydrocarbons into smaller molecules like kerosene and gasoline	The process provides various advantages compared to pyrolysis, as it effectively converts waste plastic into liquid fuels of exceptional quality	Hydrocracking is a costly process requiring expensive equipment due to the high pressure required and is more expensive than other gases like nitrogen used in the pyrolysis process	Kasar et al. (2020) and Iqbal et al. (2023)
Thermal methods	Waste to energy	By transforming waste into valuable/useful items it refers to upcycling and valorizing waste Plastic to liquid fuel gasoline and diesel etc. Plastic to solid fuel e.g. RDF and plastic to gas fuel e.g. methane and hydrogen etc.	Waste-to-energy methods can be used to overcome problems with plastic recycling, such as the necessity for polymer separation and competitiveness with virgin plastics Higher energy savings than recycling	The disadvantages of this method include pollution, particulate production, material destruction, long-term investment, and perpetuation of the linear economy	Mohanraj et al. (2017)
Chemical methods	Chemical oxidation treatment	Chemical oxidation relies on the decomposition of MPs into smaller molecules or completely converting them into mineralization	A high degradation rate is the main advantage of this process	Several drawbacks, including high energy consumption and intricate reaction conditions	Ariza-Tarazona et al. (2019)

(continued)

Table 9.1 (continued)

Types of methods	Process	Definition	Advantages	Disadvantages	References
Chemical methods	Coagulation	Coagulation involves the destabilization and aggregate of MPs particles by coagulant addition, resulting in large flocs	Simple operation, low cost	Some of the disadvantages such as consumption of flocculants	Larue et al. (2003) and Ma et al. (2019)
Chemical method	Electrocoagulation	The process of electrocoagulation utilizes an electric current to destabilize and aggregate microplastic particles, enabling their elimination through sedimentation or filtration	Electrocoagulation offers various benefits, such as high removal efficiency, versatility, the ability to remove multiple pollutants simultaneously, scalability, reduced chemical usage, and lower energy consumption	It also has disadvantages, such as energy consumption, complex equipment, limited particle removal efficiency, and electrode fouling, necessitating regular cleaning and maintenance for optimal performance	Liu et al. (2023), Sun et al. (2019) and Perren et al. (2018)
Biological methods	Bio-methanation	By employing microbial intervention, solid waste like organic polymers or organic waste can be transformed into methane and manure. This process, referred to as anaerobic digestion, requires a lack of oxygen	The plastic polymers are broken down in this method without releasing any harmful by-products. Utilising solid waste from agro-based industries, this method creates beneficial byproducts including biogas and soil manures	Some disadvantages like producing an unpleasant smell, Methane is highly explosive, and more capital intensive compared to composting and landfills	Shah et al. (2008)

(continued)

Table 9.1 (continued)

Types of methods	Process	Definition	Advantages	Disadvantages	References
Biological methods	Bioremediation	Biological remediation involves the interactions of microbial activity with environmental factors, altering the physical and chemical conditions of hazardous waste reducing its toxicity	Microorganisms are being used to biodegrade natural substances, reducing their toxicity and promoting environmental friendliness	Since these methods are costly and harmful to the environment, enzymes must be ydrocar to break down synthetic compounds	Iranzo et al. (2001) and Bernard (2021)

(continued)

Table 9.1 (continued)

Types of methods	Process	Definition	Advantages	Disadvantages	References
Membrane-based technology	Reverse Osmosis, nanofiltration, and ultrafiltration	Reverse osmosis (RO) is a technique that uses high pressure to separate water from its constituents through a semipermeable membrane. It is an effective method for filtering water, including the removal of microplastics Microfiltration (MF) is employed to reject particles larger than 1 m, while ultrafiltration (UF) targets particles larger than 0.01 m. For particles larger than 0.001 mm, nanofiltration (NF) is utilized	RO treatment in wastewater can eliminate 80% of microplastics like polyethylene, polypropylene, and polystyrene. Various filtration methods like microfiltration, ultrafiltration, and nanofiltration, each with varying particle size removal capabilities. Ultrafiltration, for example, was able to remove 99% of polyethylene from the wastewater	Membrane removal of microplastics involves considering factors like material type, pore size, thickness, ydrocarbon water source, and membrane surface qualities. However, understanding nanoplastics or microplastics effects on membrane systems is limited due to technical knowledge, data availability, and analysis methods	Poerio et al. (2019) and Ziajahromi et al. (2017)
Hybrid technology	Advanced oxidation process	Advanced oxidation processes (AOPs) have been established as highly effective approaches for the degradation of microplastics, as they generate reactive radicals that facilitate the decomposition of microplastics	Advanced oxidative processes (AOPs) are environmentally friendly, versatile methods for removing microplastics from water, targeting primary and secondary microplastics, and degrading organic pollutants in industrial and municipal facilities	AOPs offer advantages in microplastic removal but have limitations like a long degradation period of up to months and potential emission of toxic by-products, such as toxins and polycyclic aromatic ydrocarbon, due to high temperatures	Liu et al. (2019) and Shen et al. (2022)

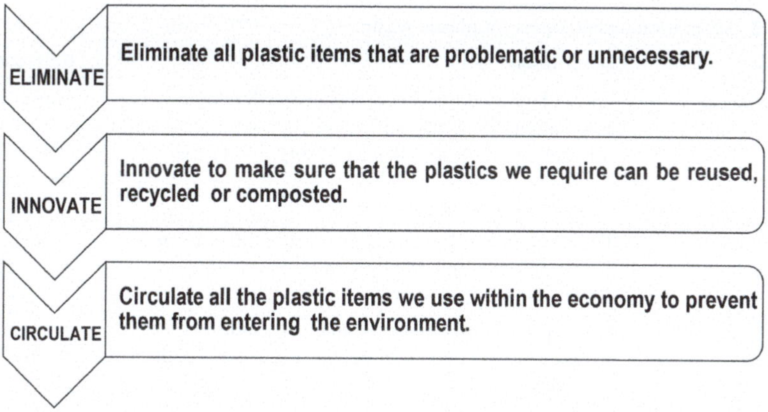

Fig. 9.2 Principles of circular economy for plastics

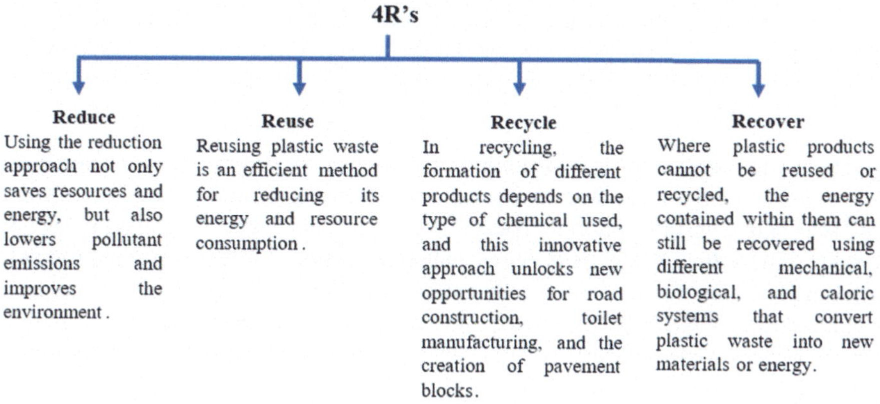

Fig. 9.3 4R's rules strategies

9.3.2.3 Regulation and Tax

Regulation of packaging (reducing packaging/material quotas, encouraging the use of alternative materials, setting standards for different types of plastics). The joint effort of implementing the EPR scheme and imposing higher taxes on plastic bags is a powerful strategy that not only reduces the flow of plastic waste into streams as well as it will discourage sellers and buyers (Ogunola et al., 2018). According to Gold et al., 2013, a prime example of technology and policy working together as effective complements can be seen in certain cities in the United States. In places like Oakland, California and Washington, District of Columbia, higher fees and taxes have been implemented on plastic products such as carrier bags, cutlery, and packaging. This approach has proven successful, with the funds generated being utilized for localized cleanup initiatives. According to Garcia et al. (2019), Governments can encourage

Table 9.2 Separation technologies for plastic waste

Methods	Description	References
Gravity-based sorting	It is widely used for plastic separation due to its automation, flexibility, and simplicity, but is slow and not suitable for all plastics due to density differences	Maisel et al. (2020), Gent et al. (2009) and Hu et al. (2013)
Electrostatic separation	The technique utilizes the triboelectric effect to charge plastic particles and employs an electric field to effectively separate them	Achouri et al. (2018)
Magnetic density separation	In this process uses a special magnetic field to create artificial gravity and separate plastic particles with close densities	Serranti and Bonifazi (2019)
Flotation processes	It involves introducing air bubbles to float hydrophobic particles and surface tension to separate types of plastics. It is widely used but has limitations in sorting plastics with similar densities	Shent et al. (1999) and Gundupalli et al. (2017)
Sensor-based sorting techniques	The techniques involve the utilization of online analytical logic and robotic units for the purpose of separation	Lubongo and Alexandridis (2022)

companies to reduce plastic consumption by offering subsidies or tax incentives to those who adopt new technologies or practices. Assessing wastewater emissions standards can help determine permitted plastic wastewater discharge amounts (Lechner & Ramler, 2015). In 2015, legislation was enacted to prohibit the use of microbeads in cosmetic items due to their significant contribution to the issue of microplastic contamination (UNEP, 2018). By focusing on hotspots of marine plastic pollution, policymakers have the opportunity to provide incentives that promote the expansion and effective implementation of these technologies.

9.3.2.4 EPR (Extended Producer Responsibility)

Extended Producer Obligation (EPR), as defined by the Organization for Economic Co-operation and Development (OECD), is a policy framework that establishes substantial obligations, either financial or physical, for producers to handle or dispose of goods after they have been used by consumers (OECD, 2021). From an environmental perspective, the EPR principle offers numerous benefits for improved plastic recycling and sustainable management of products throughout their entire lifecycle. Waste collection systems will be regulated based on Extended Producer Responsibility (EPR). To enhance recycling efforts, the State Urban Development Departments and various municipal bodies engage in collaborative and independent waste collection activities. In India, producers, importers, and brand owners typically undertake the collection of plastic sachets, pouches, and packing plastics. Norway, as an example, has successfully implemented many extended producer responsibility

(EPR) schemes in various plastic-related sectors, such as packaging waste, beverage container return systems, electrical and electronic device disposal, vehicle scrappage, and tire disposal. For example, the Norwegian government is enhancing the transition to a circular economy by strengthening existing Extended Producer Responsibility programs and also implementing an innovative EPR program for fishing gear and aquaculture waste, in addition to certain single-use products.

9.3.2.5 Bans of Plastic Bags and Single Users' Plastic Products in India and Other Countries

Governments worldwide have taken action to combat plastic marine pollution by implementing bans on certain types of single-use plastic bags (Deng et al., 2020). Denmark, Germany, North America, the UK, and Australia were the first to ban bags in 1991 and 1994, promoting eco-friendly alternatives like reusable bags. Bangladesh, the first South Asian country, has a strict ban on single-use plastics in coastal zones, hotels, and establishments until 2021. There are some strategies to prevent the sale of bags in various countries based on features like thickness and chemical properties. In the year 2016, the Indian Union government made revisions to the Plastic Waste Management Rules to impose restrictions on the utilization of plastic bags that are less than 50 μm in size (Singh & Biswas, 2023). India has taken a regulatory step to address the issue of littering caused by lightweight plastic bags. Plastic carry bags have increased in thickness from 50 to 75 μm since 30th September 2021, with a further increase to 120 μm by 31st December 2022. The Ministry of Environment, Forest, and Climate Change (MOEF&CC) issued a notification on August 12, 2021, regarding the amendment of the Plastic Waste Management (PWM) Rules. Under these amended rules, it is prohibited to produce, import, store, distribute, sell and use certain Single-Use Plastics (SUPs).

9.3.2.6 Environment-Friendly and Sustainable Strategies

Various sustainable technologies and approaches are aimed at reducing plastic consumption and preventing marine plastic pollution. To effectively gather floating debris, particularly plastic waste, from run-off water and rivers, booms, gully pots, boats, and sea bins have been identified as effective methods. Booms are logs or timbers that float on the water's surface and are anchored close to drainage banks. They have been successful in deflecting plastic waste but do not address plastics below the surface without the need for permanent structures (Tyler, 2011). Gully pots act as run-off inlet points and remove microplastics and pollutants from drainage networks (Karlsson & Luttropp, 2006). They are quick to install, reusable, and cost-effective. Skimmers or conveyor belts on boats are a cost-effective, easy-to-install, and reusable technology used to collect plastic pollution from river surfaces, such as in the successful deployment of the Chicago River (Bauer-Civiello et al., 2019). Sea bins and the Marina Trash Skimmer are trash-collecting devices powered by pumps that can collect significant amounts of debris and oil. The Seabin Project has developed the Seabin V5, a floating debris bin that can remove 1.4 tons of floating debris annually (Riggs & Naito, 2012). The Seabin V5 has undergone successful trials in New Zealand's Tutukaka Marina, where it effectively removed pieces of plastic and cigarette butts. To prevent further harm to marine ecosystems, it is crucial to decrease plastic production, enhance waste management, and encourage sustainable alternatives. Alternatives include bio-based materials such as use of made up of corn or sugarcane, straw-coated meal boxes (To, 2023), degradable plastic bags, and packaging films, non-plastic products like Jute bags, cloth bags and paper bags, waste conversion into electricity, energy, and value-added products. The utilization of plastic waste as partial aggregates in the production of concrete and as a construction material for flexible pavements not only helps in waste reduction but also contributes to employment generation, particularly in the waste collection sector (Rokade, 2012; Chavan, 2013). To effectively combat marine plastic waste, it is essential to raise awareness and actively involve individuals in beach clean-ups and recycling initiatives.

9.4 Conclusion

To confront the worldwide challenge posed by plastic pollution in aquatic ecosystems effectively, it is imperative for the scientific community to understand their responsibility to undertake appropriate corrective measures to create a sustainable environment. The detrimental impact of plastic pollution extends up to both, biotic and abiotic elements of the ecosystems. A significant amount of attention has been dedicated to tackling plastic and microplastic pollution in water. Various treatment options have been studied in recent research, exploring their physical, chemical, and biological effects. However, a universally accepted solution has not yet emerged because of the complexity and volume of plastic found in collected water samples. To successfully deal with this problem, it may be necessary to incorporate advanced oxidation and targeted biological degradation techniques into broader treatment systems. This approach can help remove microplastics from both terrestrial and inland aquatic systems along with marine systems. Educational programs spreading awareness, contribute a lot to mitigate plastic pollution. Recycling serves as a means to establish a sustainable material-cycle society, following fundamental principles of 4R's. The productivity of resources is enhanced with a high recycling rate, hence reducing the energy demand. The achievement of mitigating plastic pollution is reliant on the political will of governments, the collaboration of stakeholders and the commitment of individuals to follow regulations like plastic bag bans and restrictions on single-use plastic products, as well as the implementation of user tax etc. Encouraging people to go with alternative solutions such as the use of bio-based polymers made up of corn or sugarcane, eco-friendly materials like Jute bags, and giving some incentives to persons abiding by the rule are some good options that may lead to consumer's behavioral shifts, and as a consequence, mitigation of plastic pollution in the terrestrial and aquatic ecosystems.

References

Achouri, I. E., Zeghloul, T., Richard, G., Medles, K., Nouri, H., & Dascalescu, L. (2018). Factors that influence the performance of a two-rotating disks-type tribo-aero-electrostatic separator for micronized WEEE. *IEEE Transactions on Industry Applications, 55*(1), 802–811.

Adane, L., & Muleta, D. (2011). Survey on the usage of plastic bags, their disposal and adverse impacts on environment: A case study in Jimma City, Southwestern Ethiopia. *Journal of Toxicology and Environmental Health Sciences, 3*(8), 234–248.

Ariza-Tarazona, M. C., Villarreal-Chiu, J. F., Barbieri, V., Siligardi, C., & Cedillo-González, E. I. (2019). New strategy for microplastic degradation: Green photocatalysis using a protein-based porous N-TiO_2 semiconductor. *Ceramics International, 45*(7), 9618–9624.

Bauer-Civiello, A., Critchell, K., Hoogenboom, M., & Hamann, M. (2019). Input of plastic debris in an urban tropical river system. *Marine Pollution Bulletin, 144*, 235–242.

Bernard, P. (2021). Bioremediation technology for plastic waste. *Journal of Biochemistry and Biotechnology, 4*(6), 19.

Blettler, M. C., Garello, N., Ginon, L., Abrial, E., Espinola, L. A., & Wantzen, K. M. (2019). Massive plastic pollution in a mega-river of a developing country: Sediment deposition and ingestion by fish (*Prochilodus lineatus*). *Environmental Pollution, 255*, 113348.

Borrelle, S. B., Ringma, J., Law, K. L., Monnahan, C. C., Lebreton, L., McGivern, A., Murphy, E., Jambeck, J., Leonard, G. H., Hilleary, M. A., & Eriksen, M. (2020). Predicted growth in plastic waste exceeds efforts to mitigate plastic pollution. *Science, 369*(6510), 1515–1518.

Brems, A., Dewil, R., Baeyens, J., & Zhang, R. (2013). Gasification of plastic waste as waste-to-energy or waste-to-syngas recovery route. *Natural Science, 5*(6), 695–704.

Chavan, M. A. J. (2013). Use of plastic waste in flexible pavements. *International Journal of Application or Innovation in Engineering and Management, 2*(4), 540–552.

Conley, K., Clum, A., Deepe, J., Lane, H., & Beckingham, B. (2019). Wastewater treatment plants as a source of microplastics to an urban estuary: Removal efficiencies and loading per capita over one year. *Water Research X, 3*, 100030.

da Costa, J. P., Silva, A. L., Barcelò, D., Rocha-Santos, T., & Duarte, A. (2019). Nanoplastics in the environment. In *Plastics and the environment* (pp. 82–105). The Royal Society of Chemistry.

D'Amato, D., Droste, N., Allen, B., Kettunen, M., Lähtinen, K., Korhonen, J., Leskinen, P., Matthies, B. D., & Toppinen, A. (2017). Green, circular, bio economy: A comparative analysis of sustainability avenues. *Journal of Cleaner Production, 168*, 716–734.

Deng, Y., Zhang, J., Zhang, C., Ding, Z., Hao, C., & An, L. (2020). Countermeasures on plastic and microplastic garbage management. In *Microplastics in terrestrial environments: Emerging contaminants and major challenges* (pp. 447–469).

Duis, K., & Coors, A. (2016). Microplastics in the aquatic and terrestrial environment: Sources (with a specific focus on personal care products), fate and effects. *Environmental Sciences Europe, 28*(1), 2.

Ellis, S., Kantner, S., Saab, A., Watson, M., & Kadonaga, L. (2005). Plastic grocery bags: The ecological footprint. In *Environmental changes are spreading infectious diseases-UN study, Victoria* (pp. 1–19).

Fivga, A., & Dimitriou, I. (2018). Pyrolysis of plastic waste for production of heavy fuel substitute: A techno-economic assessment. *Energy, 149*, 865–874.

Foundation EM. (2017). The new plastics economy: rethinking the future of plastics & catalysing action. https://www.ellenmacarthurfoundation.org/publications/the-new-plastics-economy-rethinking-the-future-of-plastics-catalysing-action. Retrieved 8 Apr 2021

Galafassi, S., Nizzetto, L., & Volta, P. (2019). Plastic sources: A survey across scientific and grey literature for their inventory and relative contribution to microplastics pollution in natural environments, with an emphasis on surface water. *Science of the Total Environment, 693*, 133499.

Gall, S. C., & Thompson, R. C. (2015). The impact of debris on marine life. *Marine Pollution Bulletin, 92*(1–2), 170–179.

Garcia, B., Fang, M. M., & Lin, J. (2019). All hands on deck: Addressing the global marine plastics pollution crisis in Asia. *Forthcoming, Chinese Journal of Environmental Law, 3*(1).

Gent, M. R., Menendez, M., Toraño, J., & Diego, I. (2009). Recycling of plastic waste by density separation: Prospects for optimization. *Waste Management & Research, 27*(2), 175–187.

Gold, M., Mika, K., Horowitz, C., & Herzog, M. (2013). Stemming the tide of plastic litter: A global action agenda. *Tulane Environmental Law Journal, 27*, 165.

Groh, K. J., Backhaus, T., Carney-Almroth, B., Geueke, B., Inostroza, P. A., Lennquist, A., Leslie, H. A., Maffini, M., Slunge, D., Trasande, L., & Warhurst, A. M. (2019). Overview of known plastic packaging-associated chemicals and their hazards. *Science of the Total Environment, 651*, 3253–3268.

Gundupalli, S. P., Hait, S., & Thakur, A. (2017). A review on automated sorting of source-separated municipal solid waste for recycling. *Waste Management, 60*, 56–74.

Hofmeyr, G. G., Bester, M. N., Kirkman, S. P., Lydersen, C., & Kovacs, K. M. (2006). Entanglement of antarctic fur seals at Bouvetøya, Southern Ocean. *Marine Pollution Bulletin, 52*(9), 1077–1080.

Hu, B., Serranti, S., Fraunholcz, N., Di Maio, F., & Bonifazi, G. (2013). Recycling-oriented characterization of polyolefin package.

Iqbal, F., Shafeeq, A., Aslam, R., & Rauf, A. (2023). Comparative analysis of catalytic cracking and hydrocracking of waste plastic. *Chemical Engineering & Technology, 46*(4), 723–730.

Iranzo, M., Sainz-Pardo, I., Boluda, R., Sanchez, J., & Mormeneo, S. (2001). The use of microorganisms in environmental remediation. *Annals of Microbiology, 51*(2), 135–144.

Jambeck, J. R., Geyer, R., Wilcox, C., Siegler, T. R., Perryman, M., Andrady, A., & Law, K. L. (2015). Plastic waste inputs from land into the ocean. *Science, 347*(6223), 768–771.

Karlsson, R., & Luttropp, C. (2006). EcoDesign: What's happening? An overview of the subject area of EcoDesign and of the papers in this special issue. *Journal of Cleaner Production, 14*(15–16), 1291–1298.

Kasar, P., Sharma, D. K., & Ahmaruzzaman, M. (2020). Thermal and catalytic decomposition of waste plastics and its co-processing with petroleum residue through pyrolysis process. *Journal of Cleaner Production, 265*, 121639.

Kay, P., Hiscoe, R., Moberley, I., Bajic, L., & McKenna, N. (2018). Wastewater treatment plants as a source of microplastics in river catchments. *Environmental Science and Pollution Research, 25*, 20264–20267.

Larue, O., Vorobiev, E., Vu, C., & Durand, B. (2003). Electrocoagulation and coagulation by iron of latex particles in aqueous suspensions. *Separation and Purification Technology, 31*(2), 177–192.

Lau, W. W., Shiran, Y., Bailey, R. M., Cook, E., Stuchtey, M. R., Koskella, J., Velis, C. A., Godfrey, L., Boucher, J., Murphy, M. B., & Thompson, R. C. (2020). Evaluating scenarios toward zero plastic pollution. *Science, 369*(6510), 1455–1461.

Lebreton, L., & Andrady, A. (2019). Future scenarios of global plastic waste generation and disposal. *Palgrave Communications, 5*(1), 1–11.

Lechner, A., & Ramler, D. (2015). The discharge of certain amounts of industrial microplastic from a production plant into the River Danube is permitted by the Austrian legislation. *Environmental Pollution, 200*, 159–160.

Li, W. C., Tse, H. F., & Fok, L. (2016). Plastic waste in the marine environment: A review of sources, occurrence and effects. *Science of the Total Environment, 566*, 333–349.

Lithner, D., Damberg, J., Dave, G., & Larsson, Å. (2009). Leachates from plastic consumer products–screening for toxicity with *Daphnia magna*. *Chemosphere, 74*(9), 1195–1200.

Liu, F., Zhang, C., Li, H., Offiong, N. A. O., Bi, Y., Zhou, R., & Ren, H. (2023). A systematic review of electrocoagulation technology applied for microplastics removal in aquatic environment. *Chemical Engineering Journal, 456*, 141078.

Liu, P., Qian, L., Wang, H., Zhan, X., Lu, K., Gu, C., & Gao, S. (2019). New insights into the aging behavior of microplastics accelerated by advanced oxidation processes. *Environmental Science & Technology, 53*(7), 3579–3588.

Lubongo, C., & Alexandridis, P. (2022). Assessment of performance and challenges in use of commercial automated sorting technology for plastic waste. *Recycling, 7*(2), 11.

Ma, B., Xue, W., Hu, C., Liu, H., Qu, J., & Li, L. (2019). Characteristics of microplastic removal via coagulation and ultrafiltration during drinking water treatment. *Chemical Engineering Journal, 359*, 159–167.

Macarthur, E. L. L. E. N., & Heading, H. E. A. D. I. N. G. (2019). How the circular economy tackles climate change. *Ellen MacArthur Found, 1*, 1–71.

Maisel, F., Chancerel, P., Dimitrova, G., Emmerich, J., Nissen, N. F., & Schneider-Ramelow, M. (2020). Preparing WEEE plastics for recycling—How optimal particle sizes in pre-processing can improve the separation efficiency of high quality plastics. *Resources, Conservation and Recycling, 154*, 104619.

Masry, M., Rossignol, S., Gardette, J. L., Therias, S., Bussière, P. O., & Wong-Wah-Chung, P. (2021). Characteristics, fate, and impact of marine plastic debris exposed to sunlight: A review. *Marine Pollution Bulletin, 171*, 112701

Meijer, L. J., Van Emmerik, T., Van Der Ent, R., Schmidt, C., & Lebreton, L. (2021). More than 1000 rivers account for 80% of global riverine plastic emissions into the ocean. *Science Advances, 7*(18), eaaz5803.

Mihai, F., Ionut, M., & Grozavu, A. (2018). *Assessment of waste dumping practices in mountain creeks* (pp. 89–96).

Miranda, M., Cabrita, I., Pinto, F., & Gulyurtlu, I. (2013). Mixtures of rubber tyre and plastic wastes pyrolysis: A kinetic study. *Energy, 58*, 270–282.

Mohanraj, C., Senthilkumar, T., & Chandrasekar, M. (2017). A review on conversion techniques of liquid fuel from waste plastic materials. *International Journal of Energy Research, 41*(11), 1534–1552.

Nations, U. (2015). Transforming our world: The 2030 agenda for sustainable development. *New York: United Nations, Department of Economic and Social Affairs, 1*, 41.

Netzer, C., Li, T., & Løvås, T. (2021). Surrogate reaction mechanism for waste incineration and pollutant formation. *Energy & Fuels, 35*(9), 7030–7049.

Nielsen, T. D., Hasselbalch, J., Holmberg, K., & Stripple, J. (2020). Politics and the plastic crisis: A review throughout the plastic life cycle. *Wiley Interdisciplinary Reviews: Energy and Environment, 9*(1), e360.

Nikolaidis, P., & Poullikkas, A. (2017). A comparative overview of hydrogen production processes. *Renewable and Sustainable Energy Reviews, 67*, 597–611.

OECD. (2021). Extended producer responsibility. https://www.oecd.org/env/toolsevaluation/extendedproducerresponsibility.htm. Accessed 15 Jan 2021.

Ogunola, O. S., Onada, O. A., & Falaye, A. E. (2018). Mitigation measures to avert the impacts of plastics and microplastics in the marine environment (a review). *Environmental Science and Pollution Research, 25*, 9293–9310.

Pandey, U., Stormyr, J. A., Hassani, A., Jaiswal, R., Haugen, H. H., & Moldestad, B. M. E. (2020). Pyrolysis of plastic waste to environmentally friendly products. *Energy Produc Manage in the 21st century IV: The Quest for Sustain Energy, 246*, 61–74.

Perren, W., Wojtasik, A., & Cai, Q. (2018). Removal of microbeads from wastewater using electrocoagulation. *ACS Omega, 3*(3), 3357–3364.

Poerio, T., Piacentini, E., & Mazzei, R. (2019). Membrane processes for microplastic removal. *Molecules, 24*(22), 4148.

Pospíšil, J., Pilař, J., Billingham, N. C., Marek, A., Horak, Z., & Nešpůrek, S. (2006). Factors affecting accelerated testing of polymer photostability. *Polymer Degradation and Stability, 91*(3), 417–422.

Punyauppa-path, S., & Punyauppa-path, P. (2020). Microplastics: Origin, environmental impact, food and beverage contamination and management methods. *Naresuan University Journal: Science and Technology (NUJST), 28*(2), 72–80.

Ragaert, K., Delva, L., & Van Geem, K. (2017). Mechanical and chemical recycling of solid plastic waste. *Waste Management, 69*, 24–58.

Riggs, C. N., & Naito, C. (2012). *Debris control for bridges and culverts.* Pennsylvania Local Technical Assistance Program (LTAP) Information Sheet No. 152. United States. https://gis.penndot.gov/BPR_pdf_files/Documents/LTAP/TS_152.pdf. Accessed.

Rokade, S. (2012). Use of waste plastic and waste rubber tyres in flexible highway pavements. In *International conference on future environment and energy, IPCBEE* (Vol. 28).

Serranti, S., & Bonifazi, G. (2019). Techniques for separation of plastic wastes. In *Use of recycled plastics in eco-efficient concrete* (pp. 9–37). Woodhead Publishing.

Shah, A. A., Hasan, F., Hameed, A., & Ahmed, S. (2008). Biological degradation of plastics: A comprehensive review. *Biotechnology Advances, 26*(3), 246–265.

Shen, M., Song, B., Zhou, C., Hu, T., Zeng, G., & Zhang, Y. (2022). Advanced oxidation processes for the elimination of microplastics from aqueous systems: Assessment of efficiency, perspectives and limitations. *Science of the Total Environment, 842*, 156723.

Shent, H., Pugh, R. J., & Forssberg, E. (1999). A review of plastics waste recycling and the flotation of plastics. *Resources, Conservation and Recycling, 25*(2), 85–109.

Singh, S., & Biswas, M. K. (2023). Management strategies for single-use plastics: Lessons to learn from Indian approach of minimizing microplastic waste. *Environmental Science: Advances, 2*(12), 1680–1695.

Sun, J., Dai, X., Wang, Q., Van Loosdrecht, M. C., & Ni, B. J. (2019). Microplastics in wastewater treatment plants: Detection, occurrence and removal. *Water Research, 152*, 21–37.

Thushari, G. G. N., & Senevirathna, J. D. M. (2020). Plastic pollution in the marine environment. *Heliyon, 6*(8).

To, J. (2023). *Circular business models: Policy enablers for plastic packaging alternatives in India* (No. 8/2023). IDOS Discussion Paper.

Tulashie, S. K., Boadu, E. K., Kotoka, F., & Mensah, D. (2020). Plastic wastes to pavement blocks: A significant alternative way to reducing plastic wastes generation and accumulation in Ghana. *Construction and Building Materials, 241*, 118044.

Tyler, R. N. (2011). River debris: Causes, impacts, and mitigation techniques. *Alaska Center for Energy and Power*, 1–33.

UNEP, U. (2018). Legal limits on single-use plastics and microplastics: a global review of national laws and regulations. *United Nations Environment Programme (UNEP), Nairobi.*

Varma, P., Debsharma, C., & Sahu, P. (2022). Long-term influence of land use-land cover alteration on groundwater level in Gandhinagar taluka, India. IAH. *e-Journal of Geohydrology, 3*(1), 11–19.

Verma, R., Vinoda, K. S., Papireddy, M., & Gowda, A. N. S. (2016). Toxic pollutants from plastic waste-a review. *Procedia Environmental Sciences, 35*, 701–708.

Vollmer, I., Jenks, M. J., Roelands, M. C., White, R. J., Van Harmelen, T., De Wild, P., van Der Laan, G. P., Meirer, F., Keurentjes, J. T., & Weckhuysen, B. M. (2020). Beyond mechanical recycling: Giving new life to plastic waste. *Angewandte Chemie International Edition, 59*(36), 15402–15423.

Walker, T. R., McGuinty, E., Charlebois, S., & Music, J. (2021). Single-use plastic packaging in the Canadian food industry: Consumer behavior and perceptions. *Humanities and Social Sciences Communications, 8*(1), 1–11.

Wang, Z., Qin, Y., Li, W., Yang, W., Meng, Q., & Yang, J. (2019). Microplastic contamination in freshwater: First observation in lake ulansuhai, yellow river basin, China. *Environmental Chemistry Letters, 17*, 1821–1830.

Wong, S. L., Ngadi, N., Abdullah, T. A. T., & Inuwa, I. M. (2015). Current state and future prospects of plastic waste as source of fuel: A review. *Renewable and Sustainable Energy Reviews, 50*, 1167–1180.

Ziajahromi, S., Neale, P. A., Rintoul, L., & Leusch, F. D. (2017). Wastewater treatment plants as a pathway for microplastics: Development of a new approach to sample wastewater-based microplastics. *Water Research, 112*, 93–99.

Chapter 10
Detection and Remediation Techniques for Emerging Contaminants

Ambika Kumar, Anshu Kumar, Kumari Seema, and Karishma Kumari

Abstract Human activities have significantly enhanced the discharge of emerging contaminants (ECs) into the environment, which might have detrimental impacts on human health, even at extremely low concentrations (in the range of μg/L). The conventional wastewater detection technique and remediation methods are ineffective in removing many of these persistent emerging contaminants (ECs). Traditional methods, including Atomic Absorption Spectrometry, Mass Spectrometry, Chromatography, and other analytical techniques, are employed to measure the amounts of ECs (Emerging Contaminants) of different types. However, these approaches are subject to significant constraints and drawbacks. This chapter examines several techniques for detecting EC, such as Gas/Liquid Chromatography coupled with mass spectrometry (GC/LC–MS), Enzyme-linked immunosorbent test (ELISA), Advanced Spectroscopic Method, and Biosensors. These alternative hyphenated approaches provide several advantages compared to conventional techniques, including increased sensitivity, selectivity, reduced sample volume needs, portability, repeatability, speed, cost-effectiveness, and the ability to monitor ECs in real-time. Contemporary techniques do not need intricate solid phase extraction procedures. Current chapter provides a summary of each of the alternate techniques used to identify environmentally contaminants in various samples. The sensitivity of each approach varies based on the specific EC being tested and the corresponding matrix. Some of the current challenges encountered by these alternative methods in monitoring ECs. Further different conventional and modern remediation techniques have also been discussed.

A. Kumar (✉)
Department of Chemistry, Bhagalpur National College, Bhagalpur, Bihar 812007, India
e-mail: kumarambika.1115@gmail.com

A. Kumar (✉)
Department of Chemistry, T. N. B. College, Bhagalpur, Bihar 812007, India
e-mail: meet2anshu@gmail.com

K. Seema
Department of Chemistry, J. D. Women's College, Patna, Bihar 800022, India

K. Kumari
Department of Physics, Bhagalpur National College, Bhagalpur, Bihar 812007, India

H. Gupta et al. (eds.), *Emerging Contaminants in Water and Wastewater*, Environmental Science and Engineering, https://doi.org/10.1007/978-3-031-82579-8_10

Keywords Emerging contaminants · Adverse impact · Detection · Remediation · Quantitative analysis

10.1 Introduction

Emerging contaminants (ECs) are spread in the environment but are not yet comprehensively identified or monitored. This class of contaminants includes a wide range of materials, such as industrial chemicals, medicines, personal care items, and nanomaterials (Richardson, 2009). These are usually contaminants that have only recently been found in the environment or whose effects have only recently been observed. The adverse impact of ECs on environment or on living organism have not been comprehensively known yet. Interestingly, several such contaminants are persistent in natural environments and bioaccumulate in the food system, which might eventually endanger human health as well as ecological systems. These pollutants come from a variety of sources, including home waste, industrial emissions, and agricultural runoff. They can also reach the environment through several channels, including atmospheric deposition, wastewater discharge and leaching (Gupta, 2018; Khan et al., 2023; Kumar et al., 2014, 2017; Madhav et al., 2020, 2024). Since the anthropogenic activities, especially industrial revolution imposed many new contaminants into the environment. However, there has been a new push to classify these compounds as possible environmental hazards. The focus was mostly on conventional pollutants originated from the industrial effluents till the end of twentieth century, with a notable shift in awareness and knowledge of ECs during this time.

Sophisticated analytical techniques made it possible to identify emerging contaminants even at trace concentrations. In the beginning of twenty-first century, a fundamental change in the understanding of Emerging Contaminants was brought about by increased scientific research. It raised public awareness, leading to an increase in worries about drugs, substances that alter the endocrine system, and nanoparticles (Metcalfe, 2013). Concerns about the pervasiveness and intricate combinations of these pollutants, their cryptic nature, and the difficulties in determining and reducing their hazards are growing. A multidisciplinary strategy is required to address the issue of ECs because of their diverse character, many origins, and the difficulties associated with detecting and regulating them.

To reduce the adverse impact of these materials, regulations and concerned research have become much desirable (Satyanarayana et al., 2023). In this regard, ECs are becoming more widely acknowledged as a serious environmental as well as public health hazard. Because there are insufficient toxicological data, many adverse health effects are still poorly known, raising concerns about public safety and possible hazards to human health. Long-term exposure to these substances may have negative health impacts due to their occurrence in the atmosphere and eventually bioaccumulation in the food chains, especially when they accumulate in larger quantities in creatures, such as humans and many living organisms (Pereira et al., 2015). Also, ECs can have a negative impact on ecosystems for example, aquatic medicines can interfere with the reproductive and mental development processes of aquatic life, which has wider consequences for biodiversity and ecological services. Because many ECs have complicated chemical structures that make them difficult to identify analytically, conventional water treatment technologies are unable to remediate these chemicals. Furthermore, interactions between several ECs may have synergistic effects that make it more difficult to evaluate and manage the risks that go along with them (Tong et al., 2022).

Thus, because of the adverse impact of ECs on the environment and human health, their persistence, and their unknown consequences, it is critical to address them. Monitoring and exploring research are essential to solving these problems. The goal of this study is to present a thorough investigation of ECs, encompassing their origins, variations, adverse impact, suitable analytical techniques for their detection. By emphasizing recent developments and encountering difficulties in the identification and analysis of these pollutants, it aims to critically assess our present understanding of their effects on the environment and human health. The assessment also attempts to evaluate current EC rules and regulations, identify gaps, and suggest possible improvements. It also makes suggestions for future research, highlighting areas that require additional study in order to fully understand the ramifications of these pollutants.

The current chapter covers a broad spectrum of Emerging Contaminants (ECs), such as medicines, endocrine disruptors, nanomaterials, microplastics, and industrial chemicals. It involves looking at methods and technology that are being used now and, in the future, to identify and measure these pollutants. With the use of current research findings, it also examines the effects on the environment and human health. The review takes an interdisciplinary approach, considering the complexity of the subject and including knowledge from toxicology, public health, engineering, and environmental science.

With a comprehensive overview of the present state of knowledge surrounding ECs, this book chapter aims to be a priceless tool regarding environmental science and public health for researchers and practitioners.

10.2 Classification of ECs

ECs include a wide range of compounds found in the environment that are not extensively studied or monitored. Based on their source, purpose, and chemical makeup, these contaminants may be divided into several groups (Fig. 10.1).

10.2.1 Endocrine Disrupting Chemicals (EDCs)

Endocrine Disrupting Chemicals has ability to imitate, obstruct, or interfere with the actions of natural hormones, which can have negative consequences on development,

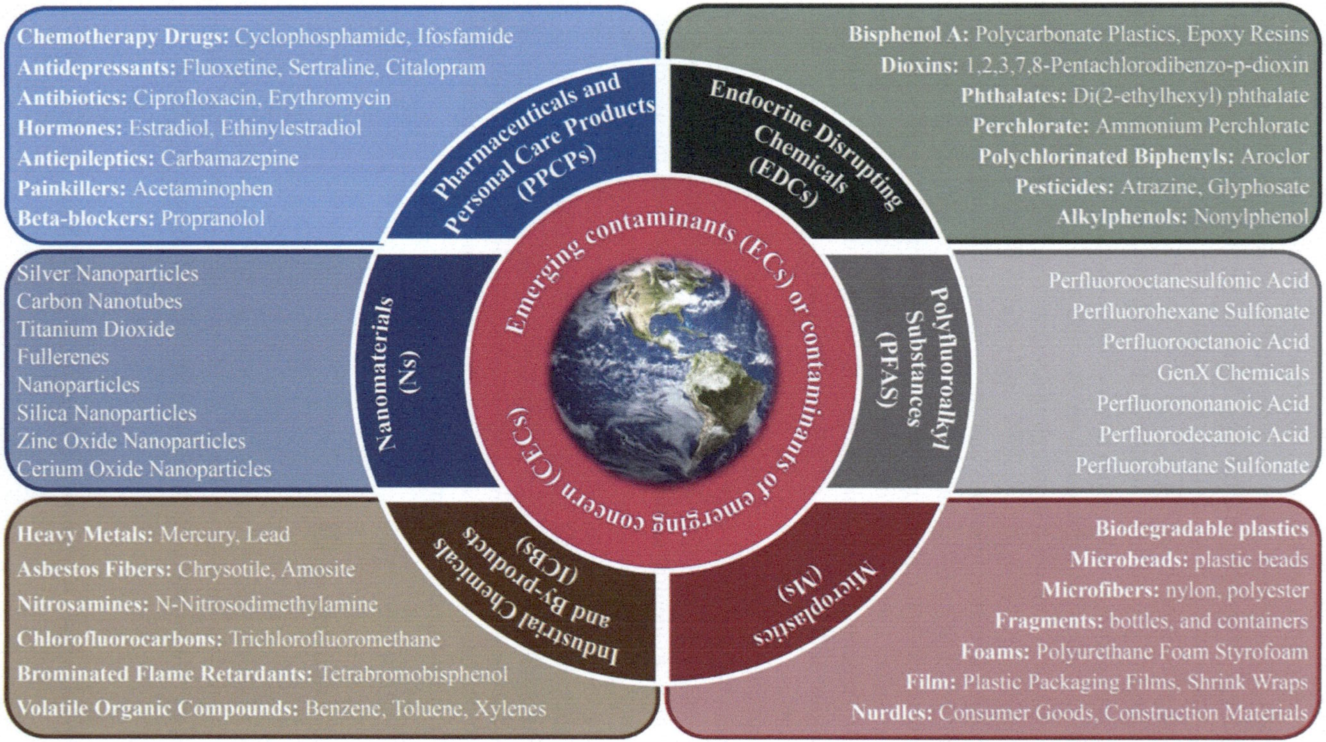

Fig. 10.1 Classification of emerging contaminants, (reprinted with permission from Li et al., 2024, Copyrights (2024) Elsevier, license no. 5797530045626)

growth, reproduction, and other essential processes. certain of the most well-known EDCs include alkylphenols, polychlorinated biphenyls (PCBs), dioxins, perchlorate and phthalates, which are being detected in certain resins, plastics and personal care items, and pesticides. Contaminated food, water, consumable material, the atmosphere, and other sources can all expose people to EDCs (Aydemir & Ulusu, 2023). The health of humans and wildlife is increasingly at risk due to the deposition of EDCs in water system. Because parabens are a prevalent kind of EDC, their disruptive effects have led to a threat for safer alternatives because of their widespread usage (Krowska et al., 2022).

10.2.2 Pharmaceuticals and Personal Care Products (PPCPs)

Pharmaceuticals and personal care products (PPCPs) are the class of chemicals that comprise cosmetics, drugs, perfumes and other consumer products. Although these products are intended to increase personal cleanliness, promote human health, there is rising concern about their occurrence in the environment. Antibiotics, birth control pills, antidepressants, painkillers, sunscreens and scents are examples of specific PPCPs. The possible effects of these compounds on ecosystems as well as human health are a subject of ongoing research. These compounds can enter the environment through a variety of channels, including agriculture runoff, inappropriate disposal, and wastewater treatment plant effluents (Gupta, 2016; Gupta & Gupta, 2016; Wang et al., 2023a). Several research has documented regarding spatiotemporally spread and environmental risk imposed by PPCPs in water bodies. These findings underline the critical necessity for extensive monitoring and detection to reduce the adverse impact of PPCPs (Ngubane et al., 2023). It will need a diverse strategy to address these issues, involving the creation of regulatory frameworks to control the discharge and environmental destiny of these new toxins, as well as better wastewater treatment technologies and enhanced analytical tools (Wang et al., 2023a; Ngubane et al., 2023).

10.2.3 Polyfluoroalkyl Substances (PFAS)

The synthetic, anthropogenic chemical class known as PFAS has attracted a lot of interest because of its ubiquitous usage and enduring presence in the environment. Multiple fluorinated carbon-fluorine linkages, which provide distinctive features including heat, water, and oil resistance, distinguish this varied collection of molecules.

Adverse health effects have been linked to certain persistent substances. Particularly, because to their toxicity and environmental persistence, long-chain PFAS, have being phased out internationally (Zheng et al., 2023). But now that replacements in the form of shorter-chain PFAS are common in the environment, adverse impact on human well-being have been raised (Gwenzi et al., 2018). It appears that PFAS, both old and new, can affect the health of male reproduction via controlling steroidogenesis in the reproductive gland. Furthermore, PFAS contaminants have been detected through raw materials, involved process, and migration from the oil-contact materials in edible oils, a vital component of the world's diet. Due to their propensity to cause obesity, particularly during periods of essential mental development, PFAS should also be taken into consideration since they raise the risk of paediatric obesity (Daugherty et al., 2023).

10.2.4 Microplastics (MPs)

Microplastics (MPs) are generally smaller plastic made particles, whose size are in the range of five millimetres. MPs are formed when bigger plastic trash or microbeads break down. Consumption of microplastics imposes adverse impact to aquatic life on both a physical and chemical level. Recent studies have exhibited that the accumulation of microplastics poses a serious hazard to ecosystems and organisms globally, making them an environmental contaminant. These microplastics have been found in soil, freshwater, marine, and the environment, among other ecosystems. According to several studies, microplastics has been emerged as a contaminant for terrestrial ecosystems by preventing plants from taking up nutrients, growing roots, or germination of seeds (Rede et al., 2023). Furthermore, toxicological research has revealed that microplastics may enter the digestive systems of a variety of creatures, making them a major worldwide concern for both terrestrial and aquatic ecosystems.

This may result in oxidative stress, histological alterations, physical harm, and intestinal obstruction. Studies on nanoparticle cleanup techniques have also linked microplastic pollution to related problems such toxic algal blooms and concerns for people and aquatic life. Worldwide, microplastic pollution of natural systems has detrimental effects on the environment and biology (Song et al., 2023).

10.2.5 Nanomaterials

Owing to the high surface-volume ratio, nanoscale materials exhibit unique properties. Metal nanoparticles, quantum dots and carbon nanotubes are a few examples. The tiny size and high surface-volume ratio of nanomaterials have, however, also drawn attention regarding possible adverse effect on the environment and human health. According to Hao et al. (2019), mesoporous carbon nanoparticles have the potential to produce phytotoxicity in rice saplings. Compared to microplastics (MPs), nanoplastics (NPs) had a more detrimental impact on photosynthetic pigments, biochemical markers, and plant development parameters. Plant health is less affected by MPs and NPs as exposure times increase. The size or concentration of MPs/NPs influences the synthesis of biological enzymes (Azeem et al., 2022). Many researchers have reported the health and environmental issues imposed by integrating nanotechnology with environmental science and agriculture. They also recommended to promote green synthesis and a variety of applications (Hendricks et al., 2023). It is vital to do research on the ambient behaviour of metal and carbon nanomaterials as well as to improve detection techniques. Agricultural chemical pollutants are the source of the demand for better food safety analysis methods (Li et al., 2023a).

10.2.6 Industry Originated Chemicals By-Products (ICBs)

Toxic and persistent in the environment, ICBs include dioxins and chemicals used in manufacturing. If not adequately handled, these substances—which are often created by numerous industrial processes—can constitute a serious hazard to both human health and the environment (Sorvari & Wahlström, 2024). For instance, the discharge of heavy metals like lead or mercury can pollute water and soil, which can have a negative impact on human populations and ecosystems. Furthermore, the incorrect disposal of pesticides and organic solvents can lead to the buildup of hazardous substances in the environment, which may disturb natural cycles as well as influence the food chain. To protect the environment and eventually the public health, industries and monitoring agencies must collaborate to develop a sustainable framework and guideline towards proper management of these hazardous materials (Tong et al., 2023).

10.3 Various Sources and Pathways of ECs

Owing to the anthropogenic activity and industrial processes, ECs routinely enter the environment through a variety of pathways (Fig. 10.2). Wastewater discharge from manufacturing plants, biosolids applied to land, landfill leaching, runoff from agriculture and cities, and air deposition are common introduction routes. Comprehending these paths of introduction is crucial in devising strategies to mitigate their impact. Reducing urban and agricultural runoff, controlling the use of biosolids, updating landfill designs, enhancing wastewater treatment technology, and limiting air emissions are some examples of mitigation techniques. Having a thorough grasp of the pathways that contaminants take to enter the environment helps to design evidence-based strategies to deal with this urgent problem.

10.3.1 Industrial Discharge

Environmental contaminants (ECs) are often released into nearby water systems as a result of industrial and manufacturing operations. Wastewater effluents can emit

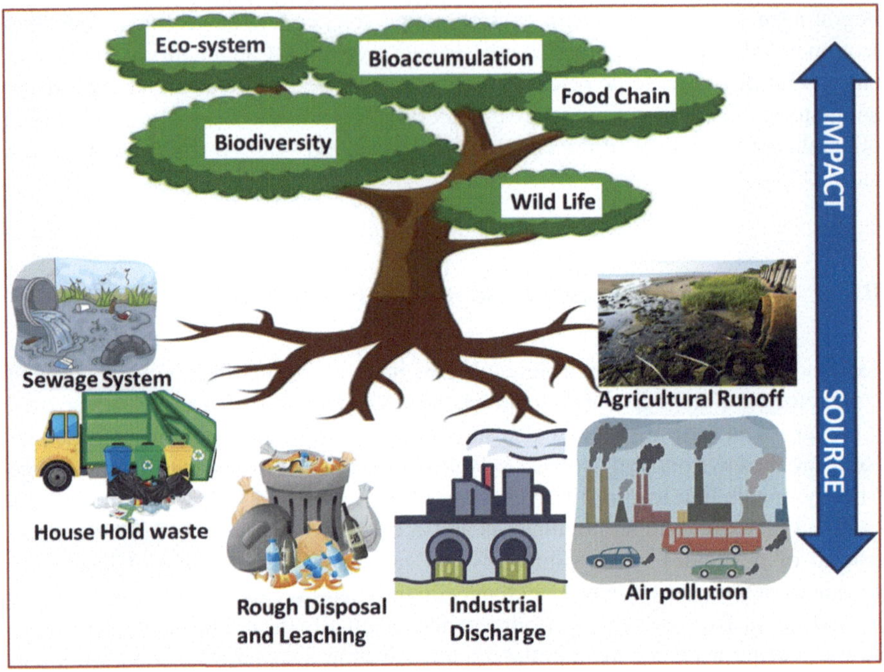

Fig. 10.2 Sources of EC's and its impact

a range of chemicals used in mining, processing, and industrial processes, posing a threat to environmental resources. The risks presented by pollutants getting into sewage networks and spreading to receiving bodies of water have been highlighted by recent studies (Jan et al., 2023). Researchers have also evaluated the amounts of developing contaminants including polycyclic aromatic hydrocarbons (PAH) in sediment in certain regions, such as the Niger Delta's Chanomi Creek region. These pollutants include benzene, toluene, ethylbenzene, and xylenes.

This risk analysis aims to enhance our understanding of the dispersion and impacts of pollutants by examining the repercussions of industrial discharge on neighboring ecosystems. Continual evaluation is still required to safeguard the water quality and habitat due to the introduction of developing pollutants into the aquatic matrix as a outcome of human activities (Onyena et al., 2023).

10.3.2 Agriculture Runoff

Agricultural runoff is the term used to describe the water that transports excessive amounts of fertilizers, pesticides, and other contaminants from agricultural fields into neighbouring bodies of water. Insufficient regulation of agricultural operations might result in inadvertent discharge of environmental contaminants (ECs). Over time, veterinary drugs, pesticides, and fertilizers used in agricultural fields may gradually infiltrate adjacent soil and eventually to water sources. A recent academic study has emphasized the potential for some agricultural methods to disseminate these pollutants (Santos et al., 2022). For example, the use of fertilizer can spread medications and personal hygiene items. Additionally, the application of nanotechnology in agricultural chemicals may help reduce pollution. In order to mitigate the inadvertent emission of pollutants from unwanted agricultural sources and their subsequent environmental impact, it is imperative to develop more effective techniques for the application, treatment, and containment of agricultural outputs. Stewardship is essential for the preservation of soil and water quality in a sustainable way (Hossain et al., 2023).

10.3.3 Sanitary Sewer Systems

Treatment of contaminated water is necessary to know the presence and nature of ECs, such as pharmaceuticals and personal hygiene waste, in water bodies. These substances often remain in the water due to inadequate removal by remediation facilities. Reports suggest that enhancing the methods of removing pollutants in man-made wetlands is of utmost importance, given the potential risks they impose to environment and human health. The research underscores the necessity for enhanced treatment protocols by underlining the consequences and challenges of insufficient sewage treatment in handling such deposits (Singh et al., 2021).

10.3.4 Atmospheric Deposition

Emerging Contaminants have the capability to traverse the atmosphere and touch down at distant locations. T3his includes particulate particles emitted by automotive exhaust and industrial pollution. This phenomenon has been elucidated by new research conducted by Barroso et al. in 2019. Tu et al. (2024) performed a study on the spatial and temporal variations and movement of air pollutants, specifically mercury, between urban areas and a seaport. The study emphasized the importance of industrial emissions and automobile exhausts in the process of atmospheric transmission.

10.3.5 Inadequate Disposal and Leaching

Inappropriate disposal of electronics and plastics containing ECs leads to environmental contamination. These products result in significant pollution, as evidenced by the existence of micro- and nanoplastics and hazardous substances in drinking water, treatment facilities, and water sources. Specifically, electronic waste emits toxic substances and heavy metals that have detrimental effects on both aquatic and terrestrial environments (Pathak & Sushil, 2022). Introduction to the hazardous chemicals found in the leachate of electronic waste poses significant health risks. Moreover, the accretion of micro- and nanoplastics disrupts the balance of ecosystems, posing a threat to biodiversity by harming both plant and animal life, as well as soil organisms.

10.3.6 Households and Businesses Product

The primary sources of ECs in water system are domestic and commercial products, such as paints, pesticides, cleaning agents, and cosmetics. These contaminants infiltrate aquatic ecosystems and provide risks to human and environmental well-being due to their inadequate removal by current techniques of treating wastewater. Their existence is linked with hazardous impact on aquatic creatures and humans. They can be discovered in both surface and drinking waters (Rathi et al., 2021).

Therefore, in order to ensure water quality and public health, it is necessary to not only create treatment procedures, but also increase surveillance to better understand how people are exposed and the resulting effects.

10.4 Effects on Human Health

Industries associated with medicines, personal care goods, and micro plastics are increasingly posing threats to human health, can be seen in Fig. 10.2. These frequently uncontrolled and little comprehended compounds have the ability to infiltrate water sources, food chains, and ecosystems, hence posing significant health risks. The potential consequences on human health encompass hormone disturbances, antibiotic resistance, and other enduring effects, necessitating the adaptation and resolution of these difficulties by public health systems (Yadav et al., 2021). The continuous investigation into the processes and impacts of these pollutants highlights the pressing requirement to review and strengthen safety standards in order to protect public well-being (Khan et al., 2023).

10.4.1 Interruption to the Endocrinal System

Endocrine-disrupting chemicals (EDCs), such as phthalates and bisphenol A (BPA), have been thorough examination due to their capacity to disrupt the hormonal systems. These substances imitate, obstruct, or otherwise disrupt the body's innate hormones, which may result in negative health consequences. More precisely, phthalates and BPA have been associated with obesity, modified reproductive habits, and reduced fertility in both people and animals (Dong et al., 2022). The substances exert their effects via linkage with hormone receptors, mostly as a result of their use in industrial and consumer products, requires thorough investigation to comprehensively comprehend their modes of operation and long-term effects on health and ecosystems. It is essential to develop safer alternatives and implement tougher regulatory procedures in order to reduce the broad impact of these substances. The citation is from Pan et al. in 2024.

10.4.2 Impact on Nervous System

Recent study emphasizes the neurotoxic properties of some contaminants, highlighting their capacity to cause neurological harm in people. These contaminating drugs have gained more recognition for their ability to disrupt neurochemical interactions, which can result in neurodevelopmental defects and increase the likelihood of developing neurodegenerative disorders including Parkinson's and Alzheimer's (Kumar et al., 2023). Prolonged exposure, even at trace levels that do not immediately cause death, has been linked to worsened toxic effects on the nervous system. Research has extensively recorded the ways in which these substances cause harm to the nervous system, such as by interfering with the functioning of neurotransmitters, causing oxidative stress, and directly damaging neurons. The available information strongly suggests the need for strict regulatory measures and more research to fully comprehend the scope of their influence on human health (Finhler et al., 2023).

10.4.3 Development of Cancer

Recent research indicates that PFAS (per- and polyfluoroalkyl materials) have been associated with cancer, namely breast cancer, as a result of their ability to disrupt the endocrine system and interfere with hormonal activity (Evich et al., 2022). Prolonged, chronic exposure to PFAS compounds elevates the likelihood of developing cancer, namely breast cancer, by imitating estrogen and stimulating the formation of tumors in breast tissue. The established correlation, supported by epidemiological studies, emphasizes the urgent requirement for further investigation into the carcinogenic processes of PFAS and strategies to mitigate contact (Seyyedsalehi and Boffetta, 2023). The presence of Micro plastics and Nanoscale materials associated with industrial processes and consumer products is posing their potential adverse impact to promote cancer disease. These tiny nanoparticles can interact with biological systems in intricate manners, thus raising worries about their impact on health (Winiarska et al., 2024).

10.4.4 Effects on the Immune System

Recent research indicates that environmental contaminants, such as heavy metals and industrial by-products, have detrimental effects on the immune system, which may lead to the development of autoimmune illnesses. These compounds have the potential to impair immune system and heighten vulnerability to illnesses. Pollutants have the ability to alter immunological tolerance, which plays a vital role in maintaining immune equilibrium (Suzuki et al., 2020). Trichloroethylene, for instance, impacts the integrity of the intestinal barrier, whereas cadmium has the potential to cause inflammatory disorders. Pollutants also modify epigenetic pathways associated with autoimmune disorders. This empirical data necessitates more investigation and more stringent policies to mitigate exposure (Lou et al., 2024).

10.4.5 Adverse Impact on Cardiovascular and Metabolic Systems

The contaminants have many methods via which they cause harm to the cardiovascular system. These mechanisms include interfering directly with cellular signalling pathways, as well as inducing oxidative stress and inflammation. Moreover, these contaminants have significant metabolic implications. Research has shown that they may play role to the onset of metabolic issues such as type II diabetes, obesity, and non-alcoholic fatty liver disease. (NAFLD). Recent research has established a connection between exposure to PFAS and the development of cardiovascular and metabolic problems (Dunder et al., 2023).

10.4.6 Effects on the Respiratory System

Industrial pollutants, particularly particulate matter, pose a significant risk to respiratory health, leading to the development of asthma and chronic obstructive pulmonary disease (COPD). The Environmental Protection Agency (EPA) cautions that these particles have the potential to trigger serious asthma episodes and necessitate hospitalization. Research indicates that prolonged exposure to these contaminants amplifies the likelihood of developing asthma in children and exacerbates chronic obstructive pulmonary disease (COPD) symptoms in adults. Furthermore, continuous exposure throughout the year adversely intensifies respiratory system in infants and worsens COPD among elder people (Josa-Culleré et al., 2024).

10.5 Detection and Analysis

Sophisticated methods are needed for the identification and measurement of the emerging contaminants, which are present at trace amounts in environmental matrices. The essential requirement is high sensitivity to make sure that ECs can be identified and measured among a wide range of other interfering chemicals. The first step is the sample preparation, which includes concentration of the ECs to minimize interference from the sample matrix, techniques such as solid-phase extraction (SPE), liquid-liquid extraction (LLE), ultrasound-assisted extraction (UAE) (Martín-Pozo et al., 2019). Some Analytical techniques are essential, such as high-resolution Mass Spectrometry (HRMS), and chromatography (gas and liquid) provide the foundation for accurate quantification and separation. The capacity to investigate ECs is further improved by sophisticated and hyphenated techniques like nuclear magnetic resonance spectroscopy and capillary electrophoresis, which meet requirements like the structural elucidation or investigation of ionic species.

10.5.1 Advanced Detection Technology

The detection and measurements of ECs is an emerging field, which is required in order to minimise the adverse impact caused by these ECs. The advancements in new contamination detection techniques are compiled in this chapter, which can be seen in Table 10.1. Several cutting-edge tools have been employed to replace the conventional methods. The innovative techniques can detect and quantify the analyte with improved accuracy. In this context, mass spectrometry and molecular techniques have been identified as the efficient tool for multiresidue analysis and to identify the unknown contaminants. Additionally, sensor technologies are still able to allow for quick on-site screening. Hence, employing innovative analytical techniques has improved our capacity to track pollutants of growing concern and obtain vital information about their existence in the atmosphere.

10.5.1.1 Gas Chromatography Hyphenated Mass Spectrometry (GC–MS)

GCMS technique is exclusively used to examine volatile and semi-volatile samples. In this technique, separation capability of gas chromatography is hyphenated with sophisticated detection ability of mass spectrometry. The usefulness of GC–MS towards applicability in analytical chemistry has been further proven by recent research. A headspace-solid phase micro-extraction-gas chromatography-mass spectrometry (HS-SPME-GC–MS) method was developed by Cincotta et al. to determine volatile ECs in fruit. The results showed that the limits of quantification (LOQ) and

Table 10.1 Represents various detection techniques and suitable contaminants

Detection techniques	Specification	Example of target emerging contaminants (ECs)	References
ELISA	Specific interaction between antibody and antigen, to detect and measure specific ECs, such as heavy metals, pesticides, and hormones	Pentachlorophenol, pesticides, herbicides, heavy metals, hormones, and many organic contaminants	Aslam et al. (2023), Silva et al. (2020) and Wang et al. (2023d)
FTIR spectroscopy	Specifically used to identify functional group of organic contaminants	Bisphenol A, microplastics, polycyclic aromatic hydrocarbons (PAH), tetracycline	Khoubi et al. (2024) and Xie et al. (2024)
GC–MS	For volatile and semi-volatile ECs	17β-estradiol, 17a-ethinylestradiol, tris(2-chloroethyl) phosphate, tris(2-chloroisopropyl) phosphate, tris (1,3-dichloro-2-propyl) phosphate, xylene, bisphenol A toluene, ethylbenzene, xylene, butylated hydroxytoluene, butylated hydroxyanisole, estriol	Albero et al. (2017), Bristow et al. (2023), Cincotta et al. (2018) and Suvar et al. (2024)
LC–MS	A sophisticated analytical method employed to separate, identify, and quantify both known and unknown samples	Bisphenol A (BPA), albendazole, caffeine, carbamazepine, lincomycin, diclofenac, carbamazepine, triclosan, phenolic compounds, dyes, paracetamol, ibuprofen, olanzapine, simvastatin	Aslam et al. (2023), He et al. (2019), Safwat et al. (2020), Soliz et al. (2024), and Wang et al. (2024)
NMR spectroscopy	Specifically useful to differentiate organic analytes with similar molecular structure	Olmesartan acid, polyfluoroalkyl substances, heparin,	Camdzic et al. (2021) and Colombo et al. (2022)

limits of detection (LOD) ranged from 0.037 ng/g to 0.098 ng/g and 0.011 ng/g to 0.033 ng/g, respectively (Cincotta et al., 2018).

Earlier, Albero et al. formulation helped in determination of several ECs and the achieved LOQ was achieved between 0.1 and 0.4 ng/g (Albero et al., 2017). In a different report, Bristow et al. created a technique for solid phase extraction hyphenated with GC–MS (SPE-GC–MS) to determine the 2-methyl isoborneol and geosmin in drinking water. The LODs for these two substances were achieved as 4.3 and 2.0 ng/g (Bristow et al., 2023). The sensitivity and selectivity are the striking features of GC–MS, but it also has significant limitations, which restricts its use only to detect volatile and semi-volatile substances. There are many ECs with large molecular weights or high polarity might not be easily detectable with GC–MS analysis. Moreover, sample preparation, introduction of samples in the system and data interpretation require highly qualified technicians/personals. Although GC–MS is an effective method for assessing some suitable ECs, these operational and technological issues limit its applicability, which eventually reduces its usefulness for extensive environmental application.

10.5.1.2 Liquid Chromatography Hyphenated Mass Spectrometry (LC–MS)

In order to detect the ultra-trace level concentration of ECs, LC–MS and LC–MS/MS have been recommended as two efficient analytical techniques. Recent advancements have demonstrated the significant effectiveness of liquid chromatography (LC), LC-mass spectrometry (LC–MS), and LC–MS/MS in analysing multiple substances in environmental samples. This enables the detection and measurement of various types of environmentally concerning compounds, such as phthalates, bisphenols, and pharmaceutical residues.

He et al. (2019) conducted a study where they quantified for several environmentally concerning substances (ECs) in three common crops (maize, wheat and tomato) using a novel and efficient freeze-centrifugation extraction technique followed by HPLC–MS/MS. The LOD range for the quantification was 0.01–2.0 ng/g. LC–MS/MS is a crucial technique in the field of analytical chemistry and environmental monitoring because to its ability to detect and identify substances in complex mixtures with high sensitivity (Magee et al., 2023). While LC–MS/MS provides high sensitivity and selectivity for detecting and analyzing ECs, it does have limitations. One of the biggest challenges in LC–MS/MS is the susceptibility to matrix effects, that can majorly influence the result. These repercussions result from the presence of additional chemicals in the sample that disrupt the ionization process of the analytes. This disruption can lead to inaccurate measurement and either suppression or enhancement of the analytes. In addition, while LC–MS/MS is generally useful for many different compounds, the chemical properties of certain ECs, such as large size or high polarity, may limit the method's efficiency and require further sophisticated analysis.

10.5.1.3 Analysis Based on Enzyme-Linked Immunosorbent Assay (ELISA)

ELISA is observed as a stablished analytical tool for analysing specific ECs. The method is based on a typical interaction between antibody and antigen, which helps to detect and measure specific ECs, such as heavy metals, pesticides, and hormones. Recently ELISA analytical technique has been used to detect potentially dangerous materials including endocrine-disrupting chemicals, pesticides, and herbicides in aquatic system. Using ELISA, Silva et al. monitored the levels of caffeine, carbamazepine, cetirizine, 17β-ethinylestradiol and 17β-estra diol in aquatic medium

(Silva et al., 2020). Commercial ELISA kits are an effective tool for onsite detection of Specific

Owing to high sensitivity and specificity, ELISA has been opted as a good analytical tool for various applications involving food safety and environmental monitoring of contaminants. But this method has also a few drawbacks, including possibility of cross-reactivity, in which the assay's antibodies attach to molecules other than to be examined analyte and provide false findings. Such issue especially troublesome with analyte having complicated matrices of molecules that may contain several related substances. The matrix effects and sample preparation techniques may have an impact on the quantitative results observed from ELISA, which might result in variations in data reliability and repeatability (Aslam et al., 2023).

10.5.1.4 Innovative Spectroscopic Methods

The analytical methods of Fourier-transform infrared spectroscopy (FTIR) and nuclear magnetic resonance (NMR) are crucial for quantification and structure determination of organic contaminants. NMR spectroscopy is specifically useful to differentiate organic analytes with similar molecular structure, like Heparin, olmesartan acid and poly-fluoroalkyl compounds (Camdzic et al., 2021; Colombo et al., 2022). In the similar pattern, FTIR spectroscopy makes it easier to identify and quantify contaminants corresponding to nanoplastics. According to Chen et al. (2020), NMR and FTIR have been found as the effective tools for characterizing complex organic pollutants found in a variety of environmental matrices. Though specific analytical ability exhibited by NMR, it also has some disadvantages. It requires larger sample amount with high concentration to get adequate signal strength.

When working with trace pollutants, NMR, which is renowned for its thorough molecular structure elucidation, can be severely limited since it needs larger amount of sample volumes and higher concentrations of the focused chemicals to generate a sufficient signal intensity. Moreover, NMR is time consuming and expensive analytical process and it also been observed to be less sensitive compared to mass spectrometry. In this regard, FTIR produces rapid result and has easy sampling procedure, but it does not provide detailed structural information other than functional group present in the analyte. For analytes consist of micro and nanoplastics, some microscopic techniques have been found very effective like, Atomic Force Microscopy, Scanning Electron Microscopy, Dissection Microscopy, Polarized Microscopy, but this instrumental analysis is unable to provide the detailed results. However, this technique when hyphenated with chemical analytical technique produces excellent result with improved selectivity and sensitivity. Many researchers can increase the precision and thoroughness the detection and characterization of micro/nanoplastics by utilizing these developments (Cai et al., 2021; Mariano et al., 2021).

10.5.1.5 Biosensors Based Analysis

Innovative biosensor technique has been developed to provide quick on-site identification of pollutants. Advanced biosensors of the future utilize biological mechanisms to precisely detect various ecological and health hazards, such as, mycotoxins, microcystins, proteins, and RNAs. They serve as a transformational tool for monitoring these hazards (Magee et al., 2023). Bobrinetskiy and Knezevic (2018) have shown that biosensors made from graphene have demonstrated potential in identifying chemical contaminants in food, suggesting that they might be used in portable devices. Moreover, advancements in electrochemical biosensing have enhanced the ability to detect and analyse contaminants (Gong et al., 2024). Despite the commendable attributes of biosensors like as mobility, rapid response, and real-time monitoring, they encounter many challenges in the context of EC detection. Their vulnerability to external influences such as pH and temperature is a significant challenge, as it can undermine the stability and performance of the biological component and lead to irratic results.

10.5.2 Challenges Encountered in Detection

Owing to the complexity present in most of the ECs, detecting them in variety of environmental matrices has always been a challenging issue for the researcher. Moreover, advanced analytical tools and expert operators are required to get desired result. Most of the ECs are present in a very trace amount, commonly reported in parts per million (ppm), parts per billion (ppb) or even parts per trillion (ppt) and at such a trace concentration only few techniques are able to detect ECs, e.g., GC–MS and UPLC–MS, in conjunction with sophisticated and effective extraction techniques (Nouri et al., 2020; Sereshti et al., 2020). Accurately identifying and measuring individual molecules within mixture of ECs is significantly hindered by their complexity. To get rid of the challenges of environmental sample investigation, Escher et al. recommend using more advanced analytical methods (Escher et al., 2020). Some research groups have systematically documented the erroneous positive and negative fluctuations in the interpretation of analytical data imposed due to these matrix effects (Noguera-Oviedo & Aga, 2016; Petrie et al., 2016). Variation corresponding to temporal and spatial changes also makes the analysis more difficult. Some researchers have reported the significance of temporal and geographical variation in the analysis. Moreover, most of the analytical techniques requires reference materials to detect the contaminant, which is a major obstacle to the validation of analytical processes for many ECs According to Richardson et al., the lack of standardization is a challenge since it causes inconsistent analytical findings, especially when mass spectrometry settings are involved (Richardson & Ternes, 2021). For accurate analysis and EC monitoring, standardized criteria and extensive reference standards are essential. Unfortunately, some novel and unknown contaminants are constantly being encountered during analysis. In this context, innovative, robust, and efficient methodologies are desperately required to identify these novel ECs (Richardson & Ternes, 2021; Zulkifli et al., 2018).

10.6 Treatment Technologies

10.6.1 Current Treatment Technologies

To eliminate or lessen ECs in the environment, a variety of techniques are used (Fig. 10.3). The efficacy and methodology of these techniques differ according on the kind of contamination.

10.6.1.1 Treatment Through Advanced Oxidation Processes (AOPs)

Application of Advanced Oxidation Processes (AOPs) is crucial for environmental management, particularly in dealing with complex organic pollutant-contaminated water and soil. Hydroxyl radicals generated by AOPs may effectively degrade these persistent pollutants, a task that traditional methods cannot effectively do (Wang et al., 2023c) Although AOPs are highly successful, they have some limitations that need to be addressed in order to enhance their use. In addition, the by-products of this oxidation processes might occasionally be as detrimental to the initial contaminants, which requires more treatment and meticulous monitoring. Engaging in research is crucial in order to rule out these obstacles. Increasing the effectiveness of Advanced Oxidation Processes (AOPs) might lead to a decrease in their energy usage, hence improving their sustainability and economic feasibility. Research might also prioritize the minimization of detrimental by-products while enhancing the safety and efficacy of these methods. Additionally, the sustainability of AOPs might be further improved by creating specific catalysts (Guo et al., 2024; Li et al., 2023b).

10.6.1.2 Treatment Approaches Towards Physisorption

Physiosorption has been emersed as a reliable and accurate tool for removal of ECs from contaminated aquatic system. This tool is based on the varying adsorbent's capability to interact and accumulate these contaminants by means of adsorption, a process in which the pollutants are attached on the surface of the adsorbing materials. Activated carbon is highly versatile and can efficiently capture a diverse array of emerging pollutants, including as pesticides, personal care chemicals, and pharmaceutical residues. The unique stratified composition and ability to exchange ions of these adsorbents derived from clay make them highly desirable towards water decontamination (Jjagwe et al., 2021). In recent times, a new category of materials known as metal-organic frameworks (MOFs) has been discovered. These materials possess

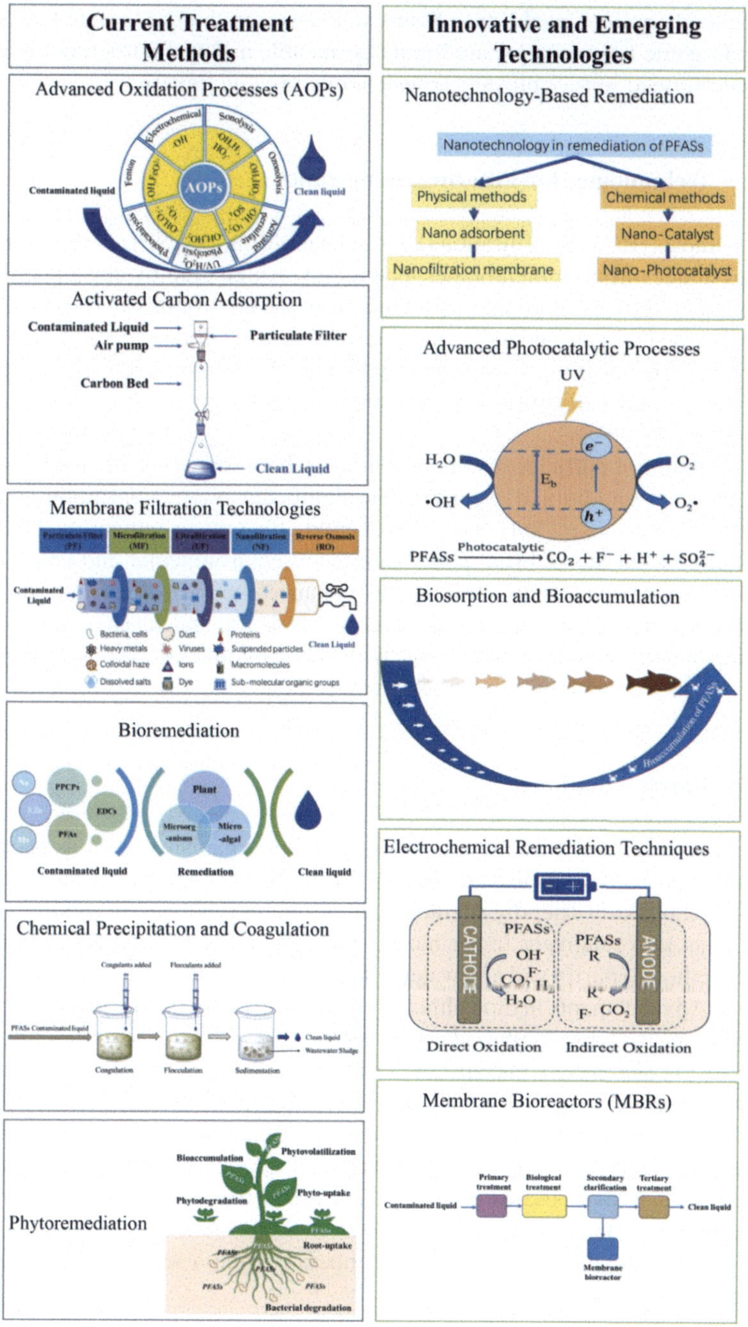

Fig. 10.3 Current and innovative treatment methods to remediate ECs (reprinted with permission from Li et al., 2024, Copyrights (2024) Elsevier, license no. 5797530045626)

exceptional adsorption capabilities. Svens son Grape et al. (2023) state that MOFs, or Metal-Organic Frameworks, are highly permeable and well-structured materials composed of metal ions or clusters connected by organic linkers.

10.6.1.3 Technologies for Filtering Membranes

Reverse osmosis (RO), ultrafiltration (UF) and nanofiltration (NF) are the techniques employed under filtrating membranes technology, which employ concepts like size exclusion to effectively eliminate pollutants from the environment. These technologies effectively segregate and refine complex mixtures of contaminants, making them very efficient for purifying wastewater and water. Membrane technologies exhibit high efficiency and versatility, yet, they are susceptible to fouling, which reduces their longevity.

Accumulation of particles results in fouling, which increases the need for maintenance and incurs additional expenses. The significant energy consumption, particularly in reverse osmosis (RO), has a detrimental effect on sustainability. Disposing of polluted concentrate or brine poses significant environmental and cost difficulties. However, advancements continue to enhance the properties and structures of membranes, hence enhancing their energy efficiency and resistance to fouling. Employing membranes in combination with other treatments has the potential to enhance production and optimize waste management.

10.6.1.4 Bioremediation

Bioremediation refers to the process of using living organisms, such as bacteria or plants, to remove or neutralize pollutants from the environment. Bioremediation is an environmentally friendly method, where microorganisms is used to convert contaminants into relatively lesser hazardous substances. It is an economical and versatile method to purify soil and water, effectively removing many types of organic pollutants. Advancements in biotechnology improve its effectiveness by customizing microorganisms to target pollutants (Kaur et al., 2023; Pokhriyal et al., 2024). But bioremediation also have some specific limitations due to external parameters such as oxygen concentration, pH, temperature, and the availability of other materials can have a substantial impact on the effectiveness of the process. These factors may restrict the applicability of bioremediation in some circumstances. In addition, as compared to other methods, the process of significant degradation may take a longer time to complete, which can be problematic in situations that require prompt repairs. In order to surmount these challenges, it is imperative that research and development in the field of bioremediation persists.

10.6.1.5 Chemical Precipitation and Coagulation

This method is often employed techniques to eliminate metal contaminants as well as heavier organic molecules from aquatic system. These methods include the employment of chemicals to generate insoluble precipitates, which are then extracted from the water. These procedures provide a straightforward approach to cleaning water and are acknowledged for their effectiveness in dealing with various contaminants. Water quality can be enhanced by employing coagulants to selectively eliminate organic molecules and heavy metals. The efficacy and scope of these processes are consistently enhanced by advancements in this field, rendering them indispensable to industrial wastewater treatment methodologies (Benalia et al., 2022). These strategies might be enhanced in certain aspects and have a few limitations. The use of chemicals also raises concerns regarding the risk of secondary contamination and the overall environmental impact of the remediation procedure.

10.6.1.6 Technology Based on Phytoremediation

Phytoremediation is a pioneering and ecofriendly technology that involves utilizing plants to eliminate or break down pollutants. This technology is highly effective in remedying damaged soils and water bodies due to its exceptional efficiency in removing organic contaminants and heavy metals. Phytoremediation operates through two primary processes: firstly phytoextraction, where suitable herbs absorb pollutants by their roots and second phytodegradation, in which the toxins are converted into less harmful material. Phytoremediation has several benefits. It is cost-effective because, in comparison to traditional cleanup methods, it often requires a lower initial investment and continuous operational costs. In addition, the process of phytoremediation can bring back plant life, so restoring ecological equilibrium and enhancing the visual appeal of the place. Furthermore, it reduces the need to dig up and transport polluted material, making it particularly suitable for on-site use (Shen et al., 2022). However, there are limitations on the use of phytoremediation. Site remediation can be a time-consuming endeavour, often spanning many growing seasons. It is important to anticipate and be prepared for a lengthy and thorough procedure. The efficacy of phytoremediation is greatly influenced by the environmental conditions in which the plants are grown, including factors such as soil type, climate, and nutritional requirements. In addition, disposing of plant debris that gets polluted may provide further challenges. Advancements in genetic engineering offer practical methods to enhance the sensitivity and usability of phytoremediation. By genetically modifying plants to enhance their abilities to absorb and metabolize substances, scientists have the potential to create plant species that are better equipped to eliminate certain toxins from the environment or survive in unfavourable conditions. To fully optimize the advantages of phytoremediation in pollution control, it is necessary to engage in ongoing research and enhancement.

10.6.2 Cutting-Edge and Nascent Technologies

Rapid advancements in novel and innovative technology are occurring in order to remediate ECs. These advancements are focused on the restriction of the present methods and many other factors like, economical, ecofriendly, easy operation mechanism. Some prominent technologies are mentioned below.

10.6.2.1 Nanotechnology

A nascent field known as "nanotechnology-based remediation" employs nanomaterials, including nanoscale materials of different shapes. Owing to their diminutive size and favourable surface-to-volume ratio, nanoscale materials exhibits significant and unique physiochemical, and biological properties that allow them to effectively interact with the contaminants of almost similar size and efficiently eliminate them from soil and water. Nano-based remedial approaches provide huge advantages compared to conventional techniques. Nanomaterials efficiently eliminate a diverse array of contaminants, such as viruses, organic compounds, and heavy metals, due to their high reactivity and extensive surface area. Moreover, synthetic nanomaterials can be modified and be made suitable for specific contaminant. Such engineering to these nanomaterials makes them more selective and sensitive towards interaction with contaminants. Furthermore, versatility exhibited by nanotechnology allows it to be customized to suit diverse environmental circumstances and contamination scenarios, such as using filters or in situ injections. Despite the advantages of remedial procedures based on nanotechnology, they also have certain downsides. Due to their ability to infiltrate and accumulate in organisms, nanomaterials pose potential risks to both the environment and human health, owing to their small size and high reactivity. In addition, the implementation of these technologies on a large scale may incur significant expenses. The general acceptance of nanoparticles employed in environmental remediation is hindered by barriers due to the ongoing development of regulatory frameworks. In conjunction with other remediation techniques, nanotechnology can enhance the development of inclusive and effective cleaning strategies.

10.6.2.2 Advanced Photoinduced Interaction

Advanced photocatalytic methods, namely those utilizing titanium dioxide (TiO_2) and other metal oxide photocatalysis, have become a potential method for breaking down organic contaminants in the environment (Wang et al., 2023b; Gupta & Gupta, 2015; Gupta et al., 2024; Gupta and Dhiman, 2024). This method utilizes solar or UV radiation to produce very reactive hydroxyl (OH·) species, which may efficiently decompose various organic pollutants. Scientists are now engaged in efforts to boost the efficacy and solar reactivity of TiO_2 photocatalysts, with the goal of harnessing a greater amount of sunlight and improving the overall effectiveness of the process. Modification in TiO_2 material and adding suitable components has ability to promote the efficiency of this technique in decontamination of environmental. The benefits of modern photocatalytic systems are in their capacity to convert organic contaminants into non-poisonous end products, such as carbon dioxide (CO_2) and water (H_2O), without the need for further chemical additives. This strategy is sustainable and ecologically favourable in reducing pollutants. Moreover, the adaptability of photocatalytic systems enables their use in many environments matrix. The citation "Constantino et al., (2022)" refers to a publication by Constantino and his colleagues in the year 2022. Nevertheless, similar to other developing technologies, the extensive implementation of sophisticated photocatalytic processes may encounter specific obstacles, including the need for improved scalability, better performance of catalyst, and economic to use. Continued research and advancements in this field are crucial to overcome these challenges and fully harness the vast potential of this innovative approach for environmental restoration (Wang et al., 2023b). The area of ecological remediation can see a substantial revolution by employing improved photocatalytic processes and upcoming technologies like nanotechnology-based remediation. This shift will lead to the creation of more efficient, sustainable, and complete solutions for efficiently tackling intricate pollution problems.

10.6.2.3 Technique Based on Bioaccumulation and Biosorption

Biosorption is a technique that utilizes both live and deceased biomass to eliminate environmental contaminants. This process has been enhanced by the application of genetic engineering and biotechnology in microbes and plants. Studies seek to improve organisms' capacity to absorb pollutants by making genetic modifications to boost their absorption ability. These advancements offer improved bioremediation techniques, particularly for the extraction of contaminants from aquatic system. Specifically modified organisms and suitable plants enhance the efficiency of biosorption and contribute to the achievement of environmental remediation (He et al., 2023; Singh et al., 2018). Current research in this area is concentrated on improving the sensitivity and selectivity of the process. Scientists genetically modify these organisms to enhance their ability to absorb certain pollutants and increase their selectivity. This has the potential to result in the creation of more effective and specialized bioremediation systems that can remove a wider variety of toxins from polluted areas. The utilization of genetically modified creatures and plants shows potential in tackling intricate pollution issues, namely in the removal of organic contaminants and heavy metals from soil. In future, these technologies may play an important role to create more sustainable solutions to remediate the contaminants. However, this method may encounter obstacles in terms of regulatory compliance and public approval, especially with regards to the utilization of genetically modified organisms. To ensure the universal acceptance of these developing technologies, it is crucial to address these concerns by conducting comprehensive risk assessments and maintaining honest communication.

10.6.2.4 Methods for Remediating Contaminants Using Electrochemical Processes

Electrochemical remediation, which involves the use of electric currents to purify water and soil, has made significant progress due to the development of improved electrodes and the use of combination treatment methods. These enhancements boost the elimination of pollutants, offering a sustainable way for cleaning up. The efficacy and use of enhanced materials and optimized procedures in pollution control have been enhanced, demonstrating their potential in comparison to conventional techniques (Nabgan et al., 2022). This integrated strategy utilizes the advantages of many technologies, resulting in enhanced overall performance and adaptability in tackling intricate contamination situations. Electrochemical remediation techniques provide several benefits due to their capacity to function without the need for toxic chemicals, which might create a secondary contamination. Such feature makes them more bearable and ecologically friendly compared to conventional remediation methods that heavily rely on chemicals. Moreover, the versatility of these systems enables their utilization in many environments, ranging from on-site remediation of groundwater and soil to off-site treatment of industrial waste and wastewater. Electrochemical techniques have shown to be a sustainable and successful alternative for addressing different environmental toxins by utilizing modifications in electrode materials, in optimized experimental condition, and towards combined treatment approaches. These techniques can have a considerable impact on the alteration towards wide-ranging and ecologically conscious pollutant management solutions.

10.6.2.5 Membrane Bioreactors (MBRs)

MBRs are systems that hyphenate biological mechanism with membrane filtration to remove contaminants from water or wastewater. Recent researchers have focused their work on improving the effectiveness of membranes by using novel designs and utilizing advanced materials to counteract fouling, hence increasing the efficiency and lifespan of the system. Enhancing parameters, utilizing materials that resist fouling, and integrating MBRs with other techniques enhance their efficiency and long-term viability. This emphasizes the necessity of continuous research on membrane technology to develop more effective solutions for water decontamination (Nadeem et al., 2022). MBRs have the benefit of efficiently eliminating various hazardous materials, including nutrients, organic contaminants, and new toxins, by combining membrane filtration with biological degradation. Moreover, the MBRs' small form and modular characteristics allow them to be easily adjusted for different water treatment purposes. Moreover, by consistently enhancing the efficiency and sustainability of this technology may have a substantial impact on meeting the increasing need for effective and ecologically conscious water decontamination. By combining MBRs with other new remediation approaches, such as adsorption-based methods, the total efficacy and adaptability of water treatment systems may be improved.

10.7 Conclusion

Owing to unknown impact and widespread effect due to presence of ECs, it complicates related environmental studies and public health related concerns. Managing medications, nano based materials, and microplastics requires innovation, relevant research, and proper monitoring. Sophisticated and sensitive detection analytical tools and suitable prevention are required to handle these dangers. It also urges more research on these emerging contaminants impacts. In this context, more funding and support for EC health and environmental research is required. To achieve the improved sensitivity, selectivity, and accuracy there must be hyphened and sophisticated detection techniques with skilled operator. A public awareness on EC sources and impacts for decision-making and regulation may be useful. By addressing these crucial matters and implementing these solutions, the detection, monitoring, and remediation of EC would be improved, resulting in a reduction of environmental and public health hazards.

References

Albero, B., Sánchez-Brunete, C., Miguel, E., & Tadeo, J. L. (2017). Application of matrix solid-phase dispersion followed by GC–MS/MS to the analysis of emerging contaminants in vegetables. *Food Chemistry, 217*, 660–667. https://doi.org/10.1016/j.foodchem.2016.09.017

Aslam, S., Zhang, Z., & Muhammad, A. (2023). A Review on validation of enzyme linked immunosorbent assay (Elisa) techniques for detection and quantification of different contaminant in aquatic environment. *Journal of Bioresource Management, 10*(3), 6. https://corescholar.libraries.wright.edu/jbm/vol10/iss3/6

Aydemir, D., & Ulusu, N. N. (2023). The impact of the endocrine-disrupting chemicals on the glucose-6-phosphate dehydrogenase enzyme activity. *Frontiers in Pharmacology, 14*, 1133741. https://doi.org/10.3389/fphar.2023.1133741

Azeem, I., Adeel, M., Ahmad, M. A., Shakoor, N., Zain, M., Yousef, N., Yinghai, Z., Azeem, K., Zhou, P., White, J. C., & Ming, X. (2022). Microplastic and nanoplastic interactions with plant species: Trends, meta-analysis, and perspectives. *Environmental Science & Technology Letters, 9*(6), 482–492. https://doi.org/10.1021/acs.estlett.2c00107

Barroso, P. J., Santos, J. L., Martín, J., Aparicio, I., & Alonso, E. (2019). Emerging contaminants in the atmosphere: Analysis, occurrence and future challenges. *Critical Reviews in Environmental Science and Technology, 49*(2), 104–171. https://doi.org/10.1080/10643389.2018.1540761

Benalia, M. C., Youcef, L., Bouaziz, M. G., Achour, S., & Menasra, H. (2022). Removal of heavy metals from industrial wastewater by chemical precipitation: Mechanisms and sludge characterization. *Arabian Journal for Science and Engineering, 47*(5), 5587–5599. https://doi.org/10.1007/s13369-021-05525-7

Bobrinetskiy, I. I., & Knezevic, N. Z. (2018). Graphene-based biosensors for on-site detection of contaminants in food. *Analytical Methods, 10*(42), 5061–5070. https://doi.org/10.1039/C8AY01913D

Bristow, R. L., Haworth-Duff, A., Young, I. S., Myers, P., Hampson, M. R., Williams, J., & Maher, S. (2023). An automated micro solid phase extraction gas chromatography–mass spectrometry (μSPE-GC–MS) detection method for geosmin and 2-methylisoborneol in drinking water. *Scientific Reports, 13*(1), 1768. https://doi.org/10.1038/s41598-023-28543-x

Cai, H., Xu, E. G., Du, F., Li, R., Liu, J., & Shi, H. (2021). Analysis of environmental nanoplastics: Progress and challenges. *Chemical Engineering Journal, 410*, 128208. https://doi.org/10.1016/j.cej.2020.128208

Camdzic, D., Dickman, R. A., & Aga, D. S. (2021). Total and class-specific analysis of per-and polyfluoroalkyl substances in environmental samples using nuclear magnetic resonance spectroscopy. *Journal of Hazardous Materials Letters, 2*, 100023. https://doi.org/10.1016/j.hazl.2021.100023

Chen, Y., Wen, D., Pei, J., Fei, Y., Ouyang, D., Zhang, H., & Luo, Y. (2020). Identification and quantification of microplastics using Fourier-transform infrared spectroscopy: Current status and future prospects. *Current Opinion in Environmental Science & Health, 18*, 14–19. https://doi.org/10.1016/j.coesh.2020.05.004

Cincotta, F., Verzera, A., Tripodi, G., & Condurso, C. (2018). Volatile emerging contaminants in melon fruits, analysed by HS-SPME-GC-MS. *Food Additives & Contaminants: Part A, 35*(3), 512–518. https://doi.org/10.1080/19440049.2017.1401738

Colombo, E., Mauri, L., Marinozzi, M., Rudd, T. R., Yates, E. A., Ballabio, D., & Guerrini, M. (2022). NMR spectroscopy and chemometric models to detect a specific non-porcine ruminant contaminant in pharmaceutical heparin. *Journal of Pharmaceutical and Biomedical Analysis, 214*, 114724. https://doi.org/10.1016/j.jpba.2022.114724

Constantino, D. S., Dias, M. M., Silva, A. M., Faria, J. L., & Silva, C. G. (2022). Intensification strategies for improving the performance of photocatalytic processes: A review. *Journal of Cleaner Production, 340*, 130800. https://doi.org/10.1016/j.jclepro.2022.130800

Daugherty, S., Mulabagal, V., Hayworth, J., & Akingbemi, B. T. (2023). Legacy and emerging perfluoroalkyl and polyfluoroalkyl substances regulate steroidogenesis in the male Gonad. *Endocrinology, 164*(12), bqad142. https://doi.org/10.1210/endocr/bqad142

Dong, Z. Y., Lin, Y. L., Zhang, T. Y., Hu, C. Y., Pan, Y., Zheng, Z. X., Tang, Y. L., Xu, B., & Gao, N. Y. (2022). Enhanced degradation of emerging contaminants by permanganate/quinone process: Case study with bisphenol A. *Water Research, 219*, 118528. https://doi.org/10.1016/j.watres.2022.118528

Dunder, L., Salihovic, S., Lind, P. M., Elmståhl, S., & Lind, L. (2023). Plasma levels of per- and polyfluoroalkyl substances (PFAS) are associated with altered levels of proteins previously linked to inflammation, metabolism and cardiovascular disease. *Environment International, 177*, 107979. https://doi.org/10.1016/j.envint.2023.107979

Escher, B. I., Stapleton, H. M., & Schymanski, E. L. (2020). Tracking complex mixtures of chemicals in our changing environment. *Science, 367*(6476), 388–392. https://doi.org/10.1126/science.aay6636

Evich, M. G., Davis, M. J., McCord, J. P., Acrey, B., Awkerman, J. A., Knappe, D. R., Lindstrom, A. B., Speth, T. F., Tebes-Stevens, C., Strynar, M. J., & Wang, Z. (2022). Per-and polyfluoroalkyl substances in the environment. *Science, 375*(6580), eabg9065. https://doi.org/10.1126/science.abg906

Finhler, S., Marchesan, G. P., Corona, C. F., Nunes, A. T., De Oliveira, K. C. S., de Moraes, A. T., Soares, L. C., Lima, F. O., Dalmolin, C., & Benvegnú, D. M. (2023). Influence of pesticide exposure on farmers' cognition: A systematic review. *Journal of Neurosciences in Rural Practice, 14*(4), 574. https://doi.org/10.25259/JNRP_58_2023

Gong, H., Bao, C., Luo, X., Yu, Y., & Yang, W. (2024). Reusable electrochemical sensor for quinine detection via β-cyclodextrin-based indicator displacement assay. *Microchemical Journal*, 110109. https://doi.org/10.1016/j.microc.2024.110109

Guo, X., Qiu, L., Li, M., Tian, F., Ren, X., Jie, S., Geng, S., Han, G., Huang, Y., Song, Y., & Yang, W. (2024). Accelerating the generation of NiOOH by in-situ surface phosphating nickel sulfide for promoting the proton-coupled electron transfer kinetics of urea electrolysis. *Chemical Engineering Journal*, 149264. https://doi.org/10.1016/j.cej.2024.149264

Gupta, B., & Gupta, H. (2015). Iron oxide mediated degradation of mutagen pyrene and determination of degradation products. *International Journal of Environmental Science and Development, 6*(12), 908. https://doi.org/10.7763/IJESD.2015.V6.720

Gupta, H. (2016). Photocatalytic degradation of phenanthrene in the presence of akaganeite nanorods and the identification of degradation products. *RSC Advances, 6*(114), 112721–112727. https://doi.org/10.1039/C6RA24602H

Gupta, H. (2018). PAH determination in effluent and sludge samples of paper industry. *Environmental Technology & Innovation, 9*, 115–121. https://doi.org/10.1016/j.eti.2017.11.009

Gupta, H., & Dhiman, S. (2024). Emerging contaminants in the environment introduction. In Occurrence, distribution and toxic effects of emerging contaminants (pp. 1-16). CRC Press. https://doi.org/10.1201/9781003335757

Gupta, H., & Gupta, B. (2016). Adsorption of polycyclic aromatic hydrocarbons on banana peel activated carbon. *Desalination and Water Treatment, 57*(20), 9498–9509.

Gupta, H., Dhiman, S., Kumar, A., Paliwal, S. K., Khan, G., Singh, H., & Mishra, A. K. (2024). Photo-and Bio-degradation of selected persistent organic pollutants. *Separation & Purification Reviews*, 1-14. https://doi.org/10.1080/15422119.2024.2332953

Gwenzi, W., Mangori, L., Danha, C., Chaukura, N., Dunjana, N., & Sanganyado, E. (2018). Sources, behaviour, and environmental and human health risks of high-technology rare earth elements as emerging contaminants. *Science of the Total Environment, 636*, 299–313. https://doi.org/10.1016/j.scitotenv.2018.04.235

Hao, Y., Xu, B., Ma, C., Shang, J., Gu, W., Li, W., & Rui, Y. (2019). Synthesis of novel mesoporous carbon nanoparticles and their phytotoxicity to rice (*Oryza sativa* L.). *Journal of Saudi Chemical Society, 23*(1), 75-82. https://doi.org/10.1016/j.jscs.2018.05.003

He, X., Xiao, W., Zeng, J., Tang, J., & Wang, L. (2023). Detoxification and removal of arsenite by Pseudomonas sp. SMS11: Oxidation, biosorption and bioaccumulation. *Journal of Environmental Management, 336*, 117641 https://doi.org/10.1016/j.jenvman.2023.117641

He, X., Zhang, H., Xue, R., Liu, W., Bagheri, M., Limmer, M. A., ... & Shi, H. (2019). Green analysis: High throughput analysis of emerging pollutants in plant sap by freeze–thaw–centrifugal membrane filtration sample preparation—HPLC-MS/MS analysis. *Journal of Agricultural and Food Chemistry, 67*(46), 12927-12935.

Hendricks, N., Olatunji, O., Ofori, I., & Gumbi, B. P. (2023). Occurrence, fate, and impact of engineered metal/carbonaceous nanomaterials in the environment, detection, and quantitation methods. *International Journal of Environmental Science and Technology, 20*(11), 12937–12954. https://doi.org/10.1007/s13762-023-04977-8

Hossain, M. B., Sultana, J., Pingki, F. H., Nur, A. A. U., Mia, M. S., Bakar, M. A., Yu, J., Paray, B. A., & Arai, T. (2023). Accumulation and contamination assessment of heavy metals in sediments of commercial aquaculture farms from a coastal area along the northern Bay of Bengal. *Frontiers in Environmental Science, 11*, 1148360. https://doi.org/10.3389/fenvs.2023.1148360

Jan, S., Mishra, A. K., Bhat, M. A., Bhat, M. A., & Jan, A. T. (2023). Pollutants in aquatic system: A frontier perspective of emerging threat and strategies to solve the crisis for safe drinking water. *Environmental Science and Pollution Research, 30*(53), 113242–113279. https://doi.org/10.1007/s11356-023-30302-4

Jjagwe, J., Olupot, P. W., Menya, E., & Kalibbala, H. M. (2021). Synthesis and application of granular activated carbon from biomass waste materials for water treatment: A review. *Journal of Bioresources and Bioproducts, 6*(4), 292–322. https://doi.org/10.1016/j.jobab.2021.03.003

Josa-Culleré, A., Basagaña, X., Koch, S., Arbillaga-Etxarri, A., Balcells, E., de Basea, M. B., Celorrio, N., Foraster, M., Rodriguez-Roisin, R., Marin, A., & Peralta, G. P. (2024). Short-term effects of air pollution and weather on physical activity in patients with chronic obstructive pulmonary disease (COPD). *Environmental Research, 247*, 118195. https://doi.org/10.1016/j.envres.2024.118195

Kaur, A., Gupta, H., & Dhiman, S. (2023). Biogenic metallic nanoparticles: Synthesis and applications using medicinal plants. In U. Shanker, C. M. Hussain, & M. Rani (Eds.), *Handbook of green and sustainable nanotechnology*. Springer. https://doi.org/10.1007/978-3-030-69023-6_101-1

Khan, W. U., Ahmed, S., Dhoble, Y., & Madhav, S. (2023). A critical review of hazardous waste generation from textile industries and associated ecological impacts. *Journal of the Indian Chemical Society, 100*(1), 100829. https://doi.org/10.1016/j.jics.2022.100829

Khoubi, J., Ghiasvand, A., Bahrami, A., Shahna, F. G., & Farhadian, M. (2024). Needle trap devices packed with an imine-based 2D COF: An innovative tool for the sampling of polycyclic aromatic hydrocarbons in air. *Microchemical Journal, 199*, 110018.

Kumar, R., Gupta, B., Gupta, H., & Rani, M. (2014). Distribution of persistent organic pollutants in urban aquatic systems. *International Journal of Scientific Research in Environmental Sciences, 2*(7), 233–243. https://doi.org/10.12983/ijsres-2014-p0233-0243

Kumar, J., Delgado, S. A., Sarma, H., & Narayan, M. (2023). Caffeic acid recarbonization: A green chemistry, sustainable carbon nano material platform to intervene in neurodegeneration induced by emerging contaminants. *Environmental Research, 237*, 116932. https://doi.org/10.1016/j.envres.2023.116932

Kumar, R., Rani, M., Gupta, H., Gupta, B., Park, D., & Jeon, B. H. (2017). Distribution of trace elements in flowing surface waters: Effect of seasons and anthropogenic practices in India. *International Journal of Environmental Analytical Chemistry, 97*(7), 637–656. https://doi.org/10.1080/03067319.2017.1339035

Kurowska, P., Mlyczyńska, E., Dawid, M., Respekta, N., Pich, K., Serra, L., Dupont, J., & Rak, A. (2022). Endocrine disruptor chemicals, adipokines and reproductive functions. *Endocrine, 78*(2), 205–218. https://doi.org/10.1007/s12020-022-03061-4

Li, H., Sheng, W., Haruna, S. A., Hassan, M. M., & Chen, Q. (2023a). Recent advances in rare earth ion-doped upconversion nanomaterials: From design to their applications in food safety analysis. *Comprehensive Reviews in Food Science and Food Safety, 22*(5), 3732–3764. https://doi.org/10.1111/1541-4337.13218

Li, L., Tian, F., Qiu, L., Wu, F., Yang, W., & Yu, Y. (2023b). Recent progress on ruthenium-based electrocatalysts towards the hydrogen evolution reaction. *Catalysts, 13*(12), 1497. https://doi.org/10.3390/catal13121497

Li, X., Shen, X., Jiang, W., Xi, Y., & Li, S. (2024). Comprehensive review of emerging contaminants: Detection technologies, environmental impact, and management strategies. *Ecotoxicology and Environmental Safety, 278*, 116420. https://doi.org/10.1016/j.ecoenv.2024.116420

Lou, H., Yang, J., Liu, Q., & Yang, Q. (2024). Hazards of trichloroethylene on the liver in animals: A systematic review and meta-analysis. *Emerging Contaminants*, 100338. https://doi.org/10.1016/j.emcon.2024.100338

Madhav, S., Kumar, A., Kushawaha, J., Ahamad, A., Singh, P., & Dwivedi, S. B. (2020). Geochemical assessment of groundwater quality in Keonjhar City, Odisha, India. *Sustainable Water Resources Management, 6*, 1–11. https://doi.org/10.1007/s40899-020-00395-7

Madhav, S., Mishra, R., Kumari, A., Srivastav, A. L., Ahamad, A., Singh, P., Ahmed, S., Mishra, P. K., & Sillanpää, M. (2024). A review on sources identification of heavy metals in soil and remediation measures by phytoremediation-induced methods. *International Journal of Environmental Science and Technology, 21*(1), 1099–1120. https://doi.org/10.1007/s13762-023-04950-5

Magee, H. Y., Maurer, M. M., Cobos, A., Adhikari, S., & Halden, R. U. (2023). Impact of LC-MS/MS quantification methods on analytical results by example of nationwide US environmental samples. *Science of the Total Environment, 892*, 164425. https://doi.org/10.1016/j.scitotenv.2023.164425

Mariano, S., Tacconi, S., Fidaleo, M., Rossi, M., & Dini, L. (2021). Micro and nanoplastics identification: Classic methods and innovative detection techniques. *Frontiers in Toxicology, 3*, 636640. https://doi.org/10.3389/ftox.2021.636640

Martín-Pozo, L., de Alarcón-Gómez, B., Rodríguez-Gómez, R., García-Córcoles, M. T., Çipa, M., & Zafra-Gómez, A. (2019). Analytical methods for the determination of emerging contaminants in sewage sludge samples. A review. *Talanta, 192*, 508–533. https://doi.org/10.1016/j.talanta.2018.09.056

Metcalfe, C. D. (2013). Pharmaceutical contaminants of emerging concern in the environment pharmaceutical contaminants of emerging concern in the environment. *Environmental Toxicology & Chemistry, 32*(8). https://doi.org/10.1002/etc.2293

Nabgan, W., Saeed, M., Jalil, A. A., Nabgan, B., Gambo, Y., Ali, M. W., Ikram, M., Fauzi, A. A., Owgi, A. H. K., Hussain, I., & Thahe, A. A. (2022). A state-of-the-art review on electrochemical technique for the remediation of pharmaceuticals containing wastewater. *Environmental Research, 210*, 112975. https://doi.org/10.1016/j.envres.2022.112975

Nadeem, K., Alliet, M., Plana, Q., Bernier, J., Azimi, S., Rocher, V., & Albasi, C. (2022). Modeling, simulation and control of biological and chemical P-removal processes for membrane bioreactors (MBRs) from lab to full-scale applications: State of the art. *Science of the Total Environment, 809*, 151109. https://doi.org/10.1016/j.scitotenv.2021.151109

Ngubane, Z., Dzwairo, B., Moodley, B., Stenström, T. A., & Sokolova, E. (2023). Quantitative assessment of human health risks from chemical pollution in the uMsunduzi River, South Africa. *Environmental Science and Pollution Research, 30*(55), 118013–118024. https://doi.org/10.1007/s11356-023-30534-4

Noguera-Oviedo, K., & Aga, D. S. (2016). Lessons learned from more than two decades of research on emerging contaminants in the environment. *Journal of Hazardous Materials, 316*, 242–251. https://doi.org/10.1016/j.jhazmat.2016.04.058

Nouri, N., Khorram, P., Duman, O., Sibel, T., & Hassan, S. (2020). Overview of nanosorbents used in solid phase extraction techniques for the monitoring of emerging organic contaminants in water and wastewater samples. *Trends in Environmental Analytical Chemistry, 25*, e00081. https://doi.org/10.1016/j.teac.2020.e00081

Onyena, A. P., Nkwoji, J. A., Chukwu, L. O., Walker, T. R., & Sam, K. (2023). Risk assessment of sediment PAH, BTEX, and emerging contaminants in Chanomi Creek Niger Delta, Nigeria. *Environmental Monitoring and Assessment, 195*(9), 1080. https://doi.org/10.1007/s10661-023-11703-x

Pan, J., Liu, P., Yu, X., Zhang, Z., & Liu, J. (2024). The adverse role of endocrine disrupting chemicals in the reproductive system. *Frontiers in Endocrinology, 14*, 1324993. https://doi.org/10.3389/fendo.2023.1324993

Pathak, P., & Sushil, K. (2022). Electronic waste: an emerging contaminant in the geo-environment. In *Emerging contaminants in the environment* (pp. 275–286). Elsevier. https://doi.org/10.1016/B978-0-323-85160-2.00018-4

Pereira, L. C., de Souza, A. O., Bernardes, M. F. F., Pazin, M., Tasso, M. J., Pereira, P. H., & Dorta, D. J. (2015). A perspective on the potential risks of emerging contaminants to human and environmental health. *Environmental Science and Pollution Research, 22*, 13800–13823. https://doi.org/10.1007/s11356-015-4896-6

Petrie, B., Youdan, J., Barden, R., & Kasprzyk-Hordern, B. (2016). Multi-residue analysis of 90 emerging contaminants in liquid and solid environmental matrices by ultra-high-performance liquid chromatography tandem mass spectrometry. *Journal of Chromatography A, 1431*, 64–78. https://doi.org/10.1016/j.chroma.2015.12.036

Pokhriyal, P., Kaur, A., Shukla, A., et al. (2024). Biocompatible zinc nanoparticles synthesis from ficus subincisa for a sustainable tomorrow: Characterization and therapeutic applications. *Russian Journal of Bioorganic Chemistry, 50*, 408–417. https://doi.org/10.1134/S1068162024020158

Rathi, B. S., Kumar, P. S., & Show, P. L. (2021). A review on effective removal of emerging contaminants from aquatic systems: Current trends and scope for further research. *Journal of Hazardous Materials, 409*, 124413. https://doi.org/10.1016/j.jhazmat.2020.124413

Rede, D., Delerue-Matos, C., & Fernandes, V. C. (2023). The microplastics iceberg: Filling gaps in our understanding. *Polymers, 15*(16), 3356. https://doi.org/10.3390/polym15163356

Richardson, S. D. (2009). Water analysis: Emerging contaminants and current issues. *Analytical Chemistry, 81*(12), 4645–4677. https://doi.org/10.1021/ac9008012

Richardson, S. D., & Ternes, T. A. (2021). Water analysis: Emerging contaminants and current issues. *Analytical Chemistry, 94*(1), 382–416. https://doi.org/10.1021/acs.analchem.1c04640

Safwat, N., Ayad, M., & Farouk, M. (2020). Development of an eco-friendly reversed-phase HPLC method for the simultaneous determination of three emerging contaminants in water samples. *Archives of Pharmaceutical Sciences Ain Shams University, 4*(1), 79-91. https://doi.org/10.21608/APS.2020.2001.1025

Santos, V. S., Anjos, J. S., de Medeiros, J. F., & Montagner, C. C. (2022). Impact of agricultural runoff and domestic sewage discharge on the spatial–temporal occurrence of emerging contaminants in an urban stream in São Paulo, Brazil. *Environmental Monitoring and Assessment, 194*(9), 637. https://doi.org/10.1007/s10661-022-10288-1

Satyanarayana, G. N. V., Kumar, A., Pandey, A. K., Sharma, M. T., Natesan, M., & Mudiam, M. K. R. (2023). Evaluating chemicals of emerging concern in the Ganga River at the two major cities Prayagraj and Varanasi through validated analytical approaches. *Environmental Science and Pollution Research, 30*(1), 1520–1539. https://doi.org/10.1007/s11356-022-22226-2

Sereshti, H., Duman, O., Tunç, S., Nouri, N., & Khorram, P. (2020). Nanosorbent-based solid phase microextraction techniques for the monitoring of emerging organic contaminants in water and wastewater samples. *Microchimica Acta, 187*, 1–35. https://doi.org/10.1007/s00604-020-04527-w

Seyyedsalehi, M. S., & Boffetta, P. (2023). Per-and poly-fluoroalkyl substances (PFAS) exposure and risk of kidney, liver, and testicular cancers: A systematic review and meta-analysis. *La Medicina del Lavoro, 114*(5). https://doi.org/10.23749/mdl.v114i5.15065

Shen, X., Dai, M., Yang, J., Sun, L., Tan, X., Peng, C., Ali, I., & Naz, I. (2022). A critical review on the phytoremediation of heavy metals from environment: Performance and challenges. *Chemosphere, 291*, 132979. https://doi.org/10.1016/j.chemosphere.2021.132979

Silva, C. P., Carvalho, T., Schneider, R. J., Esteves, V. I., & Lima, D. L. (2020). ELISA as an effective tool to determine spatial and seasonal occurrence of emerging contaminants in the aquatic environment. *Analytical Methods, 12*(19), 2517–2526. https://doi.org/10.1039/D0AY00426J

Singh, V. K., Singh, R., Kumar, A., Bhadouria, R., & Pandey, S. (2021). Sewage wastewater and sludge as source of traditional and emerging contaminants in agroecosystems. In *Sustainable agriculture reviews 50: Emerging contaminants in agriculture* (pp. 35–59). https://doi.org/10.1007/978-3-030-63249-6_2

Singh, S., Parveen, N., & Gupta, H. (2018). Adsorptive decontamination of rhodamine-B from water using banana peel powder: A biosorbent. *Environmental Technology and Innovation, 12*, 189–195. https://doi.org/10.1016/j.eti.2018.09.001

Soliz, D. L., Garcinuño, R. M., Paniagua González, G., Bravo, J. C., & Fernández Hernando, P. (2024). Assessing matrix solid-phase dispersion extraction strategies for determining bisphenols and phthalates in gilthead sea bream samples. *Foods, 13*(3), 413. https://doi.org/10.3390/foods13030413

Song, J., Xu, Z., Chen, Y., & Guo, J. (2023). Nanoparticles, an emerging control method for harmful algal blooms: Current technologies, challenges, and perspectives. *Nanomaterials, 13*(16), 2384. https://doi.org/10.3390/nano13162384

Sorvari, J., & Wahlström, M. (2024). Industrial by-products. In *Handbook of recycling* (pp. 259-285). Elsevier. https://doi.org/10.1016/B978-0-323-85514-3.00044-0

Suvar, N. S., Prodan, M., Ghicioi, E., & Toplician, A. (2024). Identification of hazardous organic substances for fire investigation with portable GC-MS. In MATEC Web of conferences (Vol. 389, p. 00020). EDP Sciences. https://doi.org/10.1051/matecconf/202438900020

Suzuki, T., Hidaka, T., Kumagai, Y., & Yamamoto, M. (2020). Environmental pollutants and the immune response. *Nature Immunology, 21*(12), 1486–1495. https://doi.org/10.1038/s41590-020-0802-6

Svensson Grape, E., Chacón-García, A. J., Rojas, S., Pérez, Y., Jaworski, A., Nero, M., Åhlén, M., Martínez-Ahumada, E., Galetsa Feindt, A. E., Pepillo, M., & Narongin-Fujikawa, M. (2023). Removal of pharmaceutical pollutants from effluent by a plant-based metal-organic framework. *Nature Water, 1*(5), 433–442. https://doi.org/10.1038/s44221-023-00070-z

Tong, X., Mohapatra, S., Zhang, J., Tran, N. H., You, L., He, Y., & Gin, K. Y. H. (2022). Source, fate, transport and modelling of selected emerging contaminants in the aquatic environment: Current status and future perspectives. *Water Research, 217*, 118418. https://doi.org/10.1016/j.watres.2022.118418

Tong, Y., Gao, J., Yue, T., Zhang, X., Liu, J., & Bai, J. (2023). Distribution, chemical fractionation, and potential environmental risks of Hg, Cr, Cd, Pb, and As in wastes from ultra-low emission coal-fired industrial boilers in China. *Journal of Hazardous Materials, 446*, 130606. https://doi.org/10.1016/j.jhazmat.2022.130606

Tu, I. C., Yuan, C. S., Tseng, Y. L., Lee, C. W., & Lin, C. (2024). Spatiotemporal variation and inter-transport of atmospheric speciated mercury between Kaohsiung Harbor and neighboring urban areas. *Environmental Pollution, 342*, 123039. https://doi.org/10.1016/j.envpol.2023.123039

Wang, N., Kang, G., Hu, G., Chen, J., Qi, D., Bi, F., Chang, N., Gao, Z., Zhang, S., & Shen, W. (2023a). Spatiotemporal distribution and ecological risk assessment of pharmaceuticals and personal care products (PPCPs) from Luoma Lake, an important node of the South-to-North Water Diversion Project. *Environmental Monitoring and Assessment, 195*(11), 1330. https://doi.org/10.1007/s10661-023-11976-2

Wang, Y., Bao, S., Liu, X., Qiu, L., Sheng, J., Yang, W., & Yu, Y. (2023b). Regulating the peroxymonosulfate activation on N doped δ-MnO_2 nanosheets for tetracycline degradation: N species as the degradation pathways switcher to convert radical to nonradical. *Chemical Engineering Journal, 477*, 147050. https://doi.org/10.1016/j.cej.2023.147050

Wang, Y., Qiu, L., Bao, S., Tian, F., Sheng, J., Yang, W., & Yu, Y. (2023c). Visible-light enhanced peroxymonosulfate activation on Co_3O_4/MnO_2 for the degradation of tetracycline: Cooperation of radical and non-radical mechanisms. *Separation and Purification Technology, 316*, 123779. https://doi.org/10.1016/j.seppur.2023.123779

Wang, Y., Wang, M., Zhou, L., Geng, X., Xu, Z., & Zhang, H. (2023d). Development of a competitive ELISA based on estrogen receptor and weak competitive molecule for the screening of potential estrogens in foods. *Food Chemistry, 401*, 134084. https://doi.org/10.1016/j.foodchem.2022.134084

Wang, F., Xiang, L., Leung, K. S. Y., Elsner, M., Zhang, Y., Guo, Y., & Tiedje, J. M. (2024). Emerging contaminants: A one health perspective. *The Innovation*. https://doi.org/10.1016/j.xinn.2024.100612

Winiarska, E., Jutel, M., & Zemelka-Wiacek, M. (2024). The potential impact of nano-and microplastics on human health: Understanding human health risks. *Environmental Research*, 118535. https://doi.org/10.1016/j.envres.2024.118535

Xie, M., Liang, M., Liu, C., Xu, Z., Yu, Y., Xu, J., & Rad, S. (2024). Peroxymonosulfate activation by CuMn-LDH for the degradation of bisphenol A: Effect, mechanism, and pathway. *Ecotoxicology and Environmental Safety, 270*, 115929. https://doi.org/10.1016/j.ecoenv.2024.115929

Yadav, D., Rangabhashiyam, S., Verma, P., Singh, P., Devi, P., Kumar, P., Hussain, C. M., Gaurav, G. K., & Kumar, K. S. (2021). Environmental and health impacts of contaminants of emerging concerns: Recent treatment challenges and approaches. *Chemosphere, 272*, 129492. https://doi.org/10.1016/j.chemosphere.2020.129492

Zheng, G., Eick, S. M., & Salamova, A. (2023). Elevated levels of ultrashort-and short-chain perfluoroalkyl acids in US homes and people. *Environmental Science & Technology, 57*(42), 15782–15793. https://doi.org/10.1021/acs.est.2c06715

Zulkifli, S. N., Rahim, H. A., & Lau, W. J. (2018). Detection of contaminants in water supply: A review on state-of-the-art monitoring technologies and their applications. *Sensors and Actuators b: Chemical, 255*, 2657–2689. https://doi.org/10.1016/j.snb.2017.09.078

Chapter 11
Biological Cleansing: Probiotic Strategies for Tackling Contaminants in Water and Wastewater

Seeta Dewali, Nagma Parveen, Neha Kathayat, Gaurav Rawat, Shikha Bora, Netra Pal Sharma, Vinay Mohan Pathak, and Satpal Singh Bisht

Abstract Increased environmental contamination leads to a continuously decline in environmental quality. Water pollution arises when detrimental substances, such as chemicals or microorganisms, infiltrate and taint bodies of water like streams, rivers, lakes, oceans, aquifers, leading to a decline in water quality and making it hazardous for both humans and the ecosystem. Water and wastewater treatment pose significant challenges due to the persistent presence of various contaminants, including pathogens, heavy metals and organic pollutants. Traditional treatment methods often fall short in effectively addressing these diverse contaminants, necessitating innovative approaches. Probiotic strategies have emerged as promising solutions, leveraging the inherent capabilities of beneficial microorganisms to combat contaminants and restore environmental balance. Furthermore, the integration of probiotics into existing treatment frameworks showcases their potential for sustainable and cost-effective remediation strategies. By harnessing natural microbial communities, probiotic-based approaches minimize reliance on synthetic chemicals and energy-intensive processes, thereby reducing environmental footprint and operational costs. This chapter highlights the multifaceted applications of probiotics in water and waste water treatment, focusing on their mechanisms of action, efficacy, and environmental implications. Continued research and technological advancements are crucial for unlocking the full potential of probiotics and integrating them into mainstream water management practices, thereby safeguarding water resources and public health in a rapidly changing world.

S. Dewali (✉) · N. Parveen · N. Kathayat · G. Rawat · S. Bora · N. P. Sharma
Department of Zoology, Kumaun University, Nainital, Uttarakhand 263002, India
e-mail: seemadeoli3@gmail.com

V. M. Pathak
Research and Development, Pritam International Private Ltd. Bhagwanpur, Roorkee, Haridwar 247667, India

S. S. Bisht
Soban Singh Jeena University, Almora, Uttarakhand 263601, India

H. Gupta et al. (eds.), *Emerging Contaminants in Water and Wastewater*, Environmental Science and Engineering, https://doi.org/10.1007/978-3-031-82579-8_11

Keywords Biological · Cleansing · Probiotic · Contaminants · Wastewater · Environmental

11.1 Introduction

Environmental contamination has been recognized as detrimental to industrial facilities' environments, especially agro-industrial ones. The population's health is declining due to the natural and agricultural systems' reaching a critical point in sustainability. The necessity to enhance the effectiveness of work on the restoration of technologically contaminated areas has grown dramatically when undertaking large-scale investment projects linked to building new facilities, modernizing current production, and implementing environmental remediation measures. Traditional techniques of cleaning industrial facilities, livestock farms, waste water, and storage waste generation are either extremely expensive or inefficient in the current context due to the increasing requirements for economic efficiency and environmental standards for food production (da Silva Vilar et al., 2019; Nițu, 2024; Saravanan et al., 2021). Conventional wastewater treatment methods often involve chemical or physical processes, which may be costly and environmentally damaging. Additionally, these methods may not effectively eliminate specific contaminants or pathogens, posing risks to aquatic ecosystems and human health. In contrast, probiotic preparations offer a more eco-friendly and holistic approach to wastewater remediation (Saiyad et al., 2024).

The discharge of untreated or inadequately treated wastewater into reservoirs by utilities, industrial companies, and agriculture, along with an increase in the frequency of oil spills into water bodies, are the primary causes of contamination of water basins (Gupta, 2018; Rai, 2024). The majority of research explains the steps involved in treating wastewater using probiotics. Simultaneously, there is no research on the potential of technology implementation in the processes under consideration. As global concerns regarding environmental pollution continue escalating, innovative and sustainable approaches are imperative to address the challenges of contaminated water bodies. Probiotics, traditionally associated with promoting gut health in humans and animals, have now found application in the remediation of waste waters. This burgeoning field explores the potential of beneficial microorganisms to enhance the efficiency of wastewater treatment processes while minimizing the environmental impact (Parker & Cruz, 2024). Probiotics, typically comprising beneficial bacteria or microbial consortia, possess inherent capabilities to degrade organic pollutants, neutralize toxins, and inhibit the growth of pathogenic microorganisms. By harnessing the natural metabolic activities of these microorganisms, probiotic-based approaches aim to enhance the biodegradation and detoxification processes within wastewater treatment systems. Moreover, using probiotics in wastewater treatment can optimize treatment efficiency, reduce energy consumption, and minimize the generation of sludge and other harmful byproducts. Furthermore, probiotic treatments may enhance wastewater treatment plants' overall resilience and stability, thereby improving their performance under fluctuating environmental conditions (Kumar et al., 2024).

This introductory exploration delves into the burgeoning field of utilizing probiotic preparations in wastewater treatment. By elucidating the mechanisms, applications, and benefits of probiotic-based approaches, this research aims to advance sustainable solutions for wastewater management. Through a comprehensive understanding of the role of probiotics in waste water cleaning, we can pave the way toward greener, more effective, and environmentally responsible wastewater treatment practices (Zhao et al., 2024). Thus, probiotic-based remediation offers a more environmentally friendly and sustainable approach than conventional remediation methods, which often involve harsh chemicals or energy-intensive processes. Probiotics are natural, non-toxic agents that can degrade contaminants under mild conditions, minimizing ecological disturbances and preserving ecosystem integrity.

This chapter aims to delve into the potential of probiotic strategies as innovative solutions for combating contaminants in water and wastewater. By dissecting the mechanisms by which probiotics interact with pollutants be it through bioaccumulation, biodegradation, or outcompeting pathogens it seeks to unveil their effectiveness. Moreover, it endeavors to scrutinize the environmental ramifications and sustainability of deploying probiotics in water treatment, considering factors such as energy efficiency, cost-effectiveness, and ecological balance. Through a thorough examination of current research, this review strives to offer profound insights into the viability, limitations and future pathways of harnessing probiotic interventions, thereby propelling the quest for cutting-edge, eco-friendly approaches to water quality enhancement.

11.2 Microbial Ecology of Water Environments: Insights into the Potential of Probiotics for Contaminant Remediation

The microbial ecology of water environments is a complex field that explores the diverse communities of microorganisms inhabiting various aquatic ecosystems. These ecosystems include marine habitats like oceans and estuaries and freshwater bodies like lakes, rivers, and wetlands. Apart from bacteria, other essential species that contribute to the microbiota structure of the water ecosystem include viruses, microbial eukaryotes, and archaea, which collectively form a diverse community of microorganisms. The health of humans, plants, and animals can all be significantly impacted by all these microbes that serve essential ecological roles (Bocanegra-García et al., 2023). In aquatic habitats, microbial communities are essential to the biogeochemical cycle (Madhav et al., 2020; Shi et al., 2022), nutrient recycling (Prasad et al., 2021), and maintaining ecosystem stability and function (Li et al., 2021). Understanding microbial ecology in water environments has practical implications for environmental management and biotechnology. Microorganisms also play vital roles in wastewater treatment, bioremediation of polluted sites, and producing biofuels and bioproducts (Ahmad et al., 2022; Chai et al., 2021; Gupta et al., 2024; Patel et al., 2022a, 2022b). Advances in molecular biology and sequencing technologies have revolutionized the study of microbial ecology, enabling researchers to explore microbial diversity and function in water environments at unprecedented scales and resolutions. Next-generation sequencing and metagenomic approaches provide insights into microbial communities' genetic potential and activity (Wani et al., 2021).

Furthermore, microbes and microbial consortia are attracting attention, with the microalgal-bacterial consortia in wastewater bioremediation, bioenergy production, and CO_2 fixation supporting sustainability and generating remarkable advancements (Khoo et al., 2021). microbial ecology is essential for creating measures to reduce pollution and remove contaminants in the context of water contamination. Probiotics—described as live microbes that, when taken in sufficient quantities, provide health benefits (Mishra & Acharya, 2021), have drawn attention due to their possible use in aquatic environments to remedy contaminants (Fig. 11.1).

Probiotics can be harnessed for bioremediation, which uses microorganisms to degrade or transform contaminants into less harmful forms. Certain probiotic strains possess metabolic capabilities to degrade organic pollutants, such as hydrocarbons, pesticides, and industrial chemicals, through enzymatic reactions. Introducing these probiotics into contaminated environments, such as soil or water, can contribute to the breakdown and detoxification of pollutants. Sarlak et al. (2021) suggested that probiotic microbes possess the functional ability to break down organophosphorus pesticides (OPPs), which are frequently employed to manage pests in agriculture. Bacillus is the most common genera among the many bacteria that have been studied as probiotics in aquaculture. In addition to adding Bacillus to feed to achieve the probiotic effect, administering this bacterial group to raise water has also been studied to enhance water quality (Kamilya & Devi, 2022).

In addition to the bioremediation methods, Probiotics are also utilized in Bioaugmentation processes. Bioaugmentation involves deliberately adding probiotic microorganisms into contaminated sites to enhance remediation processes. A study conducted reveals that the Indian white shrimp (*Penaeus indicus*) had better growth, survival, and immunity when heterotrophic bacteria (*Bacillus spp*) are augmented in the biofloc system (Panigrahi et al., 2020, 2021). Bioaugmented probiotic therapy is a viable disease mitigation technique based on emerging evidence that bacteria help plants and animals defend themselves. *Batrachochytrium dendrobatidis (Bd)*, a pathogen that causes chytridiomycosis disease in amphibians, is efficiently treated with bioaugmented probiotics (Bletz et al., 2013).

Probiotics can also be combined with other remediation strategies, such as phytoremediation, chemical oxidation, or physical containment, to achieve synergistic effects and enhance contaminant removal. Liu et al. (2020) explored tetracycline antibiotic bioremoval with probiotics *Bacillus clausii* and *Bacillus amyloliquefaciens* generating biosurfactants, and it was found that the biosurfactant removed TCs better than the nonionic surfactant. Ram explored the synergy of selenium nanoparticles (Se NPs) and lactic acid bacteria in lowering cadmium toxicity and

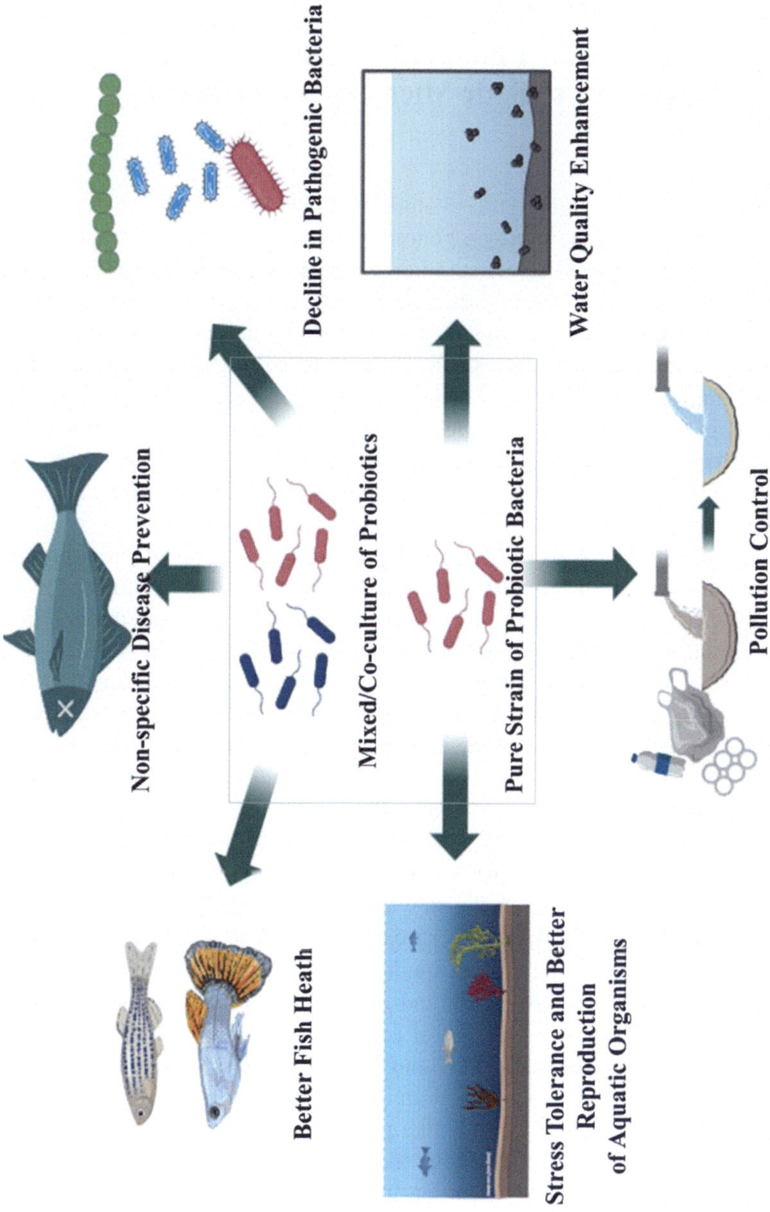

Fig. 11.1 Applications of probiotics in aquatic systems

discovered that FT-Raman spectroscopy showed greater activation of endogenous SeNPs in *L. fermentum*.

11.3 Bioengineering for Enhanced Water Purification by Harnessing Probiotic Microbes for Targeted Applications

Bioengineered probiotics signify the evolution of whole cell-mediated biotherapeutics. Progress in synthetic biology, genome engineering, and DNA sequencing and synthesis has empowered scientists to create probiotics with heightened stress resilience and the capacity to target particular pathogens and their associated toxins selectively. Furthermore, these advancements enable the precise delivery of vaccines, medications, and immunomodulators directly to host cells (Sola-Oladokun et al., 2017). Water purification is a critical process to clean and safe drinking water, and advancements in bioengineering have led to innovative approaches utilizing probiotic microbes for targeted applications in water purification. This chapter investigates how bioengineering can improve water purification by utilizing probiotic microbes for specialized purposes.

Harnessing probiotic microbes for water purification is a promising approach that leverages the beneficial properties of certain microorganisms to improve water quality. Probiotic microbes frequently influence the composition and operations of the gut microbial community in humans and animals (Hemarajata & Versalovic, 2013), exhibiting various mechanisms that can be harnessed for water purification. These mechanisms include competitive exclusion, biodegradation of pollutants using colonization, fragmentation, assimilation, and mineralization (Li et al., 2023), production of antimicrobial compounds and nutrient cycling (Rabetafika et al., 2023). When applied to water systems, probiotic microbes can outcompete pathogenic bacteria, degrade organic pollutants, produce antimicrobial compounds, and contribute to nutrient cycling, promoting a healthier aquatic environment.

Probiotic microbes play diverse roles in water purification, spanning aquaculture, wastewater treatment, and drinking water treatment. Adding probiotics to aquaculture water has proven to enhance water quality, diminish disease occurrences, and foster aquatic organisms' overall health and growth. Probiotic strains exert a positive influence on the digestive processes of aquatic animals by producing extracellular enzymes like proteases, amylases, and lipases while also supplying growth factors such as vitamins, fatty acids, and amino acids (Martínez Cruz et al., 2012; Balcázar et al., 2006). Microorganisms like probiotics (Corre & Janeo, 2002) and bacteriophages (Withey et al., 2005) ameliorated water quality. Probiotics are widely employed for remediating pathogenic microbial contamination (Zhou et al., 2023). Furthermore, the potential of probiotic microbes is being explored for point-of-use water treatment systems, offering a sustainable and eco-friendly approach to purifying drinking water.

The efficacy of probiotic microbes in reducing toxic levels in industrial wastewater enhances the removal of organic pollutants in constructed wetlands; microbial activity is thought to play a vital role in detoxifying these metals (Kosolapov et al., 2004). The integration of a gravel matrix with algal-bacteria symbiosis significantly enhances the efficacy of constructed wetlands in treating antibiotic wastewater while safeguarding the ecological community, offering an environmentally sustainable approach (Wang et al., 2022a, 2022b).

11.3.1 The Role of Probiotic Microbes in Water Purification

Recent bioengineering advancements have demonstrated probiotic microbes' potential in water purification. Probiotic microbes, such as engineered bacteria and lactic acid bacteria, have shown promise in targeted applications for degrading contaminants, removing pollutants, and improving water quality. Using probiotics as a water supplement can enhance water quality through alterations in microbial communities, inhibition of opportunistic microorganism growth, and reduced metabolic waste in the water system (Giri et al., 2018; Talpur et al., 2013). Research by Zhai and colleagues revealed that lactic acid bacteria and bifidobacteria bind to remove heavy metals both in vitro and in vivo from water and the environment (Bhakta et al., 2012; Zhai et al., 2017). Additionally, ingesting probiotics via water or feed effectively decreases mortality rates and mitigates the accumulation of heavy metals in tissues (Chizhayeva et al., 2022). The importance of probiotics in various applications, including aquaculture water purification, immunomodulatory properties and the use of probiotics in the food industry (Fig. 11.2) also highlighted the significance of engineered probiotics in aquaculture, demonstrating their role in water purification processes (Pereira et al., 2022).

11.3.2 Genetically Modified Bacteria for Plastic Degradation

The genetic engineering of marine microorganisms for breaking down plastic in salt water explicitly targets polyethylene terephthalate (PET), a commonly used plastic (Basuki et al., 2020). This application demonstrates the potential of probiotic microbes in addressing plastic pollution in aquatic environments, showcasing the real-life impact of bioengineering in water purification. Microplastic pollution significantly threatens ecosystems and impacts abiotic and biotic components. Microplastics represent a broad and intricate category of pollutants capable of transporting other contaminants and microbes. Existing approaches for microplastic removal encompass biodegradation, incineration, landfilling, and recycling. Anand et al. (2023) present biotechnological techniques to eliminate microplastics to enhance water quality.

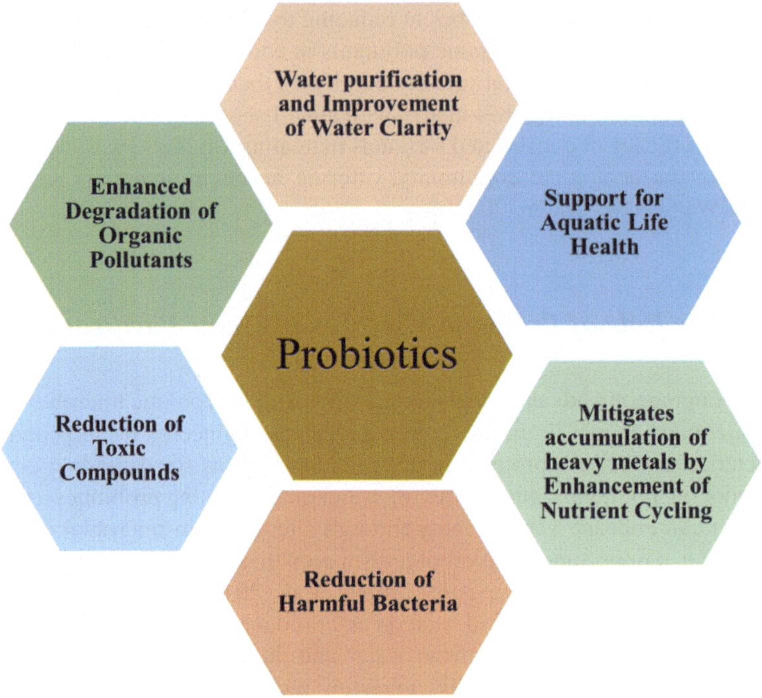

Fig. 11.2 Benefits of probiotics in maintaining water quality

11.3.3 Engineered Microbes for Biomanufacturing

Using engineered microbes as microscopic factories has revolutionized the food industry and offered sustainable solutions for valuable biomanufacturing processes (Neethirajan et al., 2018). This application highlights the versatility of probiotic microbes in diverse industrial settings, emphasizing their potential for targeted water purification applications. Genetically engineered microorganisms (GEMs) have demonstrated promise for bioremediation across soil, groundwater, and activated sludge settings. They showcase augmented degradative capacities spanning a broad spectrum of chemical contaminants (Sayler & Ripp, 2000).

Researchers have devised a 'living material' comprising natural polymers combined with genetically engineered bacteria, presenting a sustainable and eco-friendly solution for removing pollutants from water (Arif et al., 2022). These "living materials," incorporating genetically modified bacteria alongside other components, can potentially mitigate water pollutants. The material, a seaweed-based polymer combined with genetically engineered bacteria, is 3D printed into a filter. The bacteria within produce an enzyme that converts organic pollutants into harmless molecules. This innovative approach underscores the promise of bioengineering in developing environmentally friendly materials for water purification.

Bioengineering plays a crucial role in the remediation of contaminated water. However, this multidisciplinary field faces challenges in broadening the focus of scientists and engineers beyond conventional classes of pollutants towards novel, individual, and often toxic chemicals resistant to biological transformations. To effectively address these challenges, there is a need for a comprehensive understanding of the processes linking biological activity to the physical and chemical characteristics of the environment. This understanding is essential for designing systems that effectively control and enhance bioremediation for water cleanup (Sims et al., 1992).

Aquaculture effluent is primarily associated with elevated levels of organic carbon, phosphates, nitrogenous species (nitrates, nitrites, and ammonia), suspended solids, chemical oxygen demand, and biological oxygen demand (Hlordzi et al., 2020). Water becomes contaminated with the deposition of heavy metals, metalloids, dumping of fish debris (scales, offal, excreta, and dead fish), chemical fertilizers, pesticides, and plastic waste (microplastic), which results in high quantities of organic matter, phosphate, and nitrogen, which severely impact the rearing water, and reduced dissolved oxygen (DO) levels due to the decomposition of organic compounds (Roy et al., 2024; Srithongouthai & Tada, 2017; Sudan et al., 2024).

Fertilizers and commercial feeds contain higher phosphorus and nitrogen, which is the reason for eutrophication and natural ecosystem destruction. Excretory products contain high levels of phosphorus and nitrogen (Devaraja et al., 2013; Hlordzi et al., 2020). During disease outbreaks, poor water quality influences fish health and increases the rate of fish mortality; as a result, the amount of organic matter in the culture waters is increased by dead fish (Ananda Raja & Jithendran, 2015).

The deteriorating water quality affects surface and ground water for other organisms and poses a threat to the survival of aquatic life (Scanlon et al., 2023). Elevated phosphorus and nitrogen can be toxic to plants and alter their respiration, oxidative stress response, photosynthesis, enzymatic activity, membrane permeability, and protein synthesis. Pathogenic microorganisms and toxic can be transferred via the food chain. Owing to its impact on nearby rivers and groundwater, this is considered a global danger to aquatic environments (Hlordzi et al., 2020).

Frequent water exchange is the most widely utilized technique for maintaining water quality in aquaculture (Devaraja et al., 2013). In addition, a number of chemicals have been employed to improve the water quality in aquaculture Biolite plus, Aquazet Bio-tuff, Well zeolite, Geotox, Green zeolite, JV zeolite and Pontox plus (Shamsuzzaman & Biswas, 2012). The main concern with using these compounds is the bioaccumulation of these compounds in humans and other organisms.

Microbial remediation, distinctively probiotics, has recently gained popularity in improving water quality and maintaining aquatic animal health (Li et al., 2022). Probiotics are a safe supplement that aquarists use to improve feed utilization, prevent potentially harmful microbes, purify water quality, and improve host health (El-Saadony et al., 2021).

11.4 Mechanisms Used by Probiotics in Improving Water Quality

In aquaculture systems, probiotics use numerous mechanisms to remove organic wastes and pollutants (Hassan et al., 2022). Probiotics can detoxify environmental contaminants and reduce their toxicity by binding and stimulating enzymatic processes (Hu et al., 2024).

11.4.1 Probiotics and Hazardous Chemicals

Fish farmers employ a variety of chemicals and medications to cure fish illnesses and maintain ponds. The following are some typical chemicals: lime, salt, formalin, summation, potassium permanganate, malachite green, and bleaching powder (Hlordzi et al., 2020). Fish infections are treated with the antibiotics cotrimoxazole, oxytetracycline, sulphamethoxazole, oxysentin, renamox, sulphadiazine, chlorotetracycline, renamycin, amoxicillin, and orgamycine (Hlordzi et al., 2020; Okeke et al., 2022). Aquatic ecosystems are threatened by these substances and medications (Ronquillo & Hernandez, 2017).

Probiotic microbes can bind and metabolize harmful chemical substances like mycotoxins and OPPs (Srednicka et al., 2021). During bioremediation, probiotics use enzymes such as phosphatases, carboxylesterases, phosphodiesterase, and organophosphate hydrolases to break down pesticides and fulfil their carbon, energy, and nitrogen requirements (Fig. 11.3). Moreover, probiotics can also metabolize pesticides through the fermentation process (Salih et al., 2023).

According to Hu et al., (2024), probiotics can reduce the concentrations of organophosphorus pesticides (OPPs) by microbial enzyme activity and adsorption. Probiotics use adsorption, binding or degradation (*Lactobacillus plantarum* P9) as a reduction mechanism against OPPs (Sarlak et al., 2021). Probiotics enhance water quality and provide additional benefits to cultured fish. Other benefits of probiotics include immune-stimulation, which leads to pathogen resistance, direct suppression of harmful bacteria, provides nutrients, and increased utilization of food and help in growth (Hlordzi et al., 2020; Van Doan et al., 2020).

The best way to enhance water quality is to add probiotics directly; using them as feed additives has also shown some promising outcomes (Monier et al., 2023). Good quality water has lower metabolite concentration, appropriate temperature, sufficient oxygen, and relatively high transparency. *Bacillus* species have demonstrated an outstanding ability to maintain water quality in aquaculture. *Bacillus megaterium* (Hura et al., 2018), *Bacillus subtilis*, *Bacillus mojavensis*, and *Bacillus. cereus* enhance the quality of water in aquaculture ponds by maintaining biological oxygen demand, chemical oxygen demand and nitrogenous compounds in aquaculture (Haan et al., 2020; Table 11.1).

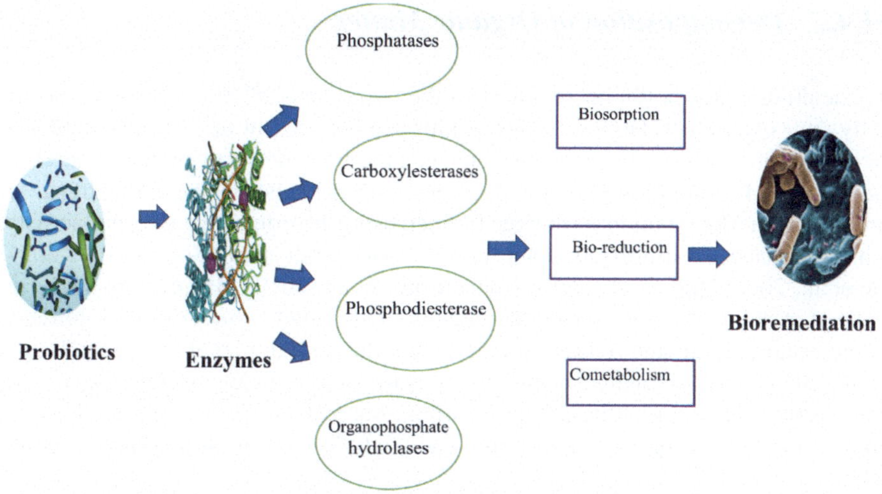

Fig. 11.3 Probiotics role in bioremediation through enzymatic activity

Table 11.1 Probiotics and their role in water quality improvement

S. No.	Probiotics (Bacillus)	Role in water quality improvement	References
1	*Bacillus subtilis* and *Bacillus cereus*	Bioaccumulate cadmium (Cd), zinc (Zn), copper (Cu), and lead (Pb) ions	Hlordzi et al. (2020)
2	*Bacillus megaterium*	Bioaccumulate copper (Cu), iron (Fe), zinc (Zn), and manganese (Mn)	Stefanescu (2015) and Hlordzi et al. (2020)
3	*Bacillus subtilis, Bacillus megaterium* and *Bacillus polymyxa*	Utilizing diesel oil by producing biosurfactants through an emulsification process	Tariq et al., (2016)
4	*Bacillus* thuringiensis and *B. cereus*	Degrade hydrocarbons	Maddela et al., (2015)
5	*Bacillus subtilis* JK-1	Crude oil utilization	Joo et al. (2013)
6	*Bacillus velezensis*	Phosphorus reduction	Thurlow et al., (2019)
7	*Bacillus velezensis* AP193	Reduced nitrate-nitrogen (NO_3^{2-}-N_2) (75%) and total nitrogen (43%) levels	Thurlow et al., (2019)
8	*Bacillus subtilis, Bacillus megaterium* and *Bacillus amyloliquefaciens*	Reduced ammonia (NH_3) levels	Hura et al., (2018) and Hlordzi et al. (2020)
9	*Bacillus subtilis*	Combat lead (Pb) poisoning in *Carassius gibelio* culture	Yin et al., (2018)

11.4.2 Decomposition of Organic Matter

Aquaculture water pollution is linked to the byproducts of fish digestion, specifically their excrement. High amounts of nitrogen are present in excretory products. According to Hai (2015), probiotics enhance the breakdown of organic materials. Leftover food is often associated with organic matter loads in aquaculture; however, probiotic *Bacil*lus stimulates hunger by increasing fish digestive enzyme activity, which enhances feed utilization and lowers waste production (Bolsan et al., 2024; He et al., 2023; Hura et al., 2018; Kuebutornye et al., 2019). Bacillus are primarily utilized in aquaculture to reduce the organic matter load, which recycles nutrients in the water column and reduces sludge accumulation (Hlordzi et al., 2020; Soltani et al., 2019). Organic matter is effectively converted into CO_2 by *Bacillus*, and β- and γ-proteobacteria use this CO_2 as a carbon source. Conversely, different bacteria convert the bulk of organic matter into bacterial biomass or slime (Hlordzi et al., 2020).

11.4.3 Hydrocarbon Degradation by Probiotics

Water pumps, aerators, outboard motors, cars, automated feeders, generators, and lawnmowers are some aquaculture equipment. These machines run on fuel, grease, or other types of oil for component maintenance or as a power source.

Microbes consume oils for food and energy, converting them into more straightforward and harmless molecules (Joo et al., 2013). According to Hlordzi et al. (2020), various bacteria use hydrocarbons as a carbon source, producing biomass, CO_2, and H_2O. Through emulsification, *Bacillus subtilis*, *Bacillus megaterium*, and *Bacillus polymyxa* effectively utilized diesel oil by generating biosurfactants (Tariq et al., 2016). *Bacillus cereus* and *Bacillus thuringiensis* are the most appropriate choices for cleaning up oil-polluted habitats because of their ability to efficiently break down hydrocarbons, according to Maddela et al. (2015). According to Joo et al. (2013) *B. subtilis* JK-1 use crude oil. These data suggest that using Bacillus in aquaculture can aid in the bioremediation of oil spills.

According to Zhao et al. (2011), low concentrations of Mg^{2+}, Fe^{2+}, and Ca^{2+} boost the capacity of *Bacillus fusiformis* to break down petroleum hydrocarbons. That means that a small concentration of metal ions also affects *Bacillus* species' capacity to enhance water quality.

11.4.4 Heavy Metal Mobilization by Probiotics

Heavy metals such as lead (Pb), cadmium (Cd), silver (Ag), chromium (Cr), mercury (Hg), cobalt (Co), zinc (Zn), iron (Fe), and copper (Cu) contaminate the environment and are mainly linked to anthropogenic activity. As a result, these heavy metals accumulate in fish tissues and are transferred to fish consumers (Roy et al., 2024; Stefanescu, 2015).

Bacteria mobilize heavy metals through metabolic processes such as producing organic and inorganic acids, excreting complexing agents, and reducing or oxidizing them (Stefanescu, 2015). Microbial cells, both dead and alive, and their products can bioaccumulate particulate and soluble metals. *Bacillus* species use heavy metals to form spores and decrease heavy metal concentration (Hlordzi et al., 2020; Stefanescu, 2015). *Bacillus subtilis* and *Bacillus cereus* can bioaccumulate cadmium (Cd), zinc (Zn), copper (Cu), and lead (Pb) ions. According to Stefanescu (2015), *Bacillus megaterium* accumulates copper (Cu), iron (Fe), zinc (Zn), and manganese (Mn).

Microbial metabolites can dissolve heavy metals and suspended particles containing heavy metal minerals. Three main processes—biosorption, bio-reduction, and cometabolism—underlie the microbial remediation of heavy metals. Lactic acid bacteria (LAB) find significant application in the bioremediation of heavy metals (Sundaramanickam et al., 2023). Jaafar (2020) adopted the use of probiotic bacteria, *Pediococcus pentosaceus* for the bioremediation of lead and cadmium and found the bacteria to be highly beneficial in eradicating the aforementioned heavy metals. In a related study, Mirza Alizadeh et al. (2021), evaluated in vitro lead nitrate and lead acetate binding capacities as well as removal potentials using five native probiotic bacterial strains (*Lactobacillus paracasei* IRBC-M 10784, *Lactobacillus acidophilus* PTCC-1932, *Lactobacillus rhamnosus* IBRC-M 10782, *Bifidobacterium bifidum* BIA-7, and *Bifidobacterium lactis* BIA-6) and inulin. It was shown that using inulin and local probiotics can be a safe and efficient way to get rid of a variety of harmful compounds, notably Pb.

According to another study by Hayati et al. (2020), probiotic supplements had a superior impact on repairing the reproductive health of fish infected with Cadmium (Cd) than other supplements. *Lactobacillus plantarum* MF042018 showed a high level of resistance against nickel and chromium in a study by Ameen et al., (2020), suggesting that it might be used as a promising biosorbent for the removal of heavy metals from industrial wastewater. Mostafidi et al. (2023) reported the synergistic effect of three LAB strains (*Enterococcus faecium, Lactiplantibacillus plantarum,* and *Limosilactobacillus fermentum*) resulted in improved metal removal (Pb, Cd, and Ni) in aqueous solutions compared to monocultures. The remediation of heavy metals, mainly Cr, Pb, Ni, Cd, Hg, Be, and As by probiotic bacterial species is provided in Table 11.1.

In this context, Probiotics might increase the production of aquaculture by reducing general illnesses and supplying clean water sources. Probiotics in aquaculture systems act through multiple methods to eliminate organic wastes and pollutants (Hassan et al., 2022).

Microorganisms can reduce hazardous contaminants in aquatic systems with minimal operational costs (Lu et al., 2012). Probiotics are a safe supplement used in aquaculture to purify water, prevent pathogenic microorganisms, improve feed utilization, and enhance host health (Olmos et al., 2011; Selim & Reda, 2015). This effectively reduced aquaculture's dependency on antibiotics (Li et al., 2022).

Table 11.2 Heavy metal bioremediation by various probiotic bacterial species

Heavy metal	Genus	Species	Reference
Cr, Pb, Ni, Cd	Bacillus	Coagulans	Goyal et al., (2019)
Cu		Siamensis	Selta et al. (2022)
Pb, Zn		Clausii	
Pb, Cd, Ni	Lactiplantibacillus	Plantarum	Mostafidi et al., (2023)
	Limosilactobacillus	Fermentum	
	Enterococcus	Faecium	
Cd, Pb	Pediococcus	pentosaceu	Jaafar (2020) and Hasan et al., (2022)
	Lactococcus	Lactis	Hasan et al., (2022)
	Lactobacillus	fermentum, plantarum	Kirillova et al. (2017), Goyal et al., (2019) and Ameen et al., (2020)
Pb		Plantarum	Shaheen et al., (2019)
		Brevis	Afraz et al., (2021)
Cd, Pb		Paracasei	
Cd, Hg, Pb, Be, As		Acidophilus, paracasei, rhamnosus, reuteri	Fang et al., (2018) and Goyal et al., (2019)
Cd, Pb		Acidophilus, delbrueckii subsp. bulgaricus	Hasan et al., (2022)
	Lacticaseibacillus	Casei subsp. casei	
Cd, Hg, Pb, Be, As	Bifidobacterium	Longum	Goyal et al., (2019)
Pb	Leuconostoc	Mesenteroides	Yi et al., (2017)

11.5 Metagenomic Insights: Unveiling the Genetic Arsenal of Probiotic Microbes in Water Quality Management

Water quality management is a critical aspect of ensuring public health and environmental sustainability (Chapman & Sullivan, 2022). Water quality degradation poses significant challenges to ecosystems and human health worldwide (UNESCO, 2021). Microbial communities play a crucial role in maintaining water quality by participating in various biogeochemical processes, including nutrient cycling and pollutant degradation (Zhang et al., 2023). Probiotic traditionally associated with health-promoting effects in humans and animals, have emerged as potential candidates for water quality improvement due to their diverse metabolic capabilities and ecological functions (Anee et al., 2021; Hassan et al., 2022; Maftei et al., 2024; Singh et al., 2022; Tabassum et al., 2021; Zorriehzahra et al., 2016). Metagenomic analysis which involves the high-throughput sequencing of DNA extracted directly from environmental samples, offers a comprehensive approach to explore the genetic repertoire of microbial communities and their functional potential (Oulas et al., 2015; Patel et al., 2022a, 2022b; Zhang et al., 2021) in various environments, including water systems

(Brar et al., 2023; Maddheshiya et al., 2024; Nam et al., 2023). By examining the genetic composition and metabolic pathways of probiotic microorganisms present in water bodies, insights can be gained into their roles in nutrient cycling, pollutant degradation, and pathogen suppression. Furthermore, understanding the interactions between probiotic microbes and indigenous microbial communities can aid in designing targeted strategies for enhancing water quality and ecosystem resilience.

Metagenomic studies have revealed a rich diversity of probiotic microbes inhabiting various aquatic environments, including freshwater bodies (Vijayan et al., 2023). By employing next-generation sequencing technologies coupled with bioinformatics tools, researchers have been able to characterize the taxonomic composition and functional potential of these microbial assemblages. For instance, studies have identified members of the genera *Lactobacillus, Bifidobacterium*, and *Bacillus* as prevalent probiotic taxa in water environments, capable of metabolizing organic matter, degrading pollutants, and inhibiting the growth of pathogenic microorganisms (Jha et al., 2020).

The genetic repertoire of probiotic microbes harbors an array of functional genes involved in nutrient metabolism, stress tolerance, and antimicrobial activity, which contribute to their ecological fitness (Mazhar et al., 2023) and potential applications in water quality management. Metagenomic analysis allows for the identification of key metabolic pathways and enzymes associated with nutrient cycling (e.g., nitrogen fixation, phosphorus metabolism) (Cremers et al., 2022; Li et al., 2018) and pollutant degradation (e.g., hydrocarbon degradation, antibiotic resistance) (Zhang et al., 2019). Moreover, probiotic microbes exhibit competitive interactions with indigenous microbial communities, modulating microbial diversity and ecosystem stability. Understanding the ecological roles of probiotic microbes in water systems is essential for harnessing their beneficial effects and mitigating potential risks associated with their introduction.

The insights gained from metagenomic studies of probiotic microbes hold significant implications for water quality management strategies. By leveraging the metabolic capabilities of probiotic microbes, it is possible to develop bioremediation approaches for mitigating water pollution caused by organic contaminants and emerging pollutants. Additionally, probiotic supplementation in water treatment processes could enhance the removal of pathogens and improve overall water quality. However, careful consideration must be given to the potential ecological impacts of introducing probiotic microbes into natural ecosystems, including unintended consequences on microbial community dynamics and ecosystem functioning. Integrating metagenomic data with ecological modeling approaches can aid in predicting the long-term effects of probiotic interventions and optimizing management strategies for sustainable water quality improvement.

11.6 Future Perspectives: Emerging Frontiers in Probiotic Research for Sustainable Water and Wastewater Management

Research in probiotics for sustainable water and wastewater management is a burgeoning area with immense potential. Here are some potential avenues for future exploration:

Bioaugmentation with Probiotics includes using probiotics to enhance microbial communities' performance in wastewater treatment plants (Balthazar et al., 2022). Probiotics could be introduced to augment specific functions, such as nutrient removal or organic matter degradation, leading to more efficient treatment processes. Exploring the integration of probiotics into microbial fuel cells (MFCs) for simultaneous wastewater treatment and energy generation (Kumar et al., 2022). Probiotic strains could be selected or engineered to maximize electron transfer efficiency, thereby enhancing electricity production while treating wastewater (Basit et al., 2021). Studying the potential of probiotics in the bioremediation of emerging contaminants, such as pharmaceuticals, personal care products, and microplastics, which are often present in wastewater (Uthra et al., 2023). Probiotic strains could be engineered to degrade or sequester these contaminants, thereby improving water quality. Investigating the use of probiotic consortia for nutrient recovery from wastewater, particularly phosphorus and nitrogen. Probiotics could be employed to selectively accumulate these nutrients from wastewater, allowing for their subsequent recovery as valuable fertilizers or agricultural amendments (Puyol et al., 2017).

Researching the potential of probiotic biofilms to control the formation of pathogenic biofilms in water distribution systems and wastewater treatment infrastructure. Probiotic biofilms could competitively exclude pathogenic species, thereby reducing the risk of biofouling and microbial contamination (Ibekwe & Murinda, 2019). Exploring the use of probiotics as biosensors for monitoring water quality parameters, such as pH, dissolved oxygen, and pollutant concentrations. Engineered probiotic strains could be designed to produce detectable signals in response to specific environmental conditions, providing real-time insights into water quality (Moraskie et al., 2021). Investigating the complex interactions between probiotic microorganisms and the native microbial communities in engineered ecosystems, such as wastewater treatment plants and constructed wetlands. Understanding these symbiotic relationships could facilitate the development of more resilient and sustainable treatment systems (Wang et al., 2022a, 2022b).

Studying the potential of probiotics to enhance the safety and quality of reclaimed wastewater for non-potable reuse applications, such as irrigation, industrial processes, and environmental restoration. Probiotic treatments could mitigate the risks associated with microbial contamination and ensure compliance with water quality standards. Exploring the application of ecological engineering principles in conjunction with probiotics to design innovative water treatment systems that mimic natural processes (Khanjani et al., 2023). This could involve the development of engineered microbial communities capable of self-regulation and adaptation to fluctuating environmental conditions. Conducting comprehensive life cycle assessments to evaluate the environmental sustainability and economic viability of probiotic-based water and wastewater treatment technologies (Tarricone et al., 2020). This holistic approach would consider the entire life cycle of these systems, from production and operation to end-of-life disposal or reuse. By pursuing research along these lines, scientists and engineers can unlock the full potential of probiotics for sustainable water and wastewater management, ultimately contributing to preserving water resources and protecting public health and ecosystems.

11.7 Conclusion

The chapters presented in this book offer a comprehensive exploration of the microbial ecology of water environments and the potential applications of probiotics in water and wastewater management. By comparing the previous data on the effectiveness of probiotics in various contexts, we can appreciate the advancements and current capabilities in this field. In earlier studies, the primary focus was on the basic understanding of microbial interactions and the general benefits of probiotics in aquatic systems. Recent chapters have expanded on these foundations by providing detailed insights into using probiotics for contaminant remediation, highlighting specific enzymes that play crucial roles in breaking down pollutants and improving water quality. The bioengineering approaches discussed in the latest chapters represent significant progress from earlier, more theoretical models to practical applications that enhance water purification processes. Additionally, the integration of metagenomic techniques has allowed for a deeper understanding of the genetic arsenal of probiotic microbes, offering new perspectives on how these organisms can be optimized for specific bioremediation tasks. Comparative analysis of previous data with recent findings shows a clear trajectory towards more sophisticated and targeted applications of probiotics. For instance, the ability to degrade hydrocarbons, solubilize phosphates, and detoxify heavy metals through specific enzymatic activities was once a promising hypothesis, but it is now supported by robust empirical data and practical examples. Looking to the future, emerging frontiers in probiotic research hold great promise for sustainable water and wastewater management. The emphasis on continued exploration and innovation in this field underscores the importance of interdisciplinary collaboration. Through ongoing research efforts, we can further advance our understanding and utilization of probiotics to benefit water quality and environmental sustainability. This progression from fundamental research to applied science marks a significant milestone in the quest for cleaner and healthier water systems.

References

Afraz, V., Younesi, H., Bolandi, M., & Hadiani, M. R. (2021). Assessment of resistance and biosorption ability of *Lactobacillus paracasei* to remove lead and cadmium from aqueous solution. *Water Environment Research, 93*(9), 1589–1599.

Ahmad, A., Banat, F., Alsafar, H., & Hasan, S. W. (2022). Algae biotechnology for industrial wastewater treatment, bioenergy production, and high-value bioproducts. *Science of the Total Environment, 806*, 150585.

Ameen, F. A., Hamdan, A. M., & El-Naggar, M. Y. (2020). Assessment of the heavy metal bioremediation efficiency of the novel marine lactic acid bacterium, *Lactobacillus plantarum* MF042018. *Scientific Reports, 10*(1), 314.

Anand, U., Dey, S., Bontempi, E., Ducoli, S., Vethaak, A. D., Dey, A., & Federici, S. (2023). Biotechnological methods to remove microplastics: A review. *Environmental Chemistry Letters, 21*(3), 1787–1810.

Ananda Raja, R., & Jithendran, K. P. (2015). Aquaculture disease diagnosis and health management. In *Advances in marine and brackishwater aquaculture* (pp. 247–255). Springer.

Anee, I. J., Alam, S., Begum, R. A., Shahjahan, R. M., & Khandaker, A. M. (2021). The role of probiotics on animal health and nutrition. *Journal of Basic and Applied Zoology, 82*, 1–16.

Arif, Z. U., Khalid, M. Y., Sheikh, M. F., Zolfagharian, A., & Bodaghi, M. (2022). Biopolymeric sustainable materials and their emerging applications. *Journal of Environmental Chemical Engineering, 10*(4), 108159.

Balcázar, J. L., De Blas, I., Ruiz-Zarzuela, I., Cunningham, D., Vendrell, D., & Múzquiz, J. L. (2006). The role of probiotics in aquaculture. *Veterinary Microbiology, 114*(3–4), 173–186.

Balthazar, C. F., Guimarães, J. F., Coutinho, N. M., Pimentel, T. C., Ranadheera, C. S., Santillo, A., Albenzio, M., Cruz, A. G., & Sant'Ana, A. S. (2022). The future of functional food: Emerging technologies application on prebiotics, probiotics and postbiotics. *Comprehensive Reviews in Food Science and Food Safety, 21*(3), 2560–2586.

Basit, A., Shah, S. T., Ullah, I., Muntha, S. T., & Mohamed, H. I. (2021). Microbe-assisted phytoremediation of environmental pollutants and energy recycling in sustainable agriculture. *Archives of Microbiology, 203*(10), 5859–5885.

Basuki, H., Chaurasia, I., Husen, K., Kasena, G., Kim, Y. E., Park, H. S., & Wanandi, R. (2020). PET-degrading cyanobacteria for marine environments. *Biotreks: Peer Reviewed High School Synthetic Biology.*

Bhakta, J. N., Ohnishi, K., Munekage, Y., Iwasaki, K., & Wei, M. Q. (2012). Characterization of lactic acid bacteria-based probiotics as potential heavy metal sorbents. *Journal of Applied Microbiology, 112*(6), 1193–1206.

Bletz, M. C., Loudon, A. H., Becker, M. H., Bell, S. C., Woodhams, D. C., Minbiole, K. P., & Harris, R. N. (2013). Mitigating amphibian chytridiomycosis with bioaugmentation: Characteristics of effective probiotics and strategies for their selection and use. *Ecology Letters, 16*(6), 807–820.

Bocanegra-García, V., Martínez-Montoya, H., Tamayo-Ordoñez, M. C., Estrada-Camacho, I., Aguilar-Salazar, A., Guaní-Sánchez, G. D., Sosa-Santillán, G. D. J., & Acosta-Cruz, E. (2023). Freshwater microbiology: Recent updates and prospects. In *Current status of fresh water microbiology* (pp. 1–27). Springer.

Bolsan, A. C., Hollas, C. E., Rodrigues, H. C., Cavaler, J. P., Lazaroto, A. C., Gabiatti, N. C., Bortoli, M., Johann, G., Antes, F. G., Kunz, A., & Edwiges, T. (2024). Challenges and operational strategies to achieve partial nitrification in biological wastewater treatment: a review. *International Journal of Environmental Research, 18*(2), 1–18.

Brar, B., Kumar, R., Sharma, D., Sharma, A. K., Thakur, K., Mahajan, D., & Kumar, R. (2023). Metagenomic analysis reveals diverse microbial community and potential functional roles in Baner rivulet, India. *Journal of Genetic Engineering and Biotechnology, 21*(1), 147.

Chai, W. S., Tan, W. G., Munawaroh, H. S. H., Gupta, V. K., Ho, S. H., & Show, P. L. (2021). Multifaceted roles of microalgae in the application of wastewater biotreatment: A review. *Environmental Pollution, 269*, 116236.

Chapman, D. V., & Sullivan, T. (2022). The role of water quality monitoring in the sustainable use of ambient waters. *One Earth, 5*(2), 132–137.

Chizhayeva, A., Amangeldi, A., Oleinikova, Y., Alybaeva, A., & Sadanov, A. (2022). Lactic acid bacteria as probiotics in sustainable development of aquaculture. *Aquatic Living Resources, 35*, 10.

Corre, V. L., & Janeo, R. L. (2002). Use of probiotics for sustainable shrimp culture in Southeast Asia. *Fisheries Science, 68*(sup1), 835–838.

Cremers, G., Jetten, M. S., Op den Camp, H. J., & Lücker, S. (2022). Metascan: METabolic analysis, SCreening and annotation of metagenomes. *Frontiers in Bioinformatics, 2*, 861505.

da Silva Vilar, D., Cruz, I. A., Torres, N. H., Figueiredo, R. T., de Melo, L., de Resende, I. T. F., Eguiluz, K. I. B., Bharagava, R. N., & Ferreira, L. F. R. (2019). Agro-industrial wastes: environmental toxicology, risks, and biological treatment approaches. *Environmental Contaminants: Ecological Implications and Management*, 1–23.

Devaraja, T., Banerjee, S., Yusoff, F., Shariff, M., & Khatoon, H. (2013). A holistic approach for selection of Bacillus spp. as a bioremediator for shrimp postlarvae culture. *Turkish Journal of Biology, 37*(1), 92–100.

Zhao, D. F., Wu, W. L., Zhang, Y. B., Liu, Q. Y., Yang, H., & Zhao, C. (2011). Study on the isolation, identification of a petroleum hydrocarbon degrading *Bacillus fusiformis* sp. bacteria and its influence of environmental factors on the degradation efficiency. *China Petroleum Processing & Petrochemical Technology, 13*(4), 74.

El-Saadony, M. T., Alagawany, M., Patra, A. K., Kar, I., Tiwari, R., Dawood, M. A., Dhama, K., & Abdel-Latif, H. M. (2021). The functionality of probiotics in aquaculture: An overview. *Fish & Shellfish Immunology, 117*, 36–52.

Fang, S., Cheng, D., Huang, Y., Hsu, T., & Lu, H. H. S. A. (2018). Pilot study of the influence of probiotics on hair toxic element levels after long-term supplement with different lactic acid bacteria strains. *Journal of Probiotics & Health, 6*, 1–8.

Giri, S. S., Yun, S., Jun, J. W., Kim, H. J., Kim, S. G., Kang, J. W., Kim, S. W., Han, S. J., Sukumaran, V., & Park, S. C. (2018). Therapeutic effect of intestinal autochthonous *Lactobacillus reuteri* P16 against waterborne lead toxicity in *Cyprinus carpio. Frontiers in Immunology, 9*, 1824.

Goyal, P., Belapurkar, P., & Kar, A. (2019). A review on in vitro and in vivo bioremediation potential of environmental and probiotic species of Bacillus and other probiotic microorganisms for two heavy metals, cadmium and nickel. *Biosciences Biotechnology Research Asia, 16*(1), 01–13.

Gupta, H. (2018). PAH determination in effluent and sludge samples of paper industry. *Environmental Technology & Innovation, 9*, 115–121. https://doi.org/10.1016/j.eti.2017.11.009

Gupta, H., Dhiman, S., Kumar, A., Paliwal, S. K., Khan, G., Singh, H., Mishra, A., & Mishra, A. K. (2024). Photo- and bio-degradation of selected persistent organic pollutants. *Separation & Purification Reviews*, 1–14. https://doi.org/10.1080/15422119.2024.2332953

Haan, T. Y., Mazuki, N. I. M., Chun, H. K., & Al-Rajabi, M. M. H. (2020). Innovation with the use of probiotics as an eco-friendly tool for sewage and palm oil mill effluent treatment. *International Journal of Nanoelectronics & Materials, 13*.

Hai, N. V. (2015). The use of probiotics in aquaculture. *Journal of Applied Microbiology, 119*(4), 917–935.

Hasan, M. S., Islam, M. Z., Liza, R. I., Sarker, M. A. H., Islam, M. A., & Harun-ur-Rashid, M. (2022). Novel probiotic lactic acid bacteria with in vitro bioremediation potential of toxic lead and cadmium. *Current Microbiology, 79*(12), 387.

Hassan, M. A., Fathallah, M. A., Elzoghby, M. A., Salem, M. G., & Helmy, M. S. (2022). Influence of probiotics on water quality in intensified *Litopenaeus vannamei* ponds under minimum-water exchange. *AMB Express, 12*(1), 22.

Hayati, A. L., Supriyanto, A. G., Suhargo, L. I., Hayaza, S. U., & Ayubu, A. (2020). Bioremediation potency of probiotics on cadmium pollution to improve fish reproductive health. *Pollution Research, 39*(4), 980–984.

He, H., Li, Y., Zhang, L., Ding, Z., & Shi, G. (2023). Understanding and application of Bacillus nitrogen regulation: A synthetic biology perspective. *Journal of Advanced Research, 49*, 1–14.

Hemarajata, P., & Versalovic, J. (2013). Effects of probiotics on gut microbiota: Mechanisms of intestinal immunomodulation and neuromodulation. *Therapeutic Advances in Gastroenterology, 6*(1), 39–51.

Hlordzi, V., Kuebutornye, F. K., Afriyie, G., Abarike, E. D., Lu, Y., Chi, S., & Anokyewaa, M. A. (2020). The use of Bacillus species in maintenance of water quality in aquaculture: A review. *Aquaculture Reports, 18*, 100503.

Hu, Y., Xiao, R., Wang, Y., Li, J., Guo, C., Bai, J., Zhang, L., Zhang, K., Jorquera, M. A., Manquian, J., & Pan, W. (2024). Distribution of organophosphorus pesticides and its potential connection with probiotics in sediments of a shallow freshwater lake. *Journal of Contaminant Hydrology, 261*, 104306.

Hura, M. U. D., Zafar, T., Borana, K., Prasad, J. R., & Iqbal, J. (2018). Effect of commercial probiotic *Bacillus megaterium* on water quality in composite culture of major carps. *International Journal of Current Agricultural Sciences, 8*(1), 268–273.

Ibekwe, A. M., & Murinda, S. E. (2019). Linking microbial community composition in treated wastewater with water quality in distribution systems and subsequent health effects. *Microorganisms, 7*(12), 660.

Jaafar, R. (2020). Bioremediation of lead and cadmium and the strive role of *Pediococcus pentosaceus* probiotic. *Iraqi Journal of Veterinary Sciences, 34*(1), 51–57.

Jha, R., Das, R., Oak, S., & Mishra, P. (2020). Probiotics (direct-fed microbials) in poultry nutrition and their effects on nutrient utilization, growth and laying performance, and gut health: A systematic review. *Animals, 10*(10), 1863.

Joo, M. H., Ki Joo, M. H., & Kim, J. Y. (2013). Characteristics of crude oil biodegradation by biosurfactant-producing bacterium *Bacillus subtilis* JK-1. *Journal of the Korean Society for Applied Biological Chemistry, 56*, 193–200.

Kamilya, D., & Devi, W. M. (2022). Bacillus probiotics and bioremediation: An aquaculture perspective. *Bacilli in agrobiotechnology: Plant stress tolerance, bioremediation, and bioprospecting* (pp. 335–347). Springer.

Khanjani, M. H., da Silva, L. O. B., Fóes, G. K., do Nascimento Vieira, F., Poli, M. A., Santos, M., & Emerenciano, M. G. C. (2023). Synbiotics and aquamimicry as alternative microbial-based approaches in intensive shrimp farming and biofloc: Novel disruptive techniques or complementary management tools? A scientific-based overview. *Aquaculture, 567*, 739273.

Khoo, K. S., Chia, W. Y., Chew, K. W., & Show, P. L. (2021). Microalgal-bacterial consortia as future prospect in wastewater bioremediation, environmental management and bioenergy production. *Indian Journal of Microbiology, 61*(3), 262–269.

Kirillova, A. V., Danilushkina, A. A., Irisov, D. S., Bruslik, N. L., Fakhrullin, R. F., Zakharov, Y. A., Bukhmin, V. S., & Yarullina, D. R. (2017). Assessment of resistance and bioremediation ability of Lactobacillus strains to lead and cadmium. *International Journal of Microbiology, 2017*.

Kosolapov, D. B., Kuschk, P., Vainshtein, M. B., Vatsourina, A. V., Wiessner, A., Kästner, M., & Müller, R. A. (2004). Microbial processes of heavy metal removal from carbon-deficient effluents in constructed wetlands. *Engineering in Life Sciences, 4*(5), 403–411. *Immunology, 9*, 346123.

Kuebutornye, F. K., Abarike, E. D., & Lu, Y. (2019). A review on the application of Bacillus as probiotics in aquaculture. *Fish & Shellfish Immunology, 87*, 820–828.

Kumar, S., Srivastava, P. K., Kumar, V., & Seth, C. S. (2024). Biofloc technology: A sustainable approach towards wastewater utilization and fish production. *Lakes & Reservoirs: Research & Management, 29*(1), e12449.

Kumar, S. D., Yasasve, M., Karthigadevi, G., Aashabharathi, M., Subbaiya, R., Karmegam, N., & Govarthanan, M. (2022). Efficiency of microbial fuel cells in the treatment and energy recovery from food wastes: Trends and applications-a review. *Chemosphere, 287*, 132439.

Li, S., Yang, Y., Yang, S., Zheng, H., Zheng, Y., Jun, M., Nagarajan, D., Varjani, S., & Chang, J. S. (2023). Recent advances in biodegradation of emerging contaminants-microplastics (MPs): Feasibility, mechanism, and future prospects. *Chemosphere*, 138776.

Li, X., Wang, A., Wan, W., Luo, X., Zheng, L., He, G., Huang, D., Chen, W., & Huang, Q. (2021). High salinity inhibits soil bacterial community mediating nitrogen cycling. *Applied and Environmental Microbiology, 87*(21), e01366-e1421.

Li, X., Wang, T., Fu, B., & Mu, X. (2022). Improvement of aquaculture water quality by mixed Bacillus and its effects on microbial community structure. *Environmental Science and Pollution Research, 29*(46), 69731–69742.

Li, Y., Jing, H., Xia, X., Cheung, S., Suzuki, K., & Liu, H. (2018). Metagenomic insights into the microbial community and nutrient cycling in the western subarctic Pacific Ocean. *Frontiers in Microbiology, 9*, 337314.

Liu, C. X., Xu, Q. M., Yu, S. C., Cheng, J. S., & Yuan, Y. J. (2020). Bio-removal of tetracycline antibiotics under the consortium with probiotics *Bacillus clausii* T and *Bacillus amyloliquefaciens* producing biosurfactants. *Science of the Total Environment, 710*, 136329.

Lu, L., Tan, H., Luo, G., & Liang, W. (2012). The effects of *Bacillus subtilis* on nitrogen recycling from aquaculture solid waste using heterotrophic nitrogen assimilation in sequencing batch reactors. *Bioresource Technology, 124*, 180–185.

Maddela, N. R., Masabanda, M., & Leiva-Mora, M. (2015). Novel diesel-oil-degrading bacteria and fungi from the Ecuadorian Amazon rainforest. *Water Science and Technology, 71*(10), 1554–1561.

Maddheshiya, R., Dhiman, S., Sahu, D. M., & Gupta, H. (2024). Chapter 16—Role of phytoremediation and metagenomics in healing the wounded (polluted) planet: Ecofriendly and sustainable approach. In I. Tyagi, V. Kumar, & R. R. Karri (Eds.), *Environmental metagenomics, water quality and suggested remediation measures of polluted waters* (pp. 201–216). Elsevier. https://doi.org/10.1016/B978-0-443-13659-7.00009-6

Madhav, S., Kumar, A., Kushawaha, J., Ahamad, A., Singh, P., & Dwivedi, S. B. (2020). Geochemical assessment of groundwater quality in Keonjhar City, Odisha, India. *Sustainable Water Resources Management, 6*, 1–11.

Maftei, N. M., Raileanu, C. R., Balta, A. A., Ambrose, L., Boev, M., Marin, D. B., & Lisa, E. L. (2024). The potential impact of probiotics on human health: An update on their health-promoting properties. *Microorganisms, 12*(2), 234.

Martínez Cruz, P., Ibáñez, A. L., Monroy Hermosillo, O. A., & Ramírez Saad, H. C. (2012). Use of probiotics in aquaculture. *International Scholarly Research Notices, 2012.*

Mazhar, S., Khokhlova, E., Colom, J., Simon, A., Deaton, J., & Rea, K. (2023). In vitro and in silico assessment of probiotic and functional properties of *Bacillus subtilis* DE111®. *Frontiers in Microbiology, 13*, 1101144.

Mirza Alizadeh, A., Hosseini, H., Mohseni, M., Eskandari, S., Sohrabvandi, S., Hosseini, M. J., Tajabadi-Ebrahimi, M., Mohammadi-Kamrood, M., & Nahavandi, S. (2021). Analytic and chemometric assessments of the native probiotic bacteria and inulin effects on bioremediation of lead salts. *Journal of the Science of Food and Agriculture, 101*(12), 5142–5153.

Mishra, S., & Acharya, S. (2021). A brief overview on probiotics: The health friendly microbes. *Biomedical and Pharmacology Journal, 14*(4), 1869–1880.

Monier, M. N., Kabary, H., Elfeky, A., Saadony, S., El-Hamed, N. N. A., Eissa, M. E., & Eissa, E. S. H. (2023). The effects of Bacillus species probiotics (*Bacillus subtilis* and *B. licheniformis*) on the water quality, immune responses, and resistance of whiteleg shrimp (*Litopenaeus vannamei*) against *Fusarium solani* infection. *Aquaculture International, 31*(6), 3437–3455.

Moraskie, M., Roshid, M. H. O., O'Connor, G., Dikici, E., Zingg, J. M., Deo, S., & Daunert, S. (2021). Microbial whole-cell biosensors: Current applications, challenges, and future perspectives. *Biosensors and Bioelectronics, 191*, 113359.

Mostafidi, M., Sanjabi, M. R., Mojgani, N., Eskandari, S., & Arbabi Bidgoli, S. (2023). Heavy metal bioremediation potential of autochthonous lactic acid bacteria for use in edible leafy vegetables. *Journal of Food Quality, 2023.*

Nam, N. N., Do, H. D. K., Loan Trinh, K. T., & Lee, N. Y. (2023). Metagenomics: An effective approach for exploring microbial diversity and functions. *Foods, 12*(11), 2140.

Neethirajan, S., Ragavan, V., Weng, X., & Chand, R. (2018). Biosensors for sustainable food engineering: Challenges and perspectives. *Biosensors, 8*(1), 23.

Novak Babič, M., Gostinčar, C., & Gunde-Cimerman, N. (2020). Microorganisms populating the water-related indoor biome. *Applied Microbiology and Biotechnology, 104*, 6443–6462.

Okeke, E. S., Chukwudozie, K. I., Nyaruaba, R., Ita, R. E., Oladipo, A., Ejeromedoghene, O., Atakpa, E. O., Agu, C. V., & Okoye, C. O. (2022). Antibiotic resistance in aquaculture and aquatic organisms: A review of current nanotechnology applications for sustainable management. *Environmental Science and Pollution Research, 29*(46), 69241–69274.

Olmos, J., Ochoa, L., Paniagua-Michel, J., & Contreras, R. (2011). Functional feed assessment on *Litopenaeus vannamei* using 100% fish meal replacement by soybean meal, high levels of complex carbohydrates and Bacillus probiotic strains. *Marine Drugs, 9*(6), 1119–1132.

Oulas, A., Pavloudi, C., Polymenakou, P., Pavlopoulos, G. A., Papanikolaou, N., Kotoulas, G., Arvanitidis, C., & Iliopoulos, L. (2015). Metagenomics: Tools and insights for analyzing next-generation sequencing data derived from biodiversity studies. *Bioinformatics and Biology Insights, 9*, BBI-S12462.

Panigrahi, A., Das, R. R., Sivakumar, M. R., Saravanan, A., Saranya, C., Sudheer, N. S., Vasagam, K. K., Mahalakshmi, P., Kannappan, S., & Gopikrishna, G. (2020). Bio-augmentation of heterotrophic bacteria in biofloc system improves growth, survival, and immunity of Indian white shrimp *Penaeus indicus. Fish & Shellfish Immunology, 98*, 477–487.

Panigrahi, A., Esakkiraj, P., Das, R. R., Saranya, C., Vinay, T. N., Otta, S. K., & Shekhar, M. S. (2021). Bioaugmentation of biofloc system with enzymatic bacterial strains for high health and production performance of *Penaeus indicus. Scientific Reports, 11*(1), 13633.

Parker, B., & Cruz, M. (2024). Augmented polycultures: Scaling up algal ecosystems and design of a biofouling aesthetic. *Critical Plant Studies*, 234.

Patel, A. K., Singhania, R. R., Albarico, F. P. J. B., Pandey, A., Chen, C. W., & Dong, C. D. (2022a). Organic wastes bioremediation and its changing prospects. *Science of the Total Environment, 824*, 153889.

Patel, T., Chaudhari, H. G., Prajapati, V., Patel, S., Mehta, V., & Soni, N. (2022b). A brief account on enzyme mining using metagenomic approach. *Frontiers in Systems Biology, 2*, 1046230.

Pereira, W. A., Piazentin, A. C. M., de Oliveira, R. C., Mendonça, C. M. N., Tabata, Y. A., Mendes, M. A., Fock, R. A., Makiyama, E. N., Corrêa, B., Vallejo, M., & Villalobos, E. F. (2022). Bacteriocinogenic probiotic bacteria isolated from an aquatic environment inhibit the growth of food and fish pathogens. *Scientific Reports, 12*(1), 5530.

Prasad, S., Malav, L. C., Choudhary, J., Kannojiya, S., Kundu, M., Kumar, S., & Yadav, A. N. (2021). Soil microbiomes for healthy nutrient recycling. *Current trends in microbial biotechnology for sustainable agriculture* (pp. 1–21).

Puyol, D., Batstone, D. J., & Krömer, J. O. (2017). Resource recovery from wastewater by biological technologies: Opportunities, challenges, and prospects. *Frontiers in Microbiology, 7*, 218193.

Rabetafika, H. N., Razafindralambo, A., Ebenso, B., & Razafindralambo, H. L. (2023). Probiotics as antibiotic alternatives for human and animal applications. *Encyclopedia, 3*(2), 561–581.

Rai, M. S. K. (2024). *Principles and practices in water and wastewater engineering*. Academic Guru Publishing House.

Nițu, R. M. (2024). The economic efficiency of waste management policies. *Theoretical and Applied Economics, 31*(1 (638), Spring), 169–178.

Ronquillo, M. G., & Hernandez, J. C. A. (2017). Antibiotic and synthetic growth promoters in animal diets: Review of impact and analytical methods. *Food Control, 72*, 255–267.

Roy, R., Samanta, S., Pandit, S., Naaz, T., Banerjee, S., Rawat, J. M., Chaubey, K. K., & Saha, R. P. (2024). An overview of bacteria-mediated heavy metal bioremediation strategies. *Applied Biochemistry and Biotechnology, 196*(3), 1712–1751.

Saiyad, M., Shah, N., Joshipura, M., Dwivedi, A., & Pillai, S. (2024). Modified biopolymers in wastewater treatment: A review. *Materials Today: Proceedings*.

Salih, A. H. M., Patra, I., Sivaraman, R., Alhamzawi, R., Khalikov, K. M., Al-Qaim, Z. H., Golgouneh, S., Jawad, M. A., Adhab, A. H., Vázquez-Cárdenas, A. L., & Abarghouei, S. (2023).

The probiotic *Lactobacillus sakei* subsp. Sakei and hawthorn extract supplements improved growth performance, digestive enzymes, immunity, and resistance to the pesticide acetamiprid in common carp (*Cyprinus carpio*). *Aquaculture Nutrition, 2023.*

Saravanan, A., Kumar, P. S., Jeevanantham, S., Karishma, S., Tajsabreen, B., Yaashikaa, P. R., & Reshma, B. (2021). Effective water/wastewater treatment methodologies for toxic pollutants removal: Processes and applications towards sustainable development. *Chemosphere, 280*, 130595.

Sarlak, Z., Khosravi-Darani, K., Rouhi, M., Garavand, F., Mohammadi, R., & Sobhiyeh, M. R. (2021). Bioremediation of organophosphorus pesticides in contaminated foodstuffs using probiotics. *Food Control, 126*, 108006.

Sayler, G. S., & Ripp, S. (2000). Field applications of genetically engineered microorganisms for bioremediation processes. *Current Opinion in Biotechnology, 11*(3), 286–289.

Scanlon, B. R., Fakhreddine, S., Rateb, A., de Graaf, I., Famiglietti, J., Gleeson, T., Grafton, R. Q., Jobbagy, E., Kebede, S., Kolusu, S. R., & Konikow, L. F. (2023). Global water resources and the role of groundwater in a resilient water future. *Nature Reviews Earth & Environment, 4*(2), 87–101.

Selim, K. M., & Reda, R. M. (2015). Improvement of immunity and disease resistance in the Nile tilapia, *Oreochromis niloticus*, by dietary supplementation with *Bacillus amyloliquefaciens. Fish & Shellfish Immunology, 44*(2), 496–503.

Selta, D. R. F., Sabna, B. S., Mahendran, R., Thankappan, B., Muthusamy, G., Balasubramanian, V., & Jayaraman, A. (2022). *Bacillus siamensis* I7B strain as heavy metal quencher and probiotics; isolated from gut of *Stolepherous commersonni.*

Shaheen, A., El-Baghdady, K. Z., & Zakaria, A. (2019). Isolation and identification of potent probiotics with high lead removal capability. *Egyptian Journal of Botany, 59*(1), 95–105.

Shamsuzzaman, M. M., & Biswas, T. K. (2012). Aqua chemicals in shrimp farm: A study from south-west coast of Bangladesh. *The Egyptian Journal of Aquatic Research, 38*(4), 275–285.

Shi, J., Zhang, B., Liu, J., Fang, Y., & Wang, A. (2022). Spatiotemporal dynamics in microbial communities mediating biogeochemical cycling of nutrients across the Xiaowan Reservoir in Lancang River. *Science of the Total Environment, 813*, 151862.

Sims, J. L., Sims, R. C., & Ginn, J. S. (1992). Bioengineering for water cleanup: State-of-the-art assessment.

Singh, R. P., Shadan, A., & Ma, Y. (2022). Biotechnological applications of probiotics: A multifarious weapon to disease and metabolic abnormality. *Probiotics and Antimicrobial Proteins, 14*(6), 1184–1210.

Sola-Oladokun, B., Culligan, E. P., & Sleator, R. D. (2017). Engineered probiotics: Applications and biological containment. *Annual Review of Food Science and Technology, 8*, 353–370.

Soltani, M., Ghosh, K., Hoseinifar, S. H., Kumar, V., Lymbery, A. J., Roy, S., & Ringø, E. (2019). Genus Bacillus, promising probiotics in aquaculture: Aquatic animal origin, bio-active components, bioremediation and efficacy in fish and shellfish. *Reviews in Fisheries Science & Aquaculture, 27*(3), 331–379.

Srednicka, P., Juszczuk-Kubiak, E., Wójcicki, M., Akimowicz, M., & Roszko, M. (2021). Probiotics as a biological detoxification tool of food chemical contamination: A review. *Food and Chemical Toxicology, 153*, 112306.

Srithongouthai, S., & Tada, K. (2017). Impacts of organic waste from a yellowtail cage farm on surface sediment and bottom water in Shido Bay (the Seto Inland Sea, Japan). *Aquaculture, 471*, 140–145.

Stefanescu, I. A. (2015). Bioaccumulation of heavy metals by *Bacillus megaterium* from phosphogypsum waste. *Scientific Study & Research. Chemistry & Chemical Engineering, Biotechnology, Food Industry, 16*(1), 93.

Sudan, A., Ghosh, A., Verma, M., & Tomar, P. (2024). Impact of water pollution & perspective techniques to mitigate it: An overview. *Handbook of water pollution* (pp. 29–64).

Sundaramanickam, A., Awasthi, S., & Patel, V. (2023). Bioremediation of heavy metals from aquatic environments by lactic acid bacteria. In *Lactic acid bacteria as cell factories* (pp. 389–399). Woodhead Publishing.

Tabassum, T., Mahamud, A. S. U., Acharjee, T. K., Hassan, R., Snigdha, T. A., Islam, T., Alam, R., Khoiam, M. U., Akter, F., Azad, M. R., & Al Mahamud, M. A. (2021). Probiotic supplementations improve growth, water quality, hematology, gut microbiota and intestinal morphology of Nile tilapia. *Aquaculture Reports, 21*, 100972.

Talpur, A. D., Memon, A. J., Khan, M. I., Ikhwanuddin, M., Abdullah, M. D. D., & Bolong, A. M. A. (2013). Gut *Lactobacillus* sp. bacteria as probiotics for *Portunus pelagicus* (Linnaeus, 1758) larviculture: Effects on survival, digestive enzyme activities and water quality. *Invertebrate Reproduction & Development, 57*(3), 173–184.

Tariq, A. L., Sudha, S., & Reyaz, A. L. (2016). Isolation and screening of Bacillus species from sediments and application in bioremediation. *International Journal of Current Microbiology and Applied Sciences, 5*(6), 916–924.

Tarricone, R., Rognoni, C., Arnoldo, L., Mazzacane, S., & Caselli, E. (2020). A probiotic-based sanitation system for the reduction of healthcare associated infections and antimicrobial resistances: A budget impact analysis. *Pathogens, 9*(6), 502.

Thurlow, C. M., Williams, M. A., Carrias, A., Ran, C., Newman, M., Tweedie, J., Allison, E., Jescovitch, L. N., Wilson, A. E., Terhune, J. S., & Liles, M. R. (2019). *Bacillus velezensis* AP193 exerts probiotic effects in channel catfish (*Ictalurus punctatus*) and reduces aquaculture pond eutrophication. *Aquaculture, 503*, 347–356.

UNESCO. (2021). International Initiative on Water Quality (IIWQ). United Nations Educational, Scientific and Cultural Organization. https://en.unesco.org/waterquality-iiwq/wq-challenge

Uthra, K. T., Chitra, V., Damodharan, N., Devadoss, A., Kuehnel, M. F., Exposito, A. J., Nagarajan, S., Pitchaimuthu, S., & Pazhani, G. P. (2023). Microplastic emerging pollutants-impact on microbiological diversity, diarrhea, antibiotic resistance, and bioremediation. *Environmental Science: Advances.*

Van Doan, H., Hoseinifar, S. H., Ringø, E., Ángeles Esteban, M., Dadar, M., Dawood, M. A., & Faggio, C. (2020). Host-associated probiotics: A key factor in sustainable aquaculture. *Reviews in Fisheries Science & Aquaculture, 28*(1), 16–42.

Vijayan, J., Nathan, V. K., Ammini, P., & Ammanamveetil, A. M. H. (2023). Bacterial diversity in the aquatic system in India based on metagenome analysis—A critical review. *Environmental Science and Pollution Research, 30*(11), 28383–28406.

Wang, D., Bai, Y., & Qu, J. (2022a). The Phragmites root-inhabiting microbiome: A critical review on its composition and environmental application. *Engineering, 9*, 42–50.

Wang, Y., Chen, P., Yu, X., & Zhang, J. (2022b). Algae-bacteria symbiotic constructed wetlands for antibiotic wastewater purification and biological response. *Frontiers in Microbiology, 13*, 1044009.

Wani, G. A., Khan, M. A., Dar, M. A., Shah, M. A., & Reshi, Z. A. (2021). Next generation high throughput sequencing to assess microbial communities: An application based on water quality. *Bulletin of Environmental Contamination and Toxicology, 106*, 727–733.

Withey, S., Cartmell, E., Avery, L. M., & Stephenson, T. (2005). Bacteriophages—Potential for application in wastewater treatment processes. *Science of the Total Environment, 339*(1–3), 1–18.

Yang, Y., Liu, W., Zhang, Z., Grossart, H. P., & Gadd, G. M. (2020). Microplastics provide new microbial niches in aquatic environments. *Applied Microbiology and Biotechnology, 104*, 6501–6511.

Yi, Y. J., Lim, J. M., Gu, S., Lee, W. K., Oh, E., Lee, S. M., & Oh, B. T. (2017). Potential use of lactic acid bacteria *Leuconostoc mesenteroides* as a probiotic for the removal of Pb (II) toxicity. *Journal of Microbiology, 55*, 296–303.

Yin, Y., Zhang, P., Yue, X., Du, X., Li, W., Yin, Y., Yi, C., & Li, Y. (2018). Effect of sub-chronic exposure to lead (Pb) and Bacillus subtilis on *Carassius auratus* gibelio: Bioaccumulation, antioxidant responses and immune responses. *Ecotoxicology and Environmental Safety, 161*, 755–762.

Zhai, Q., Yu, L., Li, T., Zhu, J., Zhang, C., Zhao, J., Zhang, H., & Chen, W. (2017). Effect of dietary probiotic supplementation on intestinal microbiota and physiological conditions of Nile tilapia (*Oreochromis niloticus*) under waterborne cadmium exposure. *Antonie Van Leeuwenhoek, 110*, 501–513.

Zhang, H., Xing, D., Wu, Y., Jin, R., Liu, D., & Deines, P. (2023). Microbial ecology and function of the aquatic systems. *Frontiers in Microbiology, 13*, 1109221.

Zhang, L., Chen, F., Zeng, Z., Xu, M., Sun, F., Yang, L., Bi, X., Lin, Y., Gao, Y., Hao, H., & Yi, W. (2021). Advances in metagenomics and its application in environmental microorganisms. *Frontiers in Microbiology, 12*, 766364.

Zhang, S., Hu, Z., & Wang, H. (2019). Metagenomic analysis exhibited the co-metabolism of polycyclic aromatic hydrocarbons by bacterial community from estuarine sediment. *Environment International, 129*, 308–319.

Zhao, N., Huang, X., Liu, Z., Gao, Y., Teng, J., Yu, T., & Yan, F. (2024). Probiotic characterization of *Bacillus smithii*: Research advances, concerns, and prospective trends. *Comprehensive Reviews in Food Science and Food Safety, 23*(2), e13308.

Zhou, Q., Wu, K., Yao, L., Chen, R., Liu, S., Xing, H., Nie, L., & Wu, Z. (2023). Bio-ecological remediation of freshwater aquaculture environments: A systematic review and bibliometric analysis. *Water Biology and Security*, 100229.

Zhu, X., Xiang, Q., Chen, L., Chen, J., Wang, L., Jiang, N., Hao, X., Zhang, H., Wang, X., Li, Y., & Omer, R. (2024). Engineered *Bacillus subtilis* Biofilm@ Biochar living materials for in-situ sensing and bioremediation of heavy metal ions pollution. *Journal of Hazardous Materials, 465*, 133119.

Zorriehzahra, M. J., Delshad, S. T., Adel, M., Tiwari, R., Karthik, K., Dhama, K., & Lazado, C. C. (2016). Probiotics as beneficial microbes in aquaculture: An update on their multiple modes of action: A review. *Veterinary Quarterly, 36*(4), 228–241.

Chapter 12
Advance and Integrated Treatment Methods of Emerging Contaminants

Leila Azaryouh, Mehdi Mennani, Amine Miftah, Khalifa El Harfi, Mounir El Achaby, Adil Aboulkas, and Zineb Kassab

Abstract The increasing levels of domestic, industrial, and agricultural wastewater pose significant challenges to water resources on a global scale. Traditional chemical and physical remediation methods, while effective in removing contaminants, often give rise to environmental and economic concerns. As a result, there is a growing interest in the development and integration of treatment processes for emerging contaminants (ECs) that are both effective and environmentally friendly. This chapter provides a comprehensive overview of recent research focusing on substances and techniques aimed at addressing various categories of ECs, such as dyes, surfactants, pesticides, personal care products, x-ray contrast media, among others. It examines the initial treatment steps, analytical techniques, and complex implementation aspects of these innovative treatment approaches. Through an analysis of the latest advancements, this study aims to contribute to the current discourse on sustainable water purification solutions. The incorporation of advanced treatment methods not only aims to enhance pollutant removal efficiency but also ensures environmental responsibility and economic feasibility. By delving into recent developments, the objective is to lay the groundwork for sustainable solutions to the pressing challenges posed by emerging contaminants in diverse environmental contexts.

12.1 Introduction

Agricultural, industrial, and domestic activities are intrinsically linked to the rise of water pollution, posing significant risks to human health, ecosystems, and biodiversity. The growing recognition of the environmental and climatic barriers associated with the availability of clean water is driving a global increase in the search for

L. Azaryouh · A. Miftah · K. El Harfi · A. Aboulkas
Chemical Processes and Applied Materials Laboratory, Polydisciplinary Faculty, Sultan Moulay Slimane University, BP 592, Beni-Mellal, Morocco

L. Azaryouh · M. Mennani · M. El Achaby · Z. Kassab (✉)
Materials Science, Energy and Nanoengineering Department (MSN), Mohammed VI Polytechnic University (UM6P), Lot 660–Hay Moulay Rachid, 43150 Ben Guerir, Morocco
e-mail: zineb.kassab@um6p.ma

H. Gupta et al. (eds.), *Emerging Contaminants in Water and Wastewater*, Environmental Science and Engineering, https://doi.org/10.1007/978-3-031-82579-8_12

sustainable and effective solutions derived from abundant, cost-efficient, natural, and renewable sources (Kasbaji et al., 2023; Zhu et al., 2022). This effort is concentrated on addressing both inorganic and organic pollutants, with a specific focus on converting these contaminants into non-harmful substances and valuable by-products (https://doi.org/10.1007/s42250-022-00391-8; https://doi.org/10.1007/s00128-020-03036-z). The key aspect lies in creating ecological strategies capable of mitigating water pollution while harnessing the potential of readily available resources to promote a healthier, more sustainable environment (Ablouh et al., 2022; Ayouch et al., 2020; Azaryouh et al., 2023). The presence of emerging contaminants in water systems has become a pressing global concern, necessitating advanced and holistic approaches for treatment to ensure the protection of human health and the environment.

Emerging contaminants (ECs) refer to synthetic or natural chemicals and microorganisms that are not regularly monitored in the environment but have the potential to enter the ecosystem and have harmful effects on ecology and human health (Richardson & Ternes, 2011; Rosenfeld & Feng, 2011). This diverse group includes pesticides, industrial chemicals, surfactants, personal care products, endocrine disruptors, painkillers, antibiotics, hormones, and various pharmaceuticals (Ali et al., 2022; Khan et al., 2023). Often found in groundwater, surface water, wastewater, drinking water, and food sources, many of these substances' environmental and human health impacts are not fully understood. Additionally, conventional water treatment processes may produce by-products with unknown chemical properties during purification. The detection of these pollutants in environmental settings previously overlooked in monitoring or regulation has raised concerns. Factors contributing to their presence include advancements in analytical methods, understanding of their persistence, bioaccumulation, and potential effects on ecosystems and health (Bodus et al., 2024; Preethi et al., 2024). Numerous ECs evade traditional water treatment methods, posing risks to drinking water sources and ecosystems. Their presence has significant environmental implications as they persist in water sources, soil, sediments, impact aquatic life, and enter the food chain (Thomas et al., 2017; Khan et al., 2023). Addressing these challenges necessitates an interdisciplinary approach that integrates knowledge from environmental sciences, analytical chemistry, toxicology, and water treatment technologies to ensure the sustainability and safety of water resources for current and future generations.

Conventional water treatment methods frequently prove inadequate in efficiently eliminating ECs due to their distinctive chemical characteristics and environmental persistence. Consequently, there is an increasing demand for sophisticated and combined treatment strategies capable of comprehensively addressing the challenges posed by these pollutants. These approaches encompass the incorporation of diverse physical, chemical, and biological techniques tailored to the specific attributes of ECs. The integration of various advanced oxidation processes (AOPs), membrane filtration, activated carbon adsorption, and biological treatment methods constitute essential elements of these holistic strategies (Richardson & Ternes, 2011; Kaur et al., 2022; Preethi et al., 2024). AOPs, such as photocatalysis and ozonation, harness the power of reactive oxygen species to break down complex organic compounds

(Sandoval et al., 2024). Membrane technologies, including reverse osmosis and nanofiltration, offer efficient separation of contaminants based on molecular size and charge (Zahmatkesh et al., 2023). The efficacy of activated carbon in adsorbing a wide range of organic pollutants is well recognized. In addition to these physicochemical methods, biological treatment approaches incorporating the use of microorganisms to break down or transform contaminants are gaining prominence. Microbial mechanisms like biodegradation and bioaugmentation play a crucial role in breaking down complex organic pollutants into less harmful forms (Dubey et al., 2023a, 2023b).

The emergence of nanotechnology in water treatment has presented a promising avenue for development and application. The distinct characteristics of nanomaterials have the potential to improve the efficiency of eliminating contaminants. Current exploration focuses on nanostructured catalysts, nanoparticles, and nanocomposites for advanced treatment methods. Nanomaterials serve as advanced adsorbents, catalysts, or membrane constituents, showcasing high effectiveness in capturing and breaking down contaminants. Additionally, nanotechnology significantly contributes to sensor advancement for the swift identification and monitoring of emerging contaminants. Nanosensors utilize the unique attributes of nanomaterials to detect minute levels of pollutants, providing real-time data for enhanced evaluation of water quality. Tailored nanoparticles such as activated carbon, metal oxides, and zeolites can selectively capture specific ECs due to their extensive surface area and distinct properties. Magnetic nanoparticles can be easily extracted from treated water using magnets, facilitating efficient contaminant removal and reuse of adsorbents.

The integration of various technologies into a cohesive approach is crucial for addressing the challenges posed by emerging contaminants in water and wastewater. This requires a comprehensive understanding of the synergies among different treatment methods and optimizing their combination for maximum efficacy. Furthermore, real-time monitoring and analytical techniques are indispensable for promptly detecting emerging contaminants and evaluating the performance of integrated treatment systems. The diverse characteristics of emerging contaminants present difficulties in accurately assessing the risks they pose to both humans and the environment. Despite advancements, the ability to monitor and quantify these compounds exceeds our knowledge of their toxic impacts on humans and wildlife. Mere observation of the simultaneous occurrence of a contaminant and adverse effects is inadequate to establish a causal relationship in toxicology. Conversely, thorough investigation into the absorption, mechanism of action, and biological impacts of each emerging contaminant is vital to establish a significant correlation between the contaminant and its effects.

As research progresses in this field, the objective of this chapter is to adopt a multidimensional strategy to address the challenges associated with ECs in water and wastewater. Initially, the focus is on evaluating the efficacy of holistic treatment approaches in eradicating ECs, emphasizing the need for comprehensive solutions. The aim is to explore the synergistic interactions among various treatment techniques, with the goal of optimizing their fusion to achieve maximum efficiency in ECs elimination. Additionally, assessing the importance of monitoring and analytical methods in immediate identification, providing insights into the functioning of holistic treatment systems. The objective is to offer a comprehensive analysis of sustainable and cost-effective strategies for EC management, underscoring the importance of scalable approaches to safeguard water resources and ensure the availability of safe drinking water for communities worldwide.

12.2 Emerging Contaminants: Types and Characteristics

Over the past three decades, there has been a notable increase in the concern for the protection of the environment. This heightened level of concern has spurred various initiatives aimed at converting agricultural waste and other natural resource-derived byproducts into valuable biosorbent materials (Noor Syuhadah & Rohasliney, 2012; Bhattacharjee et al., 2020). For an extended period, a wide range of organic and inorganic pollutants has been a significant contributor to the contamination of ground and surface water, resulting in environmental issues. Typically, industries release hazardous chemicals and contaminants into natural water systems, thereby posing environmental risks (Azaryouh et al., 2023; He et al., 2021). Water pollutants can be broadly categorized into groups such as organic pollutants, inorganic pollutants, suspended pathogens, thermal contaminants, radioactive contaminants, and others, with organic and inorganic effluents constituting the predominant industrial effluent content (Fig. 12.1). Organic compounds consist of elements like nitrogen, hydrogen, and oxygen bonded to a chain of carbon molecules through a covalent bond and are present in a significant proportion of industrial effluents. Table 12.1 presents a variety of emerging contaminants, including examples, sources, and associated challenges. Emerging contaminants can be classified into different groups, encompassing pharmaceuticals, personal care products, pesticides, Endocrine-disrupting compounds, and textile dyes.

A multitude of emerging contaminants, whether intentionally or accidentally discharged into the natural environment, have the potential to jeopardize human health and impact overall living conditions, regardless of whether they undergo controlled treatment. Pharmaceutical and personal care products encompass a diverse range of chemical compounds, ranging from common medications to personal hygiene items like perfumes and sunscreens (Intisar et al., 2023; Khan et al., 2022). While intended to enhance human health, hygiene, and aesthetics, their presence in the environment has raised significant concerns. This category includes substances such as antidepressants, painkillers, antibiotics, contraceptives, fragrances, and sunscreens. Their accumulation in nature exacerbates hormonal disruptions, leading to reproductive problems and an increased risk of breast cancer (Samal et al., 2022). Of particular concern are antibiotic residues, as they have the potential to facilitate the development of antibiotic-resistant genes, potentially giving rise to superbugs (Boshir et al., 2017). Pesticides encompass a diverse array of chemicals with properties that inhibit the growth of harmful insects, weeds, and microorganisms ganisms (Khan et al., 2022; Preethi et al., 2024). Various studies have highlighted the toxicity of

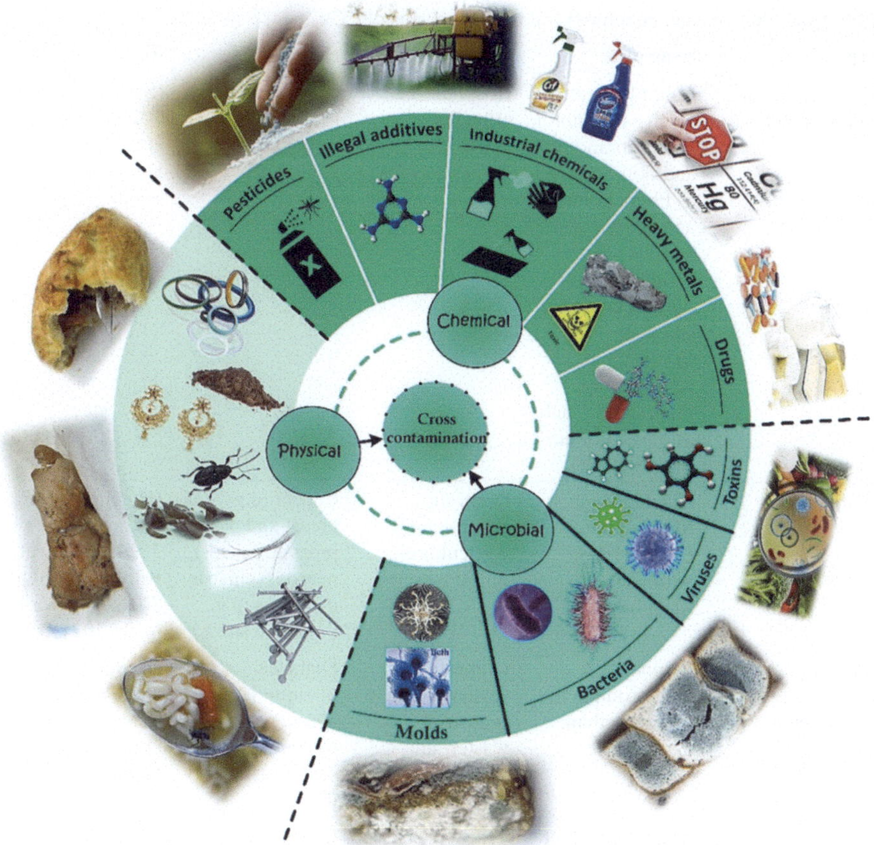

Fig. 12.1 Different contaminants and environmentally relevant threats that may be hazardous to consumers. Reproduced with permission from (Akhavan-Mahdavi et al., 2024)

pesticides to both humans and the environment. For example, prolonged exposure to herbicides like atrazine has been associated with muscle degeneration, retinal degeneration, cancer, and cardiovascular diseases in humans (Saleh et al., 2020; Huang & Li, 2024). Additionally, pesticides have been observed to weaken the immune systems of aquatic organisms and mammals, as well as alter the internal kidney hematopoietic tissue (Boshir et al., 2017). Despite their vital role in industries, textile dyes have emerged as environmental pollutants, leading to the depletion of valuable resources. Highly toxic and potentially carcinogenic, textile dyes have been linked to various human and animal ailments, ranging from skin disorders to central nervous system complications. The health challenges that may arise are likely attributable to the replacement of enzyme cofactors, resulting in the deactivation of enzymes. Swift exposure to such colorants can lead to irritation of the skin and eyes. Particularly, reactive colorants present a notable risk of allergies by binding to albumin in human

Table 12.1 Sources and health risks of different types of ECs

Type	Examples	Sources	Health risks	References
Pharmaceutical and personal care products	Beta-blockers (such as Atenolol, and propranolol), antibiotics (such as Flumequine and amoxicillin), and antidepressants Sunscreen agents (such as oxybenzone, octocrylene), Fragrances (such as musk xylene, musk ketone)	Hospital effluents Domestic wastewater	Cancer (such as beast and testicular cancer)	(Samal et al., 2022; Khan et al., 2022; Kumar et al., 2023)
Pesticide	Alachlor, Metolachlor, Chlorpyrifos, Atrazine, Malathion	Domestic wastewater Agricultural runoff	Cancer (e.g. breast and prostate cancer), neuropsychiatric disorders (e.g. oxidative stress)	(Hu et al., 2023; Huang & Li, 2024; Sabarwal, Kumar & Singh, 2018; Togawa et al., 2021)
Endocrine Disrupting Compounds	Bisphenol A. dichlorodiphenyl trichloroethane	Domestic wastewater Industrial wastewater	Diabetes, disruption of neuroendocrine system, obesity, cardiovascular disease	(Heindel, Newbold & Schug, 2015; Ho et al., 2022; Ma et al., 2019)
Dyes	Methylene blue, Crystal violet, Congo red	Textile factories wastewater	Genotoxicity, carcinogenesis, mutagenicity	(Al-Tohamy et al., 2022; Intisar et al., 2023)

blood serum, acting as antigens that trigger the production of immunoglobulin antibodies that bind to Histamine, causing allergic responses. Nevertheless, prolonged contact with these substances amplifies the chances of developing colorectal cancer, bladder cancer, and colon cancer (Al-Tohamy et al., 2022; Intisar et al., 2023).

Endocrine-disrupting compounds constitute a class of substances that infiltrate the endocrine system and disrupt hormonal functions, resulting in decreased fertility, developmental effects, and various reproductive disorders (Naidu et al., 2016; Yadav et al., 2021). Endocrine disruptors include androgens, estrogens, phthalates, steroids, surfactants, and polychlorinated biphenyls. The consequences associated with these substances encompass dysfunctions in the reproductive systems of marine mammals, heightened occurrences of breast, testicular, and prostate cancers, and the onset of endometriosis (Ma et al., 2019; Naidu et al., 2016). As understanding of emerging pollutants advances, the need to develop effective treatment methods becomes more pressing. The complex origins and characteristics of these pollutants pose intricate challenges for environmental management. Nonetheless, progress in treatment technologies offers promising solutions to address this pressing issue. By comprehensively grasping the types and sources of emerging pollutants, researchers and professionals can tailor treatment approaches to combat specific pollutants and their associated risks. Subsequently, we explore the innovative treatment techniques that are emerging as crucial elements in the effort to mitigate the impacts of these pollutants on environmental and human health.

12.3 Treatment Technologies Emerging Contaminants

Numerous strategies have been formulated for purifying water in treatment facilities to eliminate ECs and produce drinkable water, as well as to treat industrial water before its release into the environment. Given the wide range of methods for wastewater treatment, these strategies can be classified based on the nature of the processes involved, enhancing the systematic understanding of wastewater treatment procedures. The following sections offer detailed descriptions of the biological, physical, and chemical processes utilized in wastewater treatment.

12.3.1 Biological Treatment Methods

Biological treatment methods have been extensively employed to eliminate ECs, primarily through biodegradation. Microorganisms such as bacteria, fungi, and algae degrade large molecules through biodegradation, converting them into smaller molecules or mineralizing them into simpler inorganic molecules like carbon dioxide and water (Dhangar & Kumar, 2020). In traditional biodegradation, bacteria utilize organic molecules as their primary growth substrates and produce enzymes to facilitate their uptake. However, certain ECs can be toxic and resistant to microbial degradation, impeding their natural breakdown. In such cases, microbial growth for biodegradation, referred to as co-metabolism, requires a growth substrate. Biodegradation processes can be categorized into anaerobic and aerobic processes. Efficient aerobic processes include activated sludge, sequencing batch reactors, and membrane bioreactors. Environmentally sustainable alternatives are anaerobic technologies, such as anaerobic sludge reactors and anaerobic film reactors (Boshir et al., 2017).

Technologies built on membrane processes, such as membrane bioreactors (MBRs), have been employed to achieve high-quality effluent treatment against a variety of ECs. Membrane bioreactors have shown effectiveness in removing ECs, including those resistant to traditional treatment methods like activated sludge and constructed wetlands (Boshir et al., 2017; Dhangar & Kumar, 2020; Radjenović et al., 2009) The elimination of ECs is achieved through mechanisms such as sludge retention on the membrane surface, which facilitates microbial degradation and physical entrapment of molecules larger than the membrane's pore size (Li et al., 2015). The dual action of adsorption and biodegradation makes MBRs proficient in EC removal. Despite their efficiency, MBRs encounter challenges like clogging, fouling, operational issues, high energy consumption, and limitations in removing certain ECs. Recent efforts have focused on mitigating biofouling by developing new configurations to control membrane fouling (Hazrati & Gharibian, 2021; Nagda et al., 2022; Shanmuganathan et al., 2023) Constructed wetlands represent an innovative natural approach to wastewater treatment, mimicking the biological, chemical, and physical processes found in natural wetlands (Feng et al., 2021; Hu et al., 2018; Wu et al., 2018). This technology stands out for its eco-friendly treatment method, low operational costs, and ease of maintenance. These environmental and economic advantages have positioned constructed wetlands as a viable alternative to conventional wastewater treatment plants, especially in smaller communities (Ávila et al., 2015; Rathi et al., 2021).

12.3.2 Physical Treatment Methods

Membrane-based technologies, such as reverse osmosis and nanofiltration, have been utilized for the comprehensive removal of organic and inorganic contaminants in wastewater or seawater utilizing semipermeable membranes. The operational advantages of membrane techniques, such as operational flexibility, low energy requirements, cost effectiveness, and high pollutant removal efficiency, outperform other methods (Mastropietro et al., 2021). These membrane-centric techniques operate on a comparable foundation, hinging on a pressure differential as the principal propelling factor, while the semi-permeable membrane acts as a barrier against impurities. A portion of the influent stream, termed concentrate, is rebuffed by the membrane, while another fraction is permitted to traverse to the opposite end, designated as permeate. Reverse osmosis membranes possess minute pores, typically ranging from 0.1 to 1 nm, and are predominantly utilized for extracting low molecular weight substances and ions from a pure solvent. Operational procedures of reverse osmosis necessitate elevated pressure, resulting in significant energy consumption. Conversely, nanofiltration employs membranes with larger pores spanning from 1 to 10 nm, leading to diminished solute rejection (Khan et al., 2023). Nonetheless, nanofiltration functions under reduced pressure, enabling decreased maintenance costs. The majority of nanofiltration and reverse osmosis membranes feature a thin-film composite configuration, which consists of a supportive layer offering mechanical durability with limited permeate flux, in conjunction with a thin active layer that facilitates precise separation with heightened selectivity (Mastropietro et al., 2021).

Membrane fouling is a significant factor influencing both membrane rejection ability and transmembrane flux, and it is the primary barrier to the practical implementation of membrane-based purification technologies (Mastropietro et al., 2021). The extent of membrane fouling impacts the membrane's lifetime and the frequency of cleaning, and subsequently, it significantly influences the cost, design and operation of membrane plants. To overcome this issue, several additives such as nanoparticles (e.g. titanium dioxide) and biopolymers (e.g. chitosan) have been used to improve the membrane performance (El-Aassar, 2015; García et al., 2018; Kamrani et al., 2018).

12.3.3 Chemical Treatment Methods

A broad range of Emerging Contaminants (ECs) can be efficiently eliminated through biological treatment methodologies, which are contingent upon the specific pollutants targeted, the nature of the wastewater, and the operational parameters. Consequently, novel technologies like chemical treatment methods are under scrutiny to further enhance the efficacy of pollutant removal. The primary aim of chemical treatments is the mineralization of contaminants, transforming them into carbon dioxide, nitrogen, water, and various minerals. To this end, a variety of chemical substances, such as ozone, hydrogen peroxide, chlorine, as well as metal oxides and catalysts based on transition metals, along with energy sources like ultrasound, solar radiation, gamma radiation, electric current, and UV radiation, are employed by these techniques (Boshir et al., 2017; Dhangar & Kumar, 2020). Among these methodologies, advanced oxidation processes (AOPs) emerge as pivotal strategies. Advanced oxidation processes entail the generation of highly reactive radicals to disintegrate pollutants, with approaches like ozone-based oxidation, photocatalysis, and Fenton processes serving as prime examples. These advanced oxidation processes are proving effective in breaking down emerging contaminants into less harmful by-products, and offer particular advantages for the treatment of persistent organic pollutants and compounds resistant to conventional methods. Activated carbon adsorption is a widely used chemical treatment method for emerging contaminants. It takes advantage of activated carbon's high surface area and porosity to effectively adsorb a wide range of organic compounds. This method is particularly useful for removing pharmaceuticals, endocrine disruptors and other micropollutants from water sources. In addition to activated carbon adsorption, coagulation and flocculation agents represent another important approach to chemical treatment. By introducing chemicals that induce floc formation, this method facilitates the aggregation of fine particles and contaminants, simplifying their removal by sedimentation or filtration. Common coagulants, such as aluminum sulfate and ferric chloride, have proven effective in treating organic and inorganic pollutants. Chemical precipitation is another approach in the arsenal of chemical treatment methods for emerging contaminants. This method involves the addition of chemicals that induce the formation of insoluble precipitates, effectively removing contaminants from the water matrix. Lime, sodium hydroxide and various metal salts are precipitating agents commonly used in this process.

Alginate, cellulose, lignin, and chitosan, owing to their advantageous characteristics, are essential natural biopolymers extensively utilized in the remediation of ECs (Ayouch et al., 2021; Kasbaji et al., 2022a, 2022b; Kasbaji et al., 2022a, 2022b;

Kassem et al., 2022). The conjugate base of alginic acid, alginate, possesses multiple active sites for heavy metal ion adsorption due to its abundance of carboxyl groups. Alginic acid is a copolymer linear in structure, comprising homopolymer blocks containing -D-mannuronate and -L-guluronate residues covalently linked in various blocks or sequences (Ayouch et al., 2020). Alginates are derived from various brown algae species within the phaeophycae family, collected globally and processed into a water-soluble raw material. Upon interaction with divalent cations, this biopolymer is primarily used to fabricate rigid hydrogels; indeed, the alginate chains initiate interactions with the ions, leading to cross-link formation with neighboring chains upon exposure to divalent cations (Choe et al., 2018).

Cellulose, characterized as a linear polysaccharide with beta D-glucopyranose units in a 4C1 chair configuration, is connected through beta-acetal linkages (beta-1,4-glycosidic bonds), resulting in a linear chain structure. The presence of three unbound hydroxyl groups facilitates intra- and intermolecular bond formation within cellulose chains (Kushwaha & Singh, 2023). Functioning as a primary structural element in plant cell walls, cellulose is renowned for its hydrophilic groups, making it the most abundant biopolymer on Earth and ideal for hydrogel synthesis. The extraction of cellulose commences with biomass pretreatment, aiming to disintegrate recalcitrant lignocellulosic biomass into cellulose, hemicellulose, and lignin (Azaryouh et al., 2023). Following the dissolution of isolated and refined cellulose in a specific solvent system, a three-dimensional network is cross-linked using a multifunctional molecule as a crosslinking agent. Various plant-based materials have been employed in prior studies on cellulose hydrogel fabrication. Sodium carboxymethyl cellulose, also referred to as industrial MSG, serves as a natural raw material for hydrogel production (Chen et al., 2023). Citric acid, a non-toxic cross-linking agent, exhibits potential for use in the production of cellulose derivative hydrogels, as evidenced in the research by Ikrak et al. Their investigation involved the development of hydrogel membranes for Methylene blue (MB) adsorption using citric acid cross-linked sodium carboxymethyl cellulose and hydroxyethyl cellulose, achieving a maximum adsorption capacity of 769.23 mg/g (Ayouch et al., 2021).

Lignin, the second most abundant polymer on a global scale, constitutes the cellular structures of natural lignocellulosic plants (El Bouchtaoui et al., 2022; Mennani et al., 2024). Various pulping techniques and preprocessing of biomass produce three main categories of lignin: Kraft lignin, lignosulfonate lignin, and organosolve lignin. These variations have all been utilized in the development of hydrogels based on lignin (Meng et al., 2019). Kraft lignin, distinguished by a lower content of aryl ether connections (β-O-4) and a higher level of condensed configurations (C–C connections) because of the pulp cooking process, demonstrates decreased reactivity in comparison to the other types (Froass et al., 1996). Through the incorporation of lignin into polymer matrices using interpenetrating and semi-interpenetrating methodologies, hydrogels with distinct characteristics can be synthesized (Meng et al., 2019). Larrañeta et al. applied a solvent-free polymerization technique to produce lignin hydrogels in two stages, demonstrating an eco-friendly and uncomplicated approach. The procedure entailed the evaporation of aqueous ethanol combined with dissolved lignin and polyethylene glycol to generate a film, followed by polymerization at 80 °C for 24 h (Larrañeta et al., 2018).

12.4 Integrated Bio-Derived Materials and Treatments

12.4.1 Biopolymers and Biocomposites

Biomass acts as the principal origin of biopolymers and biocomposites on a global scale, exhibiting an expected annual growth rate of 10% from agricultural residue due to increased demand for agricultural goods. The substantial amount of 150 billion tons of agricultural residue produced each year, as illustrated in Fig. 12.2a, highlights the worldwide extent of agricultural waste generation (Chong et al., 2023; Mennani et al., 2024). Nonetheless, insufficient utilization of solid agricultural residue presents environmental hazards, encompassing soil pollution, gas discharges, and dust formation, particularly when mismanaged through practices like haphazard disposal or burning in public areas. To address these concerns, ongoing research underscores the integration of bio-based polymers and composites to tackle emerging pollutants through techniques such as functionalization, copolymerization, and activation (Fig. 12.2b) (Zhu et al., 2022). These innovative treatment strategies strive to enhance the adsorption efficiency, performance, surface area, structural integrity, and potential regeneration of the materials. Through the exploitation of biomass-derived materials, there is a collective endeavor to mitigate the environmental repercussions linked to underused agricultural residue while progressing efficient remedies for the mitigation of emerging pollutants.

Biopolymers and composites derived from cellulose, lignin, chitin/chitosan, starch, alginate, pectin, hemicellulose, etc., constitute the predominant categories of bio-based materials employed to effectively address issues related to ECs (Kasbaji et al., 2023; Nasrollahzadeh et al., 2021). Cellulose fiber-reinforced composites have been utilized as a biodegradable and sustainable resource for wastewater treatment and ECs mitigation. Sirajudheen et al. (2020) presented an effective technique for eliminating hazardous ECs (azo dyes) from water. The composite integrates magnetic iron oxide (Fe_3O_4) nanoparticles, graphene oxide (GO), and carboxymethyl cellulose (CMC) to create a magnetic Fe_3O_4-GO-CMC (FGC) composite (Fig. 12.2c). This study investigates the efficacy of a magnetic composite in azo dye removal via adsorption, examining variables such as initial dye concentration, contact duration, and solution acidity. The composite achieved a removal efficiency of up to 94.5%, with regeneration experiments indicating that more than 65% of the composite could be regenerated over five consecutive cycles. Notably, these materials exhibit the potential for regeneration and the reuse of biosorbents and composites. Regeneration permits multiple usage cycles, diminishing the necessity for frequent replacement of adsorbent materials and thereby reducing overall expenses.

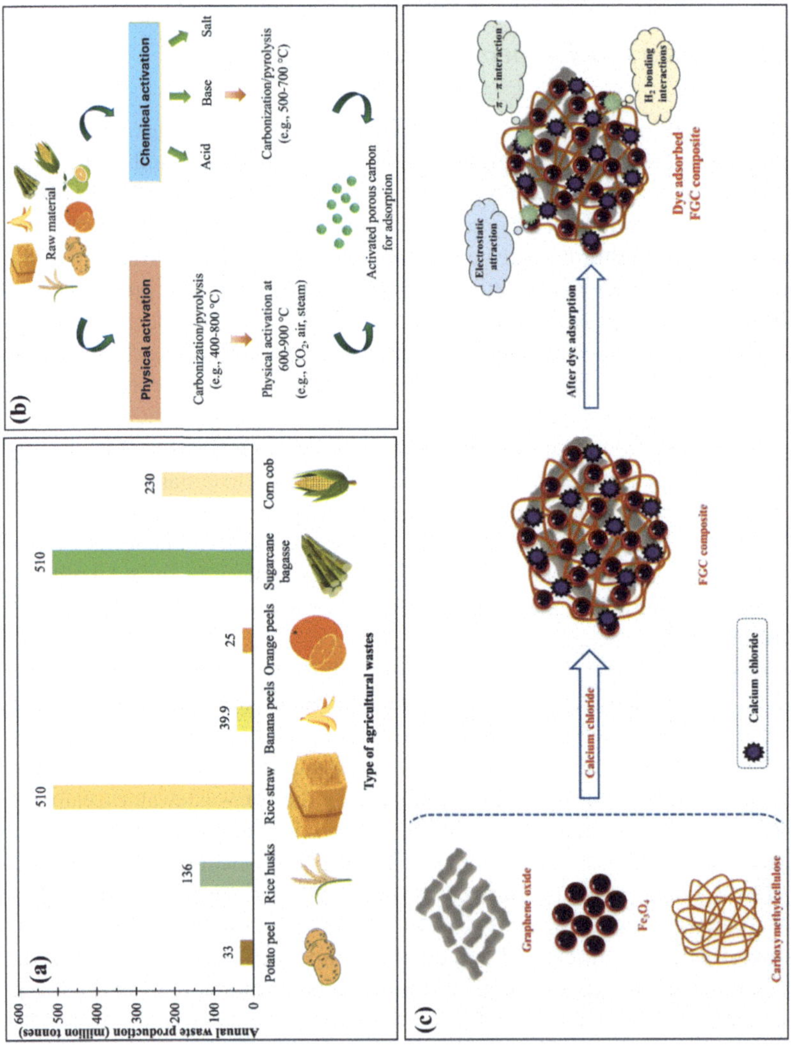

Fig. 12.2 **a** Worldwide production of agricultural waste per year. **b** Flow chart illustrating the physical and chemical activation of raw materials. Reproduced with permission from (Chong et al., 2023). **c** Preparation of FGC composites and mechanism of ECs removal. Reproduced with permission from (Zhang et al., 2022)

12.4.2 Enzyme and Nanozyme-Based Processes

Enzymes and nanozymes have emerged as promising tools for addressing emerging contaminants, providing innovative and effective solutions to environmental challenges. Enzymes, functioning as biological catalysts, play a vital role in hastening chemical reactions, especially in the degradation or alteration of specific pollutants linked to ECs. Their usage spans biodegradation, the conversion of pollutants into less hazardous substances, and bioremediation, the employment of microorganisms equipped with specific enzymes for the remediation of contaminated environments either in situ or ex-situ (Fig. 12.3a) (Amaro Bittencourt et al., 2023; Martínez-Zamudio et al., 2024). Exemplars like laccase and cytochrome P450 exemplify their efficiency, characterized by a high specificity towards particular pollutants and eco-friendly processes. Nanozymes, nanomaterials exhibiting enzyme-like catalytic activities, also complement these endeavors by emulating natural enzymes while providing benefits such as adjustable catalytic properties and improved stability. Their applications encompass the catalytic oxidation of diverse pollutants, including organic contaminants and heavy metals, along with integration into filtration systems for effective water purification. Metal oxide nanoparticles like titanium dioxide and manganese oxide, in addition to carbon-based nanozymes such as graphene and carbon nanotubes, demonstrate their adaptability and durability. The amalgamation of enzymes and nanozymes holds the potential for synergistic effects in remediating emerging contaminants. Overall efficacy can be heightened, and technical nanozymes offer precise delivery capabilities, enhancing the precision of remediation activities. Hybrid systems that combine enzymatic and nanoenzymatic approaches with other advanced treatment methods, such as adsorption or membrane filtration, are also contributing to comprehensive remediation solutions (Liang et al., 2022; Su et al., 2022).

Enzymes face challenges in bioremediation due to their higher cost in comparison to conventional wastewater treatment methods, as well as concerns regarding reusability and stability. Immobilization techniques offer a solution by securing or entrapping enzymes within a matrix, modifying their configuration, enhancing stability, and enabling numerous reaction cycles and recovery. This proves beneficial for wastewater treatment, particularly in scenarios where toxic substances may hinder enzyme function, as the applied matrix can shield the molecules. The selection of matrix and immobilization method is critical, as each approach impacts enzyme structure, function, and stability. A thorough understanding of the enzyme's attributes, encompassing its structure and optimal reaction conditions, is imperative. Furthermore, optimizing process parameters like temperature, pH, and enzyme concentration is vital to achieve maximum activity, recovery, and output (Fig. 12.3b) (Amaro Bittencourt et al., 2023). This strategy addresses cost concerns and enhances enzyme effectiveness in bioremediation systems.

The fabrication of photocatalytic nanozyme OVs-TiO_2/Pd, characterized by abundant oxygen vacancies (OVs) and adjustable Pd elements, was conducted to eliminate emerging pollutants and produce hydrogen (Fig. 12.3c) (Li et al., 2023). The

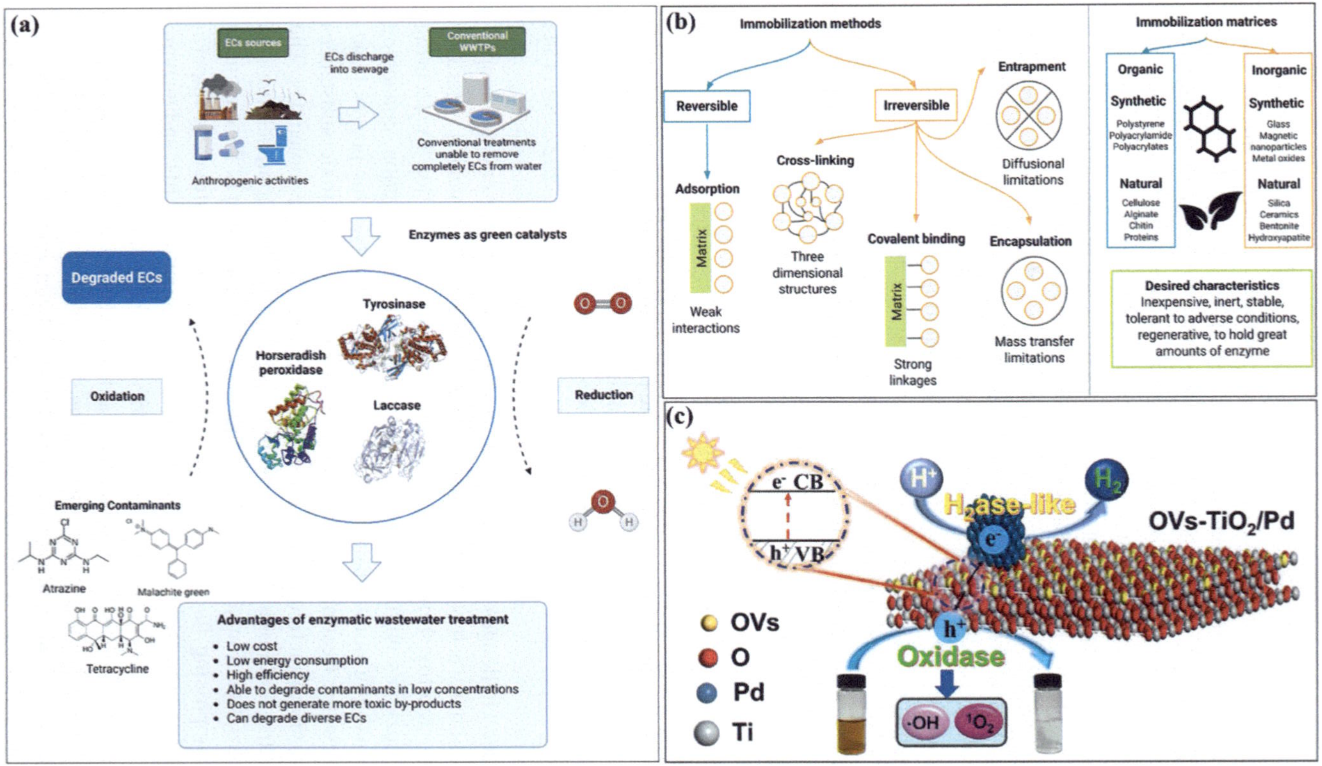

Fig. 12.3 **a** Sources of emerging contaminants and enzymatic degradation pathways: This degradation process reveal reliable, energy enduring and cost effective depolution of ECs (Martínez-Zamudio et al., 2024). **b** Diagram of the different enzyme immobilisation techniques applied to ECs (Amaro Bittencourt et al., 2023). **c** The OVs-TiO_2/Pd nanozymes exhibit exceptional photo-enhanced enzyme-like activity (Li et al., 2023)

degradation efficiency of tetracycline (TC) on OVs-TiO_2/5%Pd achieved a remarkable 87.3%, surpassing that of TiO_2 by 1.56 times. Particularly notable was the significant enhancement observed under visible light exposure, with H_2 generation reaching approximately 22.15 mmol/h/g. This demonstrates a remarkable 124.2-fold increase over the efficiency of TiO_2. The introduction of oxygen vacancies and the precise control of Pd elements in the photocatalytic nanozyme emphasize its efficacy in superior ECs removal and enhanced hydrogen production, underscoring its potential for environmentally sustainable applications. Despite these advancements, challenges persist, such as the cost linked to enzymes and the continuous necessity to develop cost-efficient nanozyme materials. Furthermore, comprehending the potential environmental implications of nanozyme materials remains essential for ensuring sustainable sanitation practices. Current research efforts are focused on boosting the utilization of enzymes and nanozymes in various environmental remediation scenarios. Their specificity, catalytic effectiveness, and potential for targeted applications make them valuable instruments in the pursuit of sustainable and efficient strategies to address emerging contaminants.

12.4.3 Monitoring and Sensor Technologies

Monitoring and detection systems are crucial in promptly identifying and analyzing new contaminants, thereby safeguarding the environment and public health. These technologies provide data in real-time or near real-time, facilitating informed decision-making. They assist in early detection across diverse environmental settings, including water, air, and soil, aiding in the identification of pollutants such as pharmaceuticals, personal care products, and industrial chemicals. Water quality monitoring utilizes sensors to detect microplastics, pharmaceutical residues, and endocrine-disrupting compounds in rivers, lakes, and drinking water (Ghanam et al., 2023; Han et al., 2023). Similarly, air quality monitoring assesses parameters like volatile organic compounds (VOCs), particulates, and airborne pathogens. Soil monitoring technologies identify contaminants affecting agricultural and ecological systems, including pesticides, heavy metals, and persistent organic pollutants. Remote sensing enables comprehensive monitoring of environmental changes, simplifying the identification of contaminant sources and assessment of land use changes (Yuan et al., 2023; Mangotra & Singh, 2024). Biosensors combine biological sensing elements for specific contaminant detection, while smart sensing platforms integrate multiple sensing techniques for thorough monitoring. Advanced data analysis, including machine learning, processes extensive datasets to discern patterns and trends, aiding in predictive modeling of contaminant behavior and risk assessment. Additionally, real-time monitoring networks and autonomous detection systems, equipped with interconnected sensors and unmanned capabilities, ensure continuous monitoring and prompt response to contamination incidents, bolstering our capacity to effectively tackle new environmental challenges.

The initiation of industrial processes has resulted in the emergence of novel pollutants in the environment, posing a risk to human health and ecosystems. Conventional techniques for monitoring water contamination have limitations, such as high expenses and environmental considerations. There is a pressing demand for the development of detection methods that are uncomplicated, cost-efficient, and highly sensitive. Fiber optic microfluidic sensors present a viable solution with benefits like controlled microfluidics, precise detection, and stability (Fig. 12.4a) (Yuan et al., 2023). The application of electrospun nanofibers (NFs) is seen as a promising strategy due to their unique three-dimensional structure, significant specific surface area, and straightforward preparation process. Moreover, NFs exhibit excellent biocompatibility, degradability, and versatility, enhancing the practicality and environmental friendliness of monitoring techniques. These characteristics are pivotal in enhancing the identification of pollutants. NF-based sensors demonstrate the ability to identify a wide range of biological, chemical, and physical hazards, including pathogens, pesticides, drug residues, toxins, allergens, and heavy metals (Fig. 12.4b) (Akhavan-Mahdavi et al., 2024). A biosensor with high electroactivity has been developed, incorporating polarity reversal after optimized inoculation. This biosensor accurately measures specific ECs and provides timely and effective alerts regarding the toxicity of unknown contaminants in real water samples. Particularly, the biosensor effectively evaluated the toxicity of recycled water with minimal levels of aerobic organic substances, enabling rapid responses and immediate alerts regarding the presence of ECs (Fig. 12.4c) (Han et al., 2023).

12.4.4 Integrated Nanomaterials

The introduction of nanotechnology in the domain of emerging contaminant remediation signifies a significant shift, leveraging the unique properties of nanoscale materials. This innovative approach addresses the crucial need to efficiently tackle ECs, which include pharmaceuticals, personal care products, pesticides, and industrial chemicals, often posing challenges for traditional remediation methods (Fig. 12.4d) (Hejji et al., 2023; Thamarai et al., 2024). The benefits of nanotechnology in this context are diverse. Nanomaterials, such as nanocomposites and nano-adsorbents, have a remarkably high surface-to-volume ratio, enhancing reactivity and increasing adsorption capacity and catalytic efficiency in contaminant removal. Nano-adsorbents have been widely used as effective agents for eliminating ECs, and this process can be facilitated through chemical (ionic or covalent) or physical (Van der Waals forces) interactions (Fig. 12.4e) (Thamarai et al., 2024). The ability to customize the design of nanomaterials allows for precise targeting of contaminants, improving the selectivity and effectiveness of the remediation process. Nanofiltration membranes, composed of nanomaterials like carbon nanotubes or graphene oxide, provide selective separation of contaminants based on their size and charge, ensuring efficient removal while retaining essential components in the water.

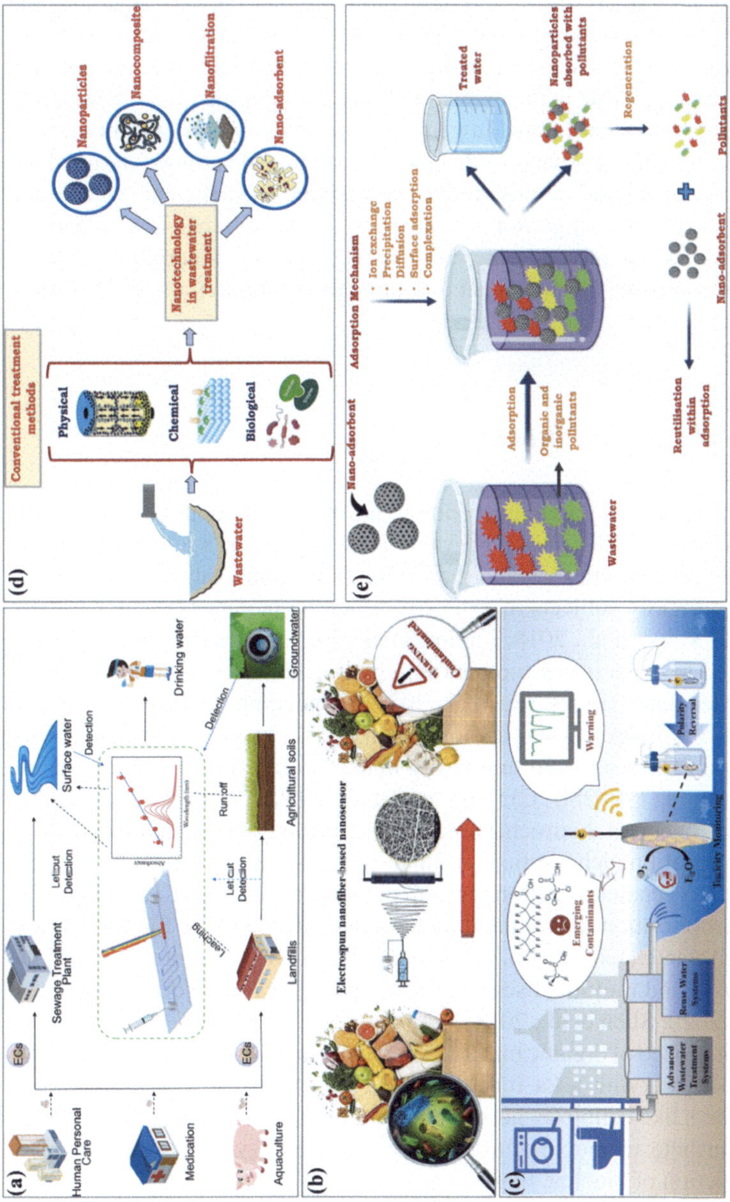

Fig. 12.4 **a** Development of highly sensitive methods for detecting toxic environmental contaminants (Yuan et al., 2023). **b** Nanosensors based on electrospun nanofibers for the detection of chemical and biological ECs in the food processing industry (Akhavan-Mahdavi et al., 2024). **c** Emerging contaminants in reuse water using biocathode sensors (Han et al., 2023). **d** The integrated potential of nanotechnology for ECs remediation. **e** Functionality of nano-adsorbents in wastewater purification (Thamarai et al., 2024)

Nanocatalysts, such as metal and metal oxide nanoparticles, catalyze the breakdown of various types of contaminants, expediting advanced oxidation processes for converting pollutants into less harmful by-products. Additionally, nanotechnologies support the creation of smart delivery systems utilizing nanocarriers, guaranteeing controlled and targeted release of sanitizing agents and enhancing treatment efficiency. The utilization of nanosensors, incorporating materials like quantum dots and nanowires, plays a pivotal role in the precise and sensitive identification of emerging pollutants at minimal concentrations. In general, nanotechnologies present a versatile and all-encompassing approach to diverse environmental settings, diminishing the environmental impact and contributing to sustainable and efficient remediation tactics. Essential further investigation is required to tackle apprehensions regarding the environmental destiny and enduring consequences of nanomaterials in order to guarantee their responsible advancement and utilization (Hejji et al., 2023; Rajput et al., 2024).

12.4.5 Biosorbents and Biochar

Under the principles of the circular economy, an intriguing method involves the conversion of biomass residue into high-value-added products, such as hydrochar. Progressions in hydrochar creation and the exploration of innovative and emerging applications are currently subjects of extensive scientific inquiry (Jain et al., 2016; Chen et al., 2017; Volpe et al., 2018; Enaime et al., 2022). A thermochemical technique is employed to convert wet biomass into hydrochar, a carbon-rich substance with elevated added value (Erses Yay et al., 2021). Depending on the feedstock and the intended use, the hydrothermal carbonization (HTC) process is conducted for a specified duration (ranging from minutes to hours) under autogenic pressure at a moderate temperature (170–280 °C) (Gimenez et al., 2020; Yaashikaa et al., 2020). The production process of hydrochar encompasses numerous reactions, such as hydrolysis, isomerization, dehydration, aromatization, polymerization, and condensation. The high oxygenated functional group content in hydrochar, resulting from hydrothermal carbonization, positions it as a productive precursor for the fabrication of chemically activated carbon. Besides applications in water treatment, the HTC product has found multiple uses like energy generation and soil enhancement (Gimenez et al., 2020; Shen et al., 2021).

Biomass is comprised of organic material derived from living organisms, primarily plants and animals, serving as a renewable and sustainable resource containing various components like cellulose, hemicellulose, lignin, proteins, and lipids. These components contribute to the structural strength and energy content of biomass. During the production of biochar, biomass goes through a process of thermal breakdown known as pyrolysis, in which it is heated in the absence of oxygen to produce a carbon-rich substance called biochar. Similar to charcoal, biochar has become important in addressing new pollutants due to its unique characteristics, including a high surface area, porous structure, and impressive adsorption capabilities. These qualities make biochar an effective sorbent for trapping and removing a wide range of pollutants from soil, water, and air, including organic contaminants, heavy metals, and nutrients. The importance of biochar in pollution control lies in its ability to capture contaminants through adsorption, ion exchange, and surface interactions.

12.5 Challenges and Perspectives

The diversity of emerging contaminants (ECs), encompassing a wide range of synthetic and natural compounds, poses significant challenges in terms of detection, treatment, and remediation. As the scientific community endeavors to comprehend and address these pollutants, numerous constraints, often associated with economic factors and the expansion of remediation processes, become prominent. A primary hurdle in addressing emerging contaminants is the economic ramifications of detecting and remediating them. Sophisticated analytical methods capable of detecting these pollutants with high precision often come with a substantial price tag. This financial encumbrance can restrict the extensive monitoring of emerging contaminants, particularly in regions with limited resources. Moreover, treatment technologies formulated to eliminate these pollutants from water and wastewater streams can be prohibitively expensive, hindering their broader adoption. The advancement of effective and scalable treatment strategies encounters technological challenges due to the complex nature of emerging contaminants, such as pharmaceuticals and industrial by-products. Conventional water treatment techniques may be insufficient in eradicating specific emerging contaminants, necessitating the application of advanced technologies. Nevertheless, upscaling these technologies to cater to the requirements of large-scale water treatment facilities introduces additional complications. Technical issues linked to the design, operation, and upkeep of these systems can impede their widespread implementation.

New contaminants exhibit diverse chemical characteristics and behaviors, complicating the prediction of their fate and transport in aquatic settings. The inadequate comprehension of how these pollutants migrate through water bodies, soils, and sediments hinders the formulation of targeted remediation approaches. Forecasting potential ecological and human health hazards becomes challenging without a thorough understanding of how contaminants interact in different environmental contexts. The regulatory framework concerning emerging contaminants often lags behind scientific advancements. Many of these pollutants may lack explicit regulations, creating a regulatory gap that obstructs proactive management. Establishing comprehensive policy frameworks that encompass the identification, monitoring, and regulation of emerging contaminants is imperative. Nonetheless, implementing these strategies may face delays and opposition from various stakeholders. Effectively addressing emerging contaminants necessitates an interdisciplinary approach that amalgamates knowledge from environmental sciences, analytical chemistry, toxicology, and engineering. The primary challenge lies in fostering collaboration and information exchange among professionals from diverse fields. Bridging these disciplinary gaps is crucial to formulating comprehensive and innovative strategies that consider both the environmental and human health impacts of emerging contaminants.

Despite the obstacles, there is a growing body of research dedicated to devising inventive and cost-efficient remediation technologies for emerging contaminants. Promising methodologies encompass nanotechnologies, advanced oxidation processes, and natural remedies. Integrating these technologies into current water treatment systems or creating independent systems can offer scalable and economically feasible solutions. Enhancing public awareness and education regarding emerging contaminants is a pivotal aspect of tackling these challenges. Informed and educated communities can advocate for sustainable water management practices, endorse research endeavors, and contribute to the adoption of conscientious consumption habits that minimize the release of contaminants into the environment.

12.6 Conclusions

This work provides a thorough analysis of emerging contaminants (ECs) and their potential implications on water resources. The examination highlights the complex nature of the challenges posed by ECs, encompassing environmental, public health, and regulatory aspects. Studies in this field have revealed the intricate mechanisms governing the behavior, transportation, and alteration of ECs in aquatic environments. The findings emphasize the urgent necessity to tackle these pollutants to safeguard water quality and ecosystem health. A critical evaluation of various novel strategies to mitigate EC impacts is presented, including reverse/forced osmosis, nanofiltration, advanced oxidation processes (AOPs), nanotechnology applications, biodegradation, and microbial remediation. These innovative approaches demonstrate ongoing efforts to develop sustainable and effective solutions for EC removal and degradation.

Nevertheless, the complexities associated with ECs go beyond mere technological fixes. Economic limitations often hinder the widespread implementation of monitoring and remediation efforts, creating significant barriers to holistic water quality management. Moreover, technological restrictions impede the development of scalable treatment methods capable of addressing the wide array of contaminants present in water sources. Addressing the multifaceted challenges linked to ECs demands a comprehensive strategy that integrates economic considerations, technological advancements, regulatory frameworks, and interdisciplinary collaboration. Encouraging progress in water treatment technologies, raising public awareness about EC consequences, and establishing robust regulatory measures are essential steps in mitigating the influence of these pollutants. By surmounting economic challenges, advancing technological innovations, addressing regulatory deficiencies, and fostering interdisciplinary collaboration among researchers, policymakers, and stakeholders, it is possible to effectively tackle emerging contaminants and ensure sustainable water resource management for current and future generations. This study represents a significant contribution to understanding ECs and their impact on water quality, underscoring the need for continuous research, technological progress, and policy formulation to effectively address this urgent environmental concern.

References

Ablouh, E. H. et al. (2022). Phosphorylated cellulose paper as highly efficient adsorbent for cadmium heavy metal ion removal in aqueous solutions. *RSC Advances*, *12*(2), 1084–1094. Available at: https://doi.org/10.1039/d1ra08060a

Akhavan-Mahdavi, S. et al. (2024). Electrospun nanofiber-based sensors for the detection of chemical and biological contaminants/hazards in the food industries. *Advances in Colloid and Interface Science*, *325*, 103111. Available at: https://doi.org/10.1016/j.cis.2024.103111

Ali, J. et al. (2022). Chapter 7—Advances in biodegradation and bioremediation of emerging contaminants in the environment. In S. Kumar & M. Z. B. T.-B. A. to C. P. Hashmi (Eds.), *Advances in pollution research* (pp. 121–138). Woodhead Publishing. Available at: https://doi.org/10.1016/B978-0-12-824316-9.00013-6

Al-Tohamy, R. et al. (2022). A critical review on the treatment of dye-containing wastewater: Ecotoxicological and health concerns of textile dyes and possible remediation approaches for environmental safety. *Ecotoxicology and Environmental Safety*, *231*, 113160. Available at: https://doi.org/10.1016/j.ecoenv.2021.113160

Amaro Bittencourt, G. et al. (2023). Emerging contaminants bioremediation by enzyme and nanozyme-based processes—A review. *iScience*, *26*(6), 106785. Available at: https://doi.org/10.1016/j.isci.2023.106785

Ávila, C. et al. (2015). Emerging organic contaminant removal in a full-scale hybrid constructed wetland system for wastewater treatment and reuse. *Ecological Engineering*, *80*, 108–116. Available at: https://doi.org/10.1016/j.ecoleng.2014.07.056

Ayouch, I. et al. (2020). Impact of the drying process on the efficiency of alginate beads for cadmium removal from water: Kinetic, isotherm and thermodynamic study. *Environmental Technology and Innovation*, *20*, 101157. Available at: https://doi.org/10.1016/j.eti.2020.101157

Ayouch, I. et al. (2021). Crosslinked carboxymethyl cellulose-hydroxyethyl cellulose hydrogel films for adsorption of cadmium and methylene blue from aqueous solutions. *Surfaces and Interfaces*, *24*, 101124. Available at: https://doi.org/10.1016/j.surfin.2021.101124

Azaryouh, L. et al. (2023). Hybrid carbonaceous adsorbents based on clay and cellulose for cadmium recovery from aqueous solution. *RSC Advances*, *13*(10), 6951–6965. Available at: https://doi.org/10.1039/d2ra08287j

Bhattacharjee, C., Dutta, S., & Saxena, V. K. (2020). A review on biosorptive removal of dyes and heavy metals from wastewater using watermelon rind as biosorbent. *Environmental Advances*, *2*, 100007. Available at: https://doi.org/10.1016/j.envadv.2020.100007

Bodus, B. et al. (2024). Review of emerging contaminants in green stormwater infrastructure: Antibiotic resistance genes, microplastics, tire wear particles, PFAS, and temperature. *Science of The Total Environment*, *906*, 167195. Available at: https://doi.org/10.1016/j.scitotenv.2023.167195

Boshir, M. et al. (2017). Progress in the biological and chemical treatment technologies for emerging contaminant removal from wastewater : A critical review. *Journal of Hazardous Materials*, *323*, 274–298. Available at: https://doi.org/10.1016/j.jhazmat.2016.04.045

El Bouchtaoui, F. Z. et al. (2022). Methylcellulose/lignin biocomposite as an eco-friendly and multifunctional coating material for slow-release fertilizers: Effect on nutrients management and wheat growth. *International Journal of Biological Macromolecules*, *221*, 398–415. Available at: https://doi.org/10.1016/j.ijbiomac.2022.08.194

Chen, R. et al. (2023). Eco-friendly semi-interpenetrating polymer network hydrogels of sodium carboxymethyl cellulose/gelatin for methylene blue removal. *Materials, 16*(9). Available at:https://doi.org/10.3390/ma16093385

Chen, J. et al. (2017). Preparation and characterization of activated carbon from hydrochar by phosphoric acid activation and its adsorption performance in prehydrolysis liquor. *BioResources, 12*(3), 5928–5941. Available at: https://doi.org/10.15376/biores.12.3.5928-5941

Choe, S. R. et al. (2018). Fabrication of alginate/humic acid/Fe-aminoclay hydrogel composed of a grafted-network for the efficient removal of strontium ions from aqueous solution. *Environmental Technology and Innovation, 9*, 285–293. Available at: https://doi.org/10.1016/j.eti.2017.12.008

Chong, Z. T., Soh, L. S., & Yong, W. F. (2023). Valorization of agriculture wastes as biosorbents for adsorption of emerging pollutants: Modification, remediation and industry application. *Results in Engineering, 17*, 100960. Available at: https://doi.org/10.1016/j.rineng.2023.100960

Dhangar, K., & Kumar, M. (2020). Science of the total environment tricks and tracks in removal of emerging contaminants from the wastewater through hybrid treatment systems : A review. *Science of the Total Environment, 738*(336), 140320. Available at: https://doi.org/10.1016/j.scitotenv.2020.140320

Dubey, M., Vellanki, B. P., & Kazmi, A. A. (2023a). Fate of emerging contaminants in a sequencing batch reactor and potential of biological activated carbon as tertiary treatment for the removal of persisting contaminants. *Journal of Environmental Management, 338*, 117802. Available at: https://doi.org/10.1016/j.jenvman.2023.117802

Dubey, M., Vellanki, B. P., & Kazmi, A. A. (2023b). Removal of emerging contaminants in conventional and advanced biological wastewater treatment plants in India-a comparison of treatment technologies. *Environmental Research, 218*, 115012. Available at: https://doi.org/10.1016/j.envres.2022.115012

El-Aassar, A., & Hameed M. A. (2015). Improvement of reverse osmosis performance of polyamide thin-film composite membranes using TiO_2 nanoparticles. *Desalination and Water Treatment, 55*(11), 2939–2950. Available at: https://doi.org/10.1080/19443994.2014.940206

Enaime, G. et al. (2022). Hydrothermal carbonization of the filter bed remained after filtration of olive mill wastewater on olive stones for biofuel application. *Biomass Conversion and Biorefinery, 12*(4), 1237–1247. Available at: https://doi.org/10.1007/s13399-020-00743-9

Erses Yay, A. S. et al. (2021). Hydrothermal carbonization of olive pomace and determining the environmental impacts of post-process products. *Journal of Cleaner Production, 315*, 128087. Available at: https://doi.org/10.1016/j.jclepro.2021.128087

Feng, L. et al. (2021). Simultaneous elimination of antibiotics resistance genes and dissolved organic matter in treatment wetlands : Characteristics and associated relationship. *Chemical Engineering Journal, 415*, 128966. Available at: https://doi.org/10.1016/j.cej.2021.128966

Froass, P. M., Ragauskas, A. J., & Jiang, J. E. (1996). Chemical structure of residual lignin from kraft pulp. *Journal of Wood Chemistry and Technology, 16*(4), 347–365. Available at: https://doi.org/10.1080/02773819608545820

García, A. et al. (2018). Incorporation of CuO nanoparticles into thin-film composite reverse osmosis membranes (TFC-RO) for antibiofouling properties. *Polymer Bulletin, 75*(5), 2053–2069. Available at: https://doi.org/10.1007/s00289-017-2146-4

Ghanam, A. et al. (2023). Selective sensing in microbial fuel cell biosensors: Insights from toxicity-adapted and non-adapted biofilms for Pb(II) and neomycin sulfate detection. *Micromachines*. Available at.https://doi.org/10.3390/mi14112027

Gimenez, M. et al. (2020). Two phase olive mill waste valorization. Hydrochar production and phenols extraction by hydrothermal carbonization', *Biomass and Bioenergy*, 143. Available at: https://doi.org/10.1016/j.biombioe.2020.105875

Han, Y. et al. (2023) Rapid warning of emerging contaminants in reuse water using biocathode sensors. *Journal of Hazardous Materials, 457*, 131735. Available at: https://doi.org/10.1016/j.jhazmat.2023.131735

Hazrati, S. S. H., & Gharibian, Y. J. A. J. S. (2021). Investigation of Ag and magnetite nanoparticle effect on the membrane fouling in membrane bioreactor. *International Journal of Environmental*

Science and Technology [Preprint], 0123456789. Available at:https://doi.org/10.1007/s13762-020-03053-9

He, X. et al. (2021). Enhanced adsorption of Cu(II) and Zn(II) from aqueous solution by polyethyleneimine modified straw hydrochar. *Science of the Total Environment*, *778*, 146116. Available at: https://doi.org/10.1016/j.scitotenv.2021.146116

Heindel, J. J., Newbold, R., & Schug, T. T. (2015). Endocrine disruptors and obesity. *Nature Reviews Endocrinology*, *11*(11), 653–661. Available at: https://doi.org/10.1038/nrendo.2015.163

Hejji, L. et al. (2023). Recent advancements in nanomaterials-based aptasensors for the detection of emerging contaminants in foodstuffs. *TrAC Trends in Analytical Chemistry*, *166*, 117194. Available at: https://doi.org/10.1016/j.trac.2023.117194

Ho, V. et al. (2022). Endocrine disruptors: Challenges and future directions in epidemiologic research. *Environmental Research*, *204*. Available at:https://doi.org/10.1016/j.envres.2021.111969

Hu, X. et al. (2018). Osteoglycin-induced VEGF inhibition enhances T lymphocytes infiltrating in colorectal cancer. *EBioMedicine*, *34*, 35–45. Available at: https://doi.org/10.1016/j.ebiom.2018.07.021

Hu, Y. et al. (2023). Risk assessment of human exposure to airborne pesticides in rural greenhouses. *Scientific Reports*, *13*(1), 1–6. Available at: https://doi.org/10.1038/s41598-023-32458-y

Huang, Y., & Li, Z. (2024). Assessing pesticides in the atmosphere: A global study on pollution, human health effects, monitoring network and regulatory performance. *Environment International*, *187*, 108653. Available at: https://doi.org/10.1016/j.envint.2024.108653

Intisar, A. et al. (2023). Adsorptive and photocatalytic degradation potential of porous polymeric materials for removal of pesticides, pharmaceuticals, and dyes-based emerging contaminants from water. *Chemosphere*, *336*, 139203. Available at: https://doi.org/10.1016/j.chemosphere.2023.139203

Jain, A., Balasubramanian, R., & Srinivasan, M. P. (2016). Hydrothermal conversion of biomass waste to activated carbon with high porosity: A review. *Chemical Engineering Journal*. Elsevier, pp. 789–805. Available at: https://doi.org/10.1016/j.cej.2015.08.014

Kamrani, M., Akbari, A., & Yunessnia lehi, A. (2018). Chitosan-modified acrylic nanofiltration membrane for efficient removal of pharmaceutical compounds. *Journal of Environmental Chemical Engineering*, *6*(1), 583–587. Available at: https://doi.org/10.1016/j.jece.2017.12.044

Kasbaji, M., Mennani, M., & Boussetta, A., et al. (2022). Bio-adsorption performances of methylene blue (MB) dye on terrestrial and marine natural fibers: Effect of physicochemical properties, kinetic models and thermodynamic parameters. *Separation Science and Technology*, *15*, 1–20. Available at: https://doi.org/10.1080/01496395.2022.2104733

Kasbaji, M., Mennani, M., & Grimi, N. et al. (2022). Implementation and physico-chemical characterization of new alkali-modified bio-sorbents for cadmium removal from industrial discharges: Adsorption isotherms and kinetic approaches. *Process Biochemistry*, *120*, 213–226. Available at: https://doi.org/10.1016/j.procbio.2022.06.010

Kasbaji, M. et al. (2023). Bio-based functionalized adsorptive polymers for sustainable water decontamination: A systematic review of challenges and real-world implementation. *Environmental Pollution*, *335*, 122349. Available at: https://doi.org/10.1016/j.envpol.2023.122349

Kassem, I. et al. (2022). Biodegradable all-cellulose composite hydrogel as eco-friendly and efficient coating material for slow-release MAP fertilizer. *Progress in Organic Coatings*, *162*, 106575. Available at: https://doi.org/10.1016/j.porgcoat.2021.106575

Kaur, J. et al. (2022). Ice nucleation of pharmaceutical and synthetic organic emerging contaminants: The impact of selected environmental conditions. *ACS Earth and Space Chemistry*, *6*(9), 2236–2249. Available at: https://doi.org/10.1021/acsearthspacechem.2c00187

Khan, N. A. et al. (2023). A state-of-art-review on emerging contaminants: Environmental chemistry, health effect, and modern treatment methods. *Chemosphere*, *344*, 140264. Available at: https://doi.org/10.1016/j.chemosphere.2023.140264

Khan, S. et al. (2022). Emerging contaminants of high concern for the environment: Current trends and future research. *Environmental Research, 207*, 112609. Available at: https://doi.org/10.1016/j.envres.2021.112609

Kumar, M. et al. (2023). Current research trends on emerging contaminants pharmaceutical and personal care products (PPCPs): A comprehensive review. *Science of the Total Environment, 859*, 160031. Available at: https://doi.org/10.1016/j.scitotenv.2022.160031

Kushwaha, J., & Singh, R. (2023). Cellulose hydrogel and its derivatives: A review of application in heavy metal adsorption. *Inorganic Chemistry Communications, 152*, 110721. Available at: https://doi.org/10.1016/j.inoche.2023.110721

Larrañeta, E. et al. (2018). Synthesis and characterization of lignin hydrogels for potential applications as drug eluting antimicrobial coatings for medical materials. *ACS Sustainable Chemistry and Engineering, 6*(7), 9037–9046. Available at: https://doi.org/10.1021/acssuschemeng.8b01371

Li, C., Cabassud, C., Guigui, C. (2015). Evaluation of membrane bioreactor on removal of pharmaceutical micropollutants: a review. *Desalination and Water Treatment, 55*(4), 845–858. Available at: https://doi.org/10.1080/19443994.2014.926839

Li, C. et al. (2023). Synergistic photocatalytic nanozymes to promote contaminant removal and hydrogen production. *Materials Today Sustainability, 24*, 100537. Available at: https://doi.org/10.1016/j.mtsust.2023.100537

Liang, S. et al. (2022). Multivalent Ce-MOFs as biomimetic laccase nanozyme for environmental remediation. *Chemical Engineering Journal, 450*, 138220. Available at: https://doi.org/10.1016/j.cej.2022.138220

Ma, Y. et al. (2019). The adverse health effects of bisphenol A and related toxicity mechanisms. *Environmental Research, 176*. Available at:https://doi.org/10.1016/j.envres.2019.108575

Mangotra, A., & Singh, S. K. (2024). Volatile organic compounds: A threat to the environment and health hazards to living organisms—A review. *Journal of Biotechnology, 382*, 51–69. Available at: https://doi.org/10.1016/j.jbiotec.2023.12.013

Martínez-Zamudio, L. Y. et al. (2024). Emerging pollutants removal from leachates and water bodies by nanozyme-based approaches. *Current Opinion in Environmental Science and Health, 37*, 100522. Available at: https://doi.org/10.1016/j.coesh.2023.100522

Mastropietro, T. F. et al. (2021). Reverse osmosis and nanofiltration membranes for highly efficient PFASs removal: Overview, challenges and future perspectives. *Dalton Transactions, 50*(16), 5398–5410. Available at: https://doi.org/10.1039/d1dt00360g

Meng, Y. et al. (2019). Lignin-based hydrogels: A review of preparation, properties, and application. *International Journal of Biological Macromolecules, 135*, 1006–1019. Available at: https://doi.org/10.1016/j.ijbiomac.2019.05.198

Mennani, M., Benhamou, A., & Mekkaoui, A. (2024). Probing the evolution in catalytic graphitization of biomass-based materials for enduring energetic applications. *Journal of Materials Chemistry A* [Preprint]. Available at:https://doi.org/10.1039/d3ta07449h

Mennani, M. et al. (2024). The potential of lignin-functionalized metal catalysts—A systematic review. *Renewable and Sustainable Energy Reviews, 189*(PA), 113936. Available at: https://doi.org/10.1016/j.rser.2023.113936

Nagda, A., Meena, M., & Shah, M. P. (2022). Bioremediation of industrial effluents: A synergistic approach. *Journal of Basic Microbiology, 62*(3–4), 395–414. Available at: https://doi.org/10.1002/jobm.202100225

Naidu, R. et al. (2016). Emerging contaminants in the environment: Risk-based analysis for better management. *Chemosphere, 154*, 350–357. Available at: https://doi.org/10.1016/j.chemosphere.2016.03.068

Nasrollahzadeh, M. et al. (2021). Starch, cellulose, pectin, gum, alginate, chitin and chitosan derived (nano)materials for sustainable water treatment: A review. *Carbohydrate Polymers, 251*, 116986. Available at: https://doi.org/10.1016/j.carbpol.2020.116986

Noor Syuhadah, S., & Rohasliney, H. (2012). Rice husk as biosorbent: A review. *Health and the Environment Journal, 3*(1), 89–95.

Preethi et al. (2024). Recent progress in mineralization of emerging contaminants by advanced oxidation process: A review. *Environmental Pollution, 341*, 122842. Available at: https://doi.org/10.1016/j.envpol.2023.122842

Radjenović, J., Petrović, M., & Barceló, D. (2009). Fate and distribution of pharmaceuticals in wastewater and sewage sludge of the conventional activated sludge (CAS) and advanced membrane bioreactor (MBR) treatment. *Water Research, 43*(3), 831–841. Available at: https://doi.org/10.1016/j.watres.2008.11.043

Rajput, P. et al. (2024). Nanomaterials and biochar mediated remediation of emerging contaminants. *Science of The Total Environment, 916*, 170064. Available at: https://doi.org/10.1016/j.scitotenv.2024.170064

Rathi, B. S., Kumar, P. S., & Show, P. (2021). A review on effective removal of emerging contaminants from aquatic systems : Current trends and scope for further research. *Journal of Hazardous Materials, 409*, 124413. Available at: https://doi.org/10.1016/j.jhazmat.2020.124413

Richardson, S. D., & Ternes, T. A. (2011). Water analysis: Emerging contaminants and current issues. *Analytical Chemistry, 83*(12), 4614–4648. Available at: https://doi.org/10.1021/ac200915r

Rosenfeld, P. E., & Feng, L. G. H. (2011). 16—Emerging Contaminants. in P.E. Rosenfeld & L. G. H. B. T.-R. of H. W. Feng (Eds.), (pp. 215–222). William Andrew Publishing. Available at: https://doi.org/10.1016/B978-1-4377-7842-7.00016-7

Sabarwal, A., Kumar, K., & Singh, R. P. (2018). Hazardous effects of chemical pesticides on human health–Cancer and other associated disorders. *Environmental Toxicology and Pharmacology, 63*, 103–114. Available at: https://doi.org/10.1016/j.etap.2018.08.018

Saleh, I. A., Zouari, N., & Al-Ghouti, M. A. (2020). Removal of pesticides from water and wastewater: Chemical, physical and biological treatment approaches. *Environmental Technology and Innovation, 19*, 101026. Available at: https://doi.org/10.1016/j.eti.2020.101026

Samal, K., Mahapatra, S., & Hibzur Ali, M. (2022). Pharmaceutical wastewater as emerging contaminants (EC): Treatment technologies, impact on environment and human health. *Energy Nexus, 6*, 100076. Available at: https://doi.org/10.1016/j.nexus.2022.100076

Sandoval, M. A. et al. (2024). Contaminants of emerging concern: Occurrence, analytical techniques, and removal with electrochemical advanced oxidation processes with special emphasis in Latin America. *Environmental Pollution, 345*, 123397. Available at: https://doi.org/10.1016/j.envpol.2024.123397

Shanmuganathan, R. et al. (2023). Recent innovations and challenges in the eradication of emerging contaminants from aquatic systems. *Chemosphere, 332*, 138812. Available at: https://doi.org/10.1016/j.chemosphere.2023.138812

Shen, R. et al. (2021). The hydrochar activation and biocrude upgrading from hydrothermal treatment of lignocellulosic biomass. *Bioresource Technology, 342*, 125914. Available at: https://doi.org/10.1016/j.biortech.2021.125914

Sirajudheen, P. et al. (2020). Perceptive removal of toxic azo dyes from water using magnetic Fe3O4 reinforced graphene oxide—carboxymethyl cellulose recyclable composite : Adsorption investigation of parametric studies and their mechanisms. *Surfaces and Interfaces, 21*, 100648. Available at: https://doi.org/10.1016/j.surfin.2020.100648

Su, Z. et al. (2022) Nanozymes for foodborne microbial contaminants detection: Mechanisms, recent advances, and challenges. *Food Control, 141*, 109165. Available at: https://doi.org/10.1016/j.foodcont.2022.109165

Thamarai, P. et al. (2024) Emerging trends and promising prospects in nanotechnology for improved remediation of wastewater contaminants: Present and future outlooks. *Environmental Nanotechnology, Monitoring and Management, 21*, 100913. Available at: https://doi.org/10.1016/j.enmm.2024.100913

Thomas, J. M. et al. (2017). Polyelectrolyte functional bilayers for the removal of model emerging contaminants. *Industrial and Engineering Chemistry Research, 56*(46), 13874–13884. Available at: https://doi.org/10.1021/acs.iecr.7b02915

Togawa, K. et al. (2021). Cancer incidence in agricultural workers: Findings from an international consortium of agricultural cohort studies (AGRICOH). *Environment International*, *157*, 106825. Available at: https://doi.org/10.1016/j.envint.2021.106825

Volpe, M. et al. (2018). One stage olive mill waste streams valorisation via hydrothermal carbonisation. *Waste Management*, *80*, 224–234. Available at: https://doi.org/10.1016/j.wasman.2018.09.021.

Wu, S. et al. (2018). Rethinking intensification of constructed wetlands as a green eco-technology for wastewater treatment. *Environmental Science and Technology*, *52*(4), 1693–1694. Available at: https://doi.org/10.1021/acs.est.8b00010

Yaashikaa, P. R. et al. (2020). A critical review on the biochar production techniques, characterization, stability and applications for circular bioeconomy. *Biotechnology Reports*. Elsevier B.V. Available at: https://doi.org/10.1016/j.btre.2020.e00570

Yadav, D. et al. (2021). Environmental and health impacts of contaminants of emerging concerns: Recent treatment challenges and approaches. *Chemosphere*, *272*, 129492. Available at: https://doi.org/10.1016/j.chemosphere.2020.129492

Yuan, Y. et al. (2023). Novel method in emerging environmental contaminants detection: Fiber optic sensors based on microfluidic chips. *Science of The Total Environment*, *857*, 159563. Available at: https://doi.org/10.1016/j.scitotenv.2022.159563

Zahmatkesh, S. et al. (2023). Wastewater treatment with algal based membrane bioreactor for the future: Removing emerging contaminants. *Chemosphere*, *335*, 139134. Available at: https://doi.org/10.1016/j.chemosphere.2023.139134

Zhang, S. et al. (2022). Prospecting cellulose fibre-reinforced composite membranes for sustainable remediation and mitigation of emerging contaminants. *Chemosphere*, *305*, 135291. Available at: https://doi.org/10.1016/j.chemosphere.2022.135291

Zhu, S. et al. (2022) Simultaneous production of clean water and organic dye from dyeing wastewater by reusable lignin-derived porous carbon. *Industrial Crops and Products*, *187*(PA), 115314. Available at: https://doi.org/10.1016/j.indcrop.2022.115314

Chapter 13
Low Molecular Weight Polycyclic Aromatic Hydrocarbons (LMW-PAHs) in Water: Sources, Their Effects and Treatments by Physical Methods

Naveen Kumar, Arun Ratn, and Naveen Chandra Joshi

Abstract Polycyclic aromatic hydrocarbons (PAHs) in aquatic environments are a priority concern due to their carcinogenic and teratogenic properties, bioaccumulation, and long-range transport. PAHs emanate from partial burning of various sources like coal, oil, and natural processes like forest fires. Due to adverse effects of PAHs, it is essential to detect and eliminate PAHs from water using reliable techniques. A variety of physical, chemical and biological methods for treating PAHs contaminated water have been investigated. Physical methods including adsorption and membrane filtration have excellent removal performance and considered to be the best method for PAHs remediation due to their ease of operation, affordability, safety and universal nature. Therefore, physical treatments are frequently used as a pre-treatment step prior to applying advanced technologies. This chapter investigated the presence of low molecular weight PAHs in aquatic environments, their adverse effects on organisms, and their treatment using physical methods.

Keywords Polycyclic aromatic hydrocarbons (PAHs) · Eco-toxic · Membrane filtration · Flotation · Adsorption · Waste water

---

N. Kumar (✉)
Department of Chemistry, S.D. College, Muzaffarnagar, Uttar Pradesh 251001, India
e-mail: naveenkumariitr@gmail.com

A. Ratn
Department of Zoology, S.D. College, Muzaffarnagar, Uttar Pradesh 251001, India

N. C. Joshi
Division of Research & Innovation, Uttaranchal University, Dehradun 248007, India

H. Gupta et al. (eds.), *Emerging Contaminants in Water and Wastewater*, Environmental Science and Engineering, https://doi.org/10.1007/978-3-031-82579-8_13

13.1 Introduction

Water contamination and the shortage of clean water has become a major global problem as agricultural and industrial activities have expanded. In recent decades, a growing number of new compounds such as pharmaceuticals, pesticides, and polycyclic aromatic hydrocarbons (PAHs) are being used in high amounts in industrial activities, leading to the release of them into water bodies and causing damage to the ecosystem. Recently PAHs, major environmental pollutants, have emerged as everywhere and have been declared as priority control pollutants by various Environmental Protection Agencies. These are fat-soluble, degradation-resistant, and are commonly found in aquatic environments, including water, sediments, fish, seabirds, and marine mammals. PAHs in aquatic environments mainly originate from the partial ignition of coal, oil, gas, and petroleum products. PAHs also come from natural processes such as forest fires (Gupta & Gupta, 2015; Madhav et al., 2020; Meng et al., 2019; Mojiri et al., 2019).

The USEPA and EU agencies have designated several PAHs as main concern because of their adverse effect on human health. Further, on the basis of human exposure and toxicity, the USEPA has identified 16 PAHs as a priority control pollutants (Busetti et al., 2006; Manoli & Samara, 2008). PAHs subdivided into two group i.e. low and high molecular weight PAHs. In this chapter these are denoted by LMW-PAHs and HMW-PAHs, respectively. LMW-PAHs have two or three rings while HMW-PAHs have four or more than four rings. It was found that LMW-PAHs are more easily degrade as compare to HMW-PAHs and HMW-PAHs are sturdy absorbed to sediments. PAHs have very low water solubility and therefore persist for long time in the aquatic environment. Due to strongly adsorption of PAHs in soil sediments, these are little penetrate into groundwater, therefore their concentration levels in uncontaminated groundwater are typically very low (0–5 ng L^{-1}). However, at contaminated sites, they exceed above 10 ng L^{-1} concentration. Therefore, due to their persisting nature and trace concentrations (pg L^{-1} to ng L^{-1}) in aquatic environment it is difficult to identify and treat PAHs contaminated water (Behera et al., 2018; Prince et al., 2021).

PAHs and their derivatives spread through the food chain and accumulate in organisms. They are causing a number of adverse effects such as endocrine alteration, growth reduction, DNA damage, to various aquatic organisms. Human exposure to PAHs can cause a higher risk of developing cancer in several different organs such as the pancreas, skin, stomach, lung, bladder and breast, as well, they have been related with suppression of the immune system and are suspected to be endocrine disruptors (Rengarajan et al., 2015; Siddens et al., 2012; Zhang et al., 2016). Many research revealed that due to the ability of accumulation of PAHs in organisms they are considered as carcinogenic, acutely toxic and genotoxic chemicals (Juhasz & Naidu, 2000). Therefore, suitable and efficient methods are required for the treatment of PAHs contaminated water.

Numbers of physical, chemical and biological methods for eliminating PAHs from polluted water have been investigated. The most common methods for treating

PAH-contaminated waters are extraction, coagulation, membrane filtration, adsorption (employing adsorbents such as activated carbons, carbon nanotubes, graphene, graphene oxide, mesoporous silica, clays, zeolites, various nanocomposites and many bichar based materials), chemical oxidation, photodegradation, bioremediation, and phytoremidiation. Electrokinetic remediation, advanced oxidation methods, advanced phytoremediation, green nanoremediation, and integrated approach is still in the development stage and has great potential (Adeola & Forbes, 2020; Madhav et al., 2024; Smol & Włodarczyk-makuła, 2016). However, the most of these methods have several disadvantages. In chemical process some by product are more dangerous than the PAHs. Further, their chemical degradation efficiency also decreases due to poor water solubility (Ukiwe et al., 2013). Bioremediation requires pollutants to be in a chemical state that microbes can use, but PAHs are often not soluble in water, leading to their persistence in nature. As a result, traditional biological degradation systems are not very effective. However, adding surfactants or emulsifying agents can help solubilise the compounds and speed up microbial degradation (Mueller et al., 1989).

Moreover, these processes require high investment and maintenance costs. Therefore, from the above mentioned point of views physical methods (adsorption, membrane filtration, etc.) may be the best method for PAHs treatment. Physical methods are more popular due to their ease operating procedure, safety, affordability and reliability (Akinpelu et al., 2018; Lamichhaneet al., 2016).

Based on the size of the contaminants retained on the membrane, the membrane filtration processes are classified into four types. These four membrane filtration processes are microfiltration (MF), nanofiltration (NF), ultrafiltration (UF), and reverse osmosis (RO). Out of these filtration processes NF and RO are more effective techniques which can be used independently or in integration with other conventional methods (Smol & Włodarczyk-makuła, 2016). To eliminate PAHs from public landfill leachate, Smol et al. (2014) used RO and UF membrane filtration process. In this process, the PAHs were extracted from landfill leachate using 2-propanol. The findings of research showed that the initial concentration of PAHs was found 17.5–30.4 mg L^{-1} which are reduces to 19–100% and 48–73% in RO and UF processes, respectively (Smol et al., 2014). Barman et al. (2022) prepared a mesoporous chitosan based carbon-bentonite nano composite membrane which was denoted as AC-NB-C. The AC-NB-C had 482.07 m^2 g^{-1} BET surface area and 0.251 cm^3 g^{-1} of total pore volume. For efficiency of AC-NB-C membrane for PAHs removal was optimized by varying initial concentration of PAH, pH of solution and applied pressure. They reported that the prepared membrane rejected about 99.3% of PAHs (NAP and ACE) under optimized reaction conditions. FTIR studies showed that the adsorption takes place through π–π stacking and hydrogen bonding (Barman et al., 2022).

Adsorption method is also very effective for the removal of PAHs from water and for this purpose many low cost adsorbents and bio adsorbents can be used. In this context, Researchers investigated and reviewed the effectiveness of various adsorbents for the removal of PAHs. Some popular adsorbents used for PAHs removal are activated carbon, carbon nanotubes, graphene, lignocelluloses, chitin, algae, bacterial and fungal biomass (Akinpelu et al., 2018; Lamichhane et al., 2016; Tran et al.,

2015). For this reason, there has been an increasing interest in finding low-cost, readily available adsorbents for adsorbing PAH. In this relation, Kumar and Gupta (2020) obtained sawdust based activated carbon (SDAC) which is inexpensive and exhibited high specific BET surface area of 1400 m^2 g^{-1}. The SDAC used for removal of NAP from water and its desorption efficiency was greater than 90% for up to six cycles (Kumar & Gupta, 2020). Cheng et al. (2019) developed corn cobs and rape straw based highly efficient porous carbons (PCs) by hydrothermal techniques followed by activation with $KHCO_3$. Synthesized PCs efficiently remove NAP, ACE, and PHE from water with high adsorption efficiency due to their large pore volume and surface area. The PCs illustrate 592.97, 480.27, and 692.27 mg g^{-1} adsorption efficiency for NAP, ACE, and PHE, respectively (Cheng et al., 2019). Eeshwarasinghe et al. (2019) used granular activated carbon (GAC) to remove heavy metals (Cd, Cu, Zn) and PAHs (ACE and PHE) from water. The study found that the adsorption of these contaminants together was lower than when they were present separately. The Langmuir isotherm model effectively represented the adsorption equilibrium data, with maximum adsorption capacities of 7.36 mg g^{-1} for PHE and 2.63 mg g^{-1} for ACE (Eeshwarasinghe et al., 2019). Yuan et al. (2010) was fabricated four porous carbons (specific surface area ranging from 562 to 1904 m^2 g^{-1}) using petroleum coke for the elimination of selected PAHs (NAP, FL, PHE, PYR and FLU) from water. These carbons synthesised by the chemical activation of coke using KOH at 600–1000 °C. The value of capacity parameter (K_f) for all the four fabricated porous carbon follows the order of NAP > FL > PHE > PYR with > 90% of PAHs removal in 30 min (Yuan et al., 2010). Thus, adsorptive removal of PAHs appears to be the most efficient technique owing to its simplicity and excellent removal efficiency. This chapter investigates the sources of LMW-PAHs, their impact on the environment and people, as well as their removal through physical methods i.e. membrane filtration, flotation and adsorption processes.

13.2 Physico-chemical Properties of LMW-PAHs

Chemically, in LMW-PAHs two or three aromatic rings are fused in linear, clustered, or angular configurations. They are predominantly white, pale yellow or colourless, solid compounds. PAHs have low vapour pressures and are very poorly soluble in water. The solubility and vapour pressure decreases as the molecular weight increases of PAH (Abdel-shafy & Mansour, 2015; Patel et al., 2020). Furthermore, temperature, pH, and ionic strength affect the water solubility of PAHs. PAHs have low solubility in water (ranging from 0.73×10^5 ng L^{-1} for ANT to 3.17×10^7 ng L^{-1} for NAP), but easily dissolve in fats and oils, leading to their accumulation in living organisms' fatty tissue (Behera et al., 2018; Rubio-clementeet al., 2014; Smol & Włodarczyk-makuła, 2016). Table 13.1 summarizes some important physicochemical characteristics of LMW-PAHs.

Table 13.1 Physico-chemical properties of LMW-PAHs

Name (symbol)	Chemical formula	Structure	Molecular weight (g/mole)	MP (°C)	BP (°C)	Solubility in water (ng/L)
Naphthalene, (NAP)	$C_{10}H_8$	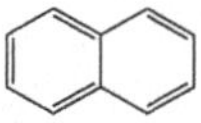	128.2	80.2	218	3.17×10^7
Acenaphthene, (ACE)	$C_{12}H_{10}$	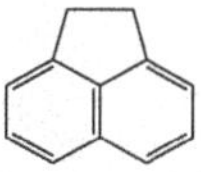	154.2	95.0	279	3.9×10^6
Acenaphthylene, (ACY)	$C_{12}H_8$	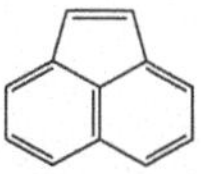	152.2	91.8	280	3.93×10^6
Fluorene, (FL)	$C_{13}H_{10}$	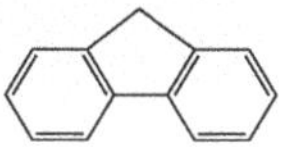	166.2	116	295	1.98×10^6
Anthracene, (ANT)	$C_{14}H_{10}$	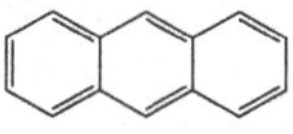	178.2	218	340	0.73×10^5
Phenanthrene, (PHE)	$C_{14}H_{10}$	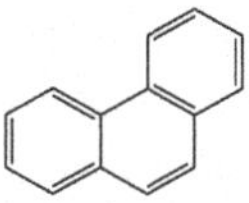	178.2	101	340	1.29×10^6

13.3 Sources and Occurrence of LMW-PAHs in Water and Wastewater

Due to both synthetic and natural processes, PAHs are found everywhere in the natural environments. Natural sources of PAH production include oil spills, volcanoes, wildfires, and forest fires, but they are not very significant. Anthropogenic activities, like vehicle exhaust and waste incineration are the main sources of PAHs. Thus, incomplete combustion is a significant contributor to the emissions of PAHs from a range of industrial processes, such as the incineration of waste, the production of iron, steel, and aluminium, cement, rubber tires, manufacturing, coal tar pitch, dyes, and electricity (Masih et al., 2010; Ravindraet al., 2008). Figure 13.1 depicts several anthropogenic and natural sources of PAHs.

Water contains PAHs, which can enter the water from soil, industrial effluents, or accidental marine spills during oil transportation. Industrial effluents, municipal runoff, contaminated groundwater runoff, spills of oils, and atmospheric particulate depositions are the primary sources of PAHs in surface waters. The sources of PAHs in groundwater include contaminated surface water, contaminated irrigation water, landfill leachate, and contaminated soil. As groundwater passes through

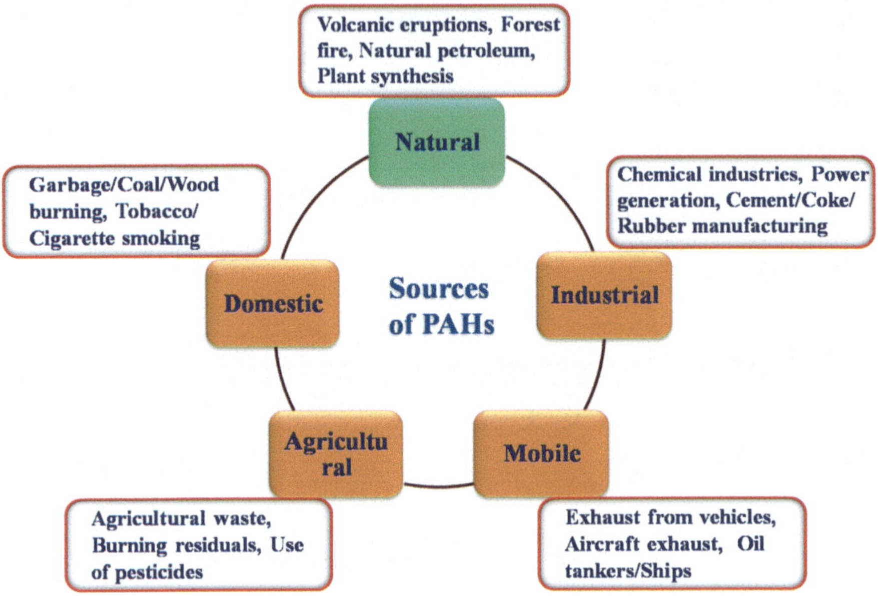

Fig. 13.1 Various natural and anthropogenic sources of PAHs

different soil matrices, it naturally filters itself, and PAHs easily attach to organic soil. Carcinogenic PAHs concentrations in groundwater in the US ranging from 0.2 to 6.9 ng L^{-1}, whereas surface water concentrations ranged from 0.1 to 800 ng L^{-1}, with the majority of cases falling between 2 and 50 ng L^{-1} (Alegbeleyeet al., 2017; Srogi, 2007). Zhang et al. (2019) investigated the geographically concentrations of PAHs in China's drinking water and reported up to 71.49 ng L^{-1} of Σ16PAHs concentration in drinking water in the Prefecture-level City (Zhang et al., 2019).

Naphthalene (NAP) is the simplest PAH, and the primary sources of environmental naphthalene are petroleum refining, coal gasification, chemical synthesis, and petrochemical wastewater. It is the most abundant (8–12%) compound in coal tar, which is obtained from during the petroleum refining process. It is used in the preparation of 2-naphthol, dyes, chemical solvents, resins, mothballs, sanitary balls and insect repellents. The environment is now seriously threatened by NAP pollution as a result of wastewater discharges from the coal tar extraction process and everyday mothball after-sales treatment (Sharma & Lee, 2015; Xu et al., 2022).

ACE and ACY are mainly released into the environment by wood combustion, burning of coal and oil, natural fires, and emissions and effluents from different industries. In many different industries, ACE and ACY are widely used in the manufacturing of plastics, dyes, fungicides, insecticides, and medications (Mallick, 2019; Shimada et al., 2015).

ANT and PHE are found mainly at oil-contaminated sites. ANT exists in exhaust fumes, crude oil, and fossil fuels (Nzila et al., 2018; Qu et al., 2021). In most of the cases the concentrations of PHE in natural water were found in the range of pg L^{-1}

to ng L^{-1}. However, the concentration of PHE in water at highly polluted site may be exceed up to 7.6–9.9 μg L^{-1} (Wang et al., 2019).

13.4 Routes of Exposure and Health Effects of LMW-PAHs

PAHs are found everywhere (air, water, soil, plant and animal) in the natural environments.

The general population is exposing to various PAHs through inhalation of cigarette smoke, fireplace smoke, and consumption of PAH-containing foods. Occupational exposure of workers to PAHs also occurs by breathing exhaust fumes during mining, oil refining, coke production, and coal gasification processes. Therefore, we can say that people are exposed to dangerous PAHs through the air, water, and contaminated foods (Lawal, 2017; Rubio-clementeet al., 2014). Fish are a key source of essential nutrients for many people, but they also contain high levels of PAH-based pollutants due to biotransformation and biomagnification processes, as well as increased human activity. Thus, the eventual ingestion of these tainted fish is now a major way that people are exposed to PAHs (Nwaichi & Ntorgbo, 2016). In aquatic organisms, exposure is thought to occur through respiration, food ingestion, sediments, airborne particles, and dermal uptake from the surrounding environment. PAHs are absorbed by mammals through ingestion, skin contact, and inhalation, while plants absorb PAHs from the soil via their roots and move them within the plant that varies depending on concentration, solubility, and soil type. PAHs are hydrophobic and accumulate in aquatic ecosystems, becoming bioavailable to organisms and accumulating in higher-trophic level biota (Abdel-shafy & Mansour, 2015; Gupta, 2016; Gupta & Kumar, 2020; Honda & Suzuki, 2020; Zhao et al., 2014).

PAHs have been found in living organisms throughout history, particularly in humans, and numerous cancers, including those of the lung, bladder, breast, oesophagus, kidney, skin, and pancreas, have been connected to PAHs. They may also lead to endocrine disruption and immune system suppression (Rengarajan et al., 2015; Siddens et al., 2012; Zhang et al., 2016). Behera et al. (2018) reported that exposure of fish to 4 mg L^{-1} NAP caused osmoregulatory imbalance, while exposure to other PAHs increased bile and gastric PAH metabolite levels and caused liver dysfunction and CYP1A activity (Behera et al., 2018). Naphthalene's toxicity increases when it forms covalent bonds with molecules found in the tissues of liver, lungs and kidney. It has also been linked to visual issues and haemolytic anaemia. NAP exposure causes skin irritation and dermatitis in sensitive individuals (Alegbeleye et al., 2017). Naphthalene derivatives such as acenaphthene and acenaphthylene are known carcinogens. Short and long-term ACE exposure damages the skin, body fluids and immune systems of animals (Kim et al., 2013; Shimada et al., 2015). Anthracene (ANT) is a highly lipophilic compound found in surface water, sewage sludge, and food, and is resistant to photodegradation. It can easily enter into the food chain via water and is associated with serious diseases (Zango et al., 2019). It has been

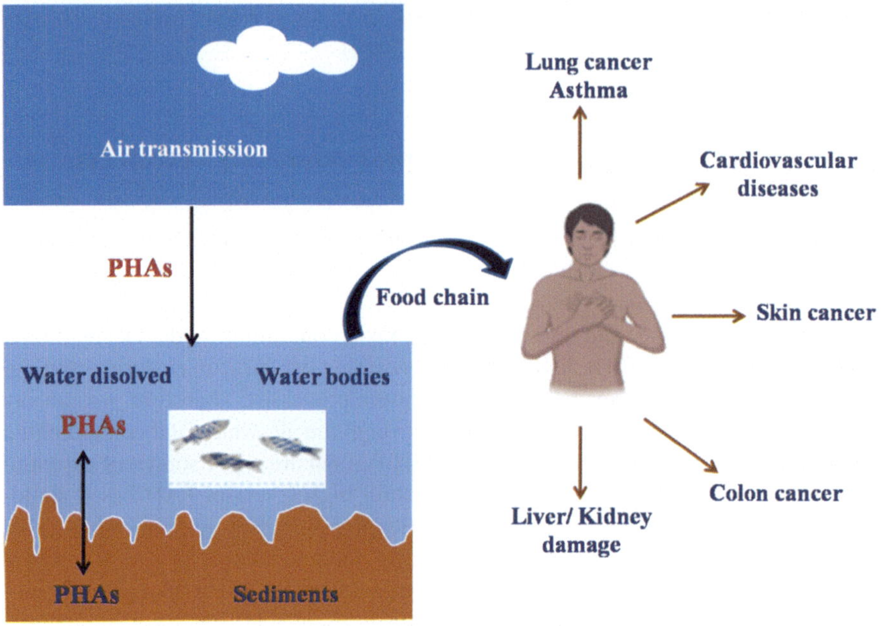

Fig. 13.2 Routes of exposure of PAHs and their outcome on human health

demonstrated that phenanthrene (PHE), a significant component of urban air pollution, causes mutations and compromises immunological function (Liu et al., 2013). Ram et al. (2009) found that PHE levels are higher in children with bronchial asthma, causing oxidative stress by affecting GSH and MDA levels (Ram et al., 2009). PHE causes swelling and oxidative stress in female rat lungs and liver. This inflammation may be linked to suppressed Treg cell function (Ma et al., 2019). According to Huang et al. (2016), exposure to PHE may cause disruptions to important regulators of cardiac hypertrophy, decrease the expression of miR-133a through DNA methylation, and ultimately cause cardiomyocyte hypertrophy in rats (Huang et al., 2016). Figure 13.2 depicts the various routes of PAHs exposure and their detrimental effects on human health.

13.5 Physical Methods of Removal of LMW-PAHs

Methods for removing PAHs include physical, chemical, biological, and advanced processes such as ultrasound treatment, activated sludge, adsorption, membrane filtration and many integrated processes (Fig. 13.3) (Fatone et al., 2011; Gong et al., 2017; Rani & Karthikeyan, 2019; Sher et al., 2023).

Physical approaches such as filtration and adsorption are the most excellent methods at the moment because of their ease of operation, safety, affordability, and

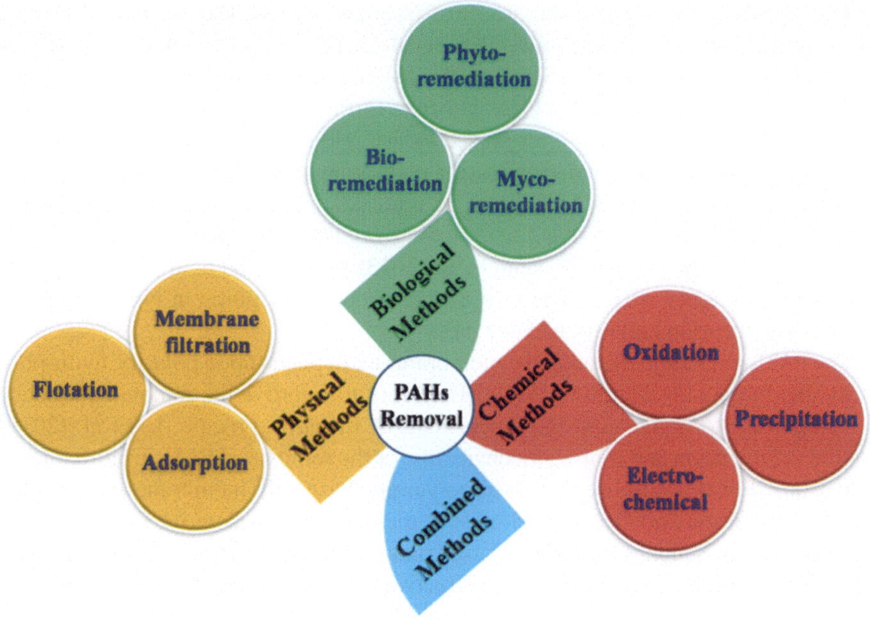

Fig. 13.3 Various methods of removal of PAHs from water

universality (Akinpelu et al., 2018; Lamichhane et al., 2016). Both organic and inorganic pollutants, as well as coarse particles, can be eliminated from wastewater by physical treatment techniques which are discussed below.

13.5.1 Membrane Filtration

The process of treating water with porous filter medium is called membrane filtration. The filter media catches impurities and let's only water pass through. Hydrostatic pressure is used to filter high molecular weight solutes through sand, polymers, ceramics, and zeolites as filter media. A membrane separation technology is suggested for on-site treatment of oily wastewater from sea oil spills. In this technology the removal process can occur via three main mechanisms: size exclusion, charge repulsion, or physicochemical interactions. The filtration characteristics, as well as the efficiency of contaminant removal and rejection, are influenced by the type of membrane used, the compounds, and the conditions to which they are exposed. Some of the complex factors influencing removal include membrane material and contaminant properties such as pore size, hydrophobicity, functional groups, and p^{Ka}, as well as the quality of the treated water (Sher et al., 2023).

The membrane processes that are most frequently employed are microfiltration (MF), ultra filtration (UF) (González-Pérez et al., 2012; Smol & Włodarczyk-Makuła, 2012), reverse osmosis (RO) (Smol et al., 2015) and nanofiltration (NF) (Barman et al., 2022; Sanches et al., 2011). Membrane filtration systems have a small footprint, are highly efficient and emit no secondary pollutants. However, PAH particle sizes are much smaller than microfiltration membrane pore sizes. As a result, microfiltration membranes are not effective in removing these contaminants from the water. Nanofiltration is popular for removing organic micropollutants from water through its small pore size (Hosseini et al., 2023; Mohammad et al., 2014; Smol & Włodarczyk-makuła, 2016). The size and distribution of pores in NF membranes affect the adsorption and removal of PAHs.

The PAHs exchanges through the membrane are depends upon polarity, hydrophobicity and molecular weight of PAHs. Therefore to improve the removal efficiency it is important to understand the mechanism of filtration process. Li et al. (2019) studied the rejection and fouling mechanisms for the removal of PAHs from synthetic wastewater. For this purpose they used polyamide based NF membrane and reported that the adsorption and size exclusion were the main mechanism for PAHs removal (Li et al., 2019). Sanches et al. (2011) used nanofiltration flat-sheet membrane with a polysulphone layers to understand the effectiveness and rejection mechanism of NF to remove PAHs from various drinking water bodies. They reported that PAHs were eliminated with 99.4% (NAP) and 94.2% (ANT) removal efficiency via hydrophobic interactions mechanism (Sanches et al., 2011). Celebioglu et al. (2019) fabricated an electrospun polycyclodextrin (poly-CD) based membrane by the electrospinning and followed by heat treatment of HP-β-CD solutions. The prepared poly-CD was used to eliminate various PAHs including ACE, FL, FLU, PHE and PYR and heavy metals from water. The removal efficiency for PAHs was reported approximately 92.6% with 0.43 ± 0.045 mg/g of the adsorption capacity (qe) of poly-CD. The poly-CD membrane recycled by using CH_3CN or 2% HNO_3 solution and reuse up to four times without losing its performance (Celebioglu et al., 2019). Gong et al. (2017) applied electrocoagulation (EC) and membrane filtration techniques in the integration mode for the treatment of paper-making water. They introduced EC technique in combination with six types of flat-sheet membranes. In this study more than 90% of PAHs removal was achieved (Gong et al., 2017).

Table 13.2 outlines the data for PAHs removal from membrane filtration processes.

13.5.2 Flotation

Flotation process is used to remove lighter substances such as oil and fatty material from water. It is defined as the upward motion of lighter particles, flakes or aggregates due to buoyancy effect (Wang et al., 2012). Because of its low operating costs and high rate of separation in relatively short residence periods, flotation is now being used for a wider range of water and wastewater treatment applications. It effectively removes organic and inorganic pollutants using gas bubbles for separation (Kyzas &

Table 13.2 A summary for the removal of various LMW-PAHs from membrane filtration procedure

Name of technique	Contaminants	Removal efficiency	Additional information	References
UF	NAP, ACE, ACY, ANT, FL, PHE	~ 100%	Polytetrafluoroethylene (PTFE) based membrane	(Hosseini et al., 2023)
NF	NAP	–	Polyamide (PA) nanofiltration membranes	(Li et al., 2019)
NF/UF/RO	NAP, ACE, ACY, ANT, FL, PHE	> 90%	Combined approach using electro-coagulation and membrane filtration	(Gong et al., 2017)
NF	ACE, FL, PHE	~ 92%	Electrospunpolycyclodextrin (poly-CD) nanofibers-based membrane	(Celebioglu et al., 2019)
UF	NAP, ACE, ACY, ANT, FL, PHE	19.9–70.8%	–	(Smol & Włodarczyk-Makuła, 2012)
UF	PHE	82%	Polyvinylidenefluoride (PVDF) based UF Membrane bioreactor (MBR)	(González-Pérez et al., 2012)
UF	NAP	92.8%	TiO_2/UV-C photocatalysis coupled with ultrafiltration (UF) membrane	(Rani & Karthikeyan, 2019)
RO	NAP, ACE, ACY, ANT, FL, PHE	43–67%	Polyamide (PA) based membrane	(Smol et al., 2015)
NF	NAP, ACE	~ 99%	Activated carbon-nano-bentonite composite linked with chitosan	(Barman et al., 2022)
NF	NAP, ANT	99.4% (NAP), 94.2% (ANT)	The nanofiltration flat-sheet membrane with a polysulphone support layer	(Sanches et al., 2011)
UF/RO	NAP, ACE, ACY, ANT, FL, PHE	19 to 77%	–	(Smol et al., 2014)
UF	NAP, ACE, ACY, ANT, FL, PHE	38 to 94%	–	(Fatone et al., 2011)

Matis, 2018; Rubioet al., 2002). Moreover, this method is based on the variation in aquaphobicity between the molecules that need to be removed. It involves the subsequent actions as follows (1) bubble generation, (2) bubble-droplet dispersion and contact, (3) aggregate bubble-droplet formation, and (4) aggregate transportation (da Silva et al., 2014). PAHs are removed during flotation using induced gas flotation (IGF), dissolved gas flotation (DGF), and electro-flotation. IGF and DGF produce bubbles differently, which affects the retention time, size, surface area etc. of the bubbles (Piccioli et al., 2020). Table 13.3 compares and summarizes the approaches.

Ozerova et al. (2021) reported that the use of flocculants and coagulants enhances the flotation process of pollutants by increasing the hydrophobicity of the pollutants. Flocculants contain a hydrophilic and hydrophobic portion. They are classified as cationic, anionic, non-ionic, and amphoteric based on their hydrophilic part (Ozerova et al., 2021). For example, to increase hydrophobic nature of PAHs a chemical reagent Tween 80 (known as collectors) was used successfully by Chebbi et al. (2019). They investigated the importance of the conditioning process as well as effect of adding Tween 80 and ethanol on PAH removal efficiency. In this study 93.67% recovery of PAHs from was achieved after 30 min of conditioning and 20 min of flotation which indicating the importance of the conditioning phase before the flotation process. Moreover, the conditioning and flotation times were lowered from 30 to 10 min and 20 to 12 min, respectively, by using ethanol as a frother substance (Chebbi et al., 2019).

Graphene oxide (GO) has huge surface area and appropriate surface functionalization for PAHs adsorption. Yahya and Lau (2021) developed GO-coated microbubbles to enhance flotation. They used 1-Dodecyl-3-methylimidazolium chloride to connect microbubbles and GO. The fabricated GO-coated microbubbles had a considerably removal efficiency of 64.77% for PHE and 74.01% for PYR than uncoated microbubbles (Yahya & Lau, 2021). Recent research confirmed that flotation process with colloidal gas aphrons (CGAs) showed excellent removal efficiency for many pollutants. Pal et al. (2022) created CGAs by using various surfactants and used them for the elimination of low ring PAHs (NAP, FL, PHE) from water. They optimized removal efficiency by varying type of surfactant, and other important factors such as the flow rate, pH and temperature of PAHs and CGAs solution. In this method the CGAs derived from cationic surfactant completely removed the FL and PHE at 5.6 pH and 25 °C temperature, while NAP was removed to 88.4% under the analogous situation (Pal et al., 2022). Thus, the ability to save time, improve the efficiency of small particle removal, lower operating costs, and a small area requirement make flotation a good choice for wastewater treatment.

Table 13.3 Comparison between IGF and DGF type systems

Parameter	IGF system	DGF system
Bubble size	100–1000 μm	10–100 μm
Bubble generation method	Velocity based	Pressure based
Operating condition	Turbulent and less quiescent; multi-cell configuration	Quiescent; usually single-cell configuration
Retention time	< 5 min	5–15 min
Capital cost	Low	High

13.5.3 Adsorption

Adsorption is a mass transfer mechanism where substances accumulated at the interface between two phases. Adsorption in water treatment involves the interaction between contaminated water and a solid adsorbent at their interface (Dąbrowski, 2001). Adsorption is divided into chemical (chemisorption) and physical (physisorption) categories based on whether the adsorbate is chemically or physically bound to the surface. Physisorption occurs through weak Vander Waals forces, whereas chemisorption involves stronger forces (typically covalent chemical bonding) (Katibi et al., 2021). Due to PAHs' low water solubility and high adsorption affinity for solid media, adsorption methods have emerged as a promising technique to eliminate PAHs (Liu et al., 2014; Wang et al., 2014). Moreover, adsorption is the most simple, affordable, effective, and efficient remediation methods, and produces no unnecessary harmful by-products during the process. Many adsorbents such as activated carbon, carbon nanoparticles, graphene, bentonite, silica-based adsorbents and low-cost agricultural biomass, have been used for pollutant treatment in water (Akinpelu et al., 2018; Kumar et al., 2023; Lamichhane et al., 2016; Singh et al., 2018, 2020, 2022; Tran et al., 2015). The composition of adsorbents is principally responsible for removing the particular contaminants from water. Moreover, the adsorption efficiency depends upon the surface area, size, polarity, and morphology of adsorbent (Ali et al., 2016). Thus, in the adsorption process the removal efficiency of PAHs will depends mainly on the choice of adsorbent.

Eeshwarasinghe et al. (2018) examined the adsorptive removal of selected PAHs (NAP, ACE, ACY, FL and PHE) using granular activated carbon. In this study the adsorption efficiencies were reported as 16.2, 20.0, and 31.0 (mg g^{-1}) for NAP, ACE and ACY respectively (Eeshwarasinghe et al., 2018). Due to high surface area activated carbon is increasingly popular for adsorption. In a study vehicle tyre based high quality activated carbon (VTAC) was prepared for the removal of ANT from water. It was very effective adsorbent with respect to ANT and more than 99% of ANT was adsorbed on 14 mg of carbon in 75 min of contact time under optimized reaction conditions (Gupta, 2018). Kong et al. (2011) studied the removal of NAP, ACE, and PHE using soybean stalk-based activated carbon. Their findings revealed high removal efficiencies for NAP (100%), ACE (95.64%), and PHE (99.89%), respectively (Kong et al., 2011). In one study, microwave power was used to produce coal-based activated carbon. Under microwave treatment, activated carbon's pore volume and BET surface area increased, making it an effective adsorbent (adsorption capacity = 180.87 mg g^{-1}) for removing NAP from aqueous solution (Ge et al., 2015).

Biochar made from agricultural wastes effectively removes PAHs from water, with remarkable performance even at very low concentrations. It is cost-effective and environmentally friendly. Biochar, like activated carbon, is made by pyrolyzing agricultural, residues (such as rice husks, peanut hulls, wheat residue, pecan shells, corn stover, pine needle, apricot stones, and coniferous saw mill residue) in the absence of O_2 at 300–700 °C temperature (Lamichhane et al., 2016; Tran et al., 2015). Rashad et al. (2023) recently created three distinct biochar adsorbents from pinewood sawdust (PWSDB) by pyrolyzing it for three hours at 500 °C after chemically and physically activating it. The synthesized PWSDB biochar was used to eliminate LMW-PAHs (NAP, ANT, and PHE) from synthetic wastewater. Batch adsorption experiments demonstrated that the removal efficacy of PAHs was around 90%, providing an environmentally friendly and prompting opportunity for water treatment (Rashad et al., 2023). Gupta and co-workers developed biochar-based activated carbon from orange rind and banana peel to remove specific PAHs from water. They found that synthesized activated carbon was a cheap, effective, and recyclable adsorbent for removing PAHs from water (Gupta, 2015; Gupta & Gupta, 2016). Zhang et al. (2018) found that the *Klebsiella oxytoca* PYR-1 has a high ability for removing PAHs from wastewater in a shorter equilibrium time and with greater adsorption capacity. They reported that baking 0.2 g L^{-1} of adsorbent resulted in removal efficiencies of 13.2%, 31.1%, 34.6%, 59.7%, and 64.6% for NAP, ACE, FL, PHE, and ANT, respectively (Zhang et al., 2018). In one study, activated porous biochar (PBC_{KOH}) was created through pyrolysis of corn straw followed by activation by KOH. The PBC_{KOH} demonstrated excellent adsorption performance, with 450.43 mg g^{-1} adsorption capacity of NAP (Qu et al., 2020).

In addition to the common adsorbents mentioned above, other important adsorbents for the efficient elimination of PAHs from water were used. Hu et al. (2014) used hydrophobic mesoporous MCM-41 molecular sieves (surface area 37.2/g) to remove PHE. The removal rate of PHE reached 94% with an adsorbent dose of up to 0.4 g L^{-1}. These findings indicate that MCM-41 molecular sieves have a significant adsorption effect for PHE even at low doses (Hu et al., 2014). Various clays like bentonite, zeolite, montmorillonite, and kaolinite have been applied extensively to eliminate PAHs from water (Shang & Sun, 2019). Zhou et al. (2013) investigated the adsorptive removal of ACE, FL, and PHE from surfactant solution using organobentonite adsorbent. Different PAHs removal percentages from surfactant solution were in the following order: PHE > FL > ACE, and increased from 49.6% to 69.4% as PAH hydrophobicity increased. Hydrophilicity of surfactant also play an important role, and in this study 96.7% removal with the highest selectivity 587.9 of PHE was reported by using the most hydrophilic surfactant (Zhou et al., 2013). In a study, Hedayati and Li (2020) fabricated modified clinoptilolites adsorbent by using various surfactants. The fabricated cost-effective, modified clinoptilolites used for the removal of ANT, FL, FLU, PHE, and PYR from aqueous media. The adsorption mechanism of PAHs was based on the hydrophobicity of PAHs and π-π electron donor–acceptor interactions with surfactants. According to authors 84–94% of PAHs (ANT, PHE, and FL) removal was observed in optimized reaction conditions (Hedayati & Li, 2020). Changchaivong and Khaodhiar (2009) prepared eight modified organoclays using sodium-rich bentonite and dodecylpyridinium chloride. The as prepared dodecylpyridinium-modified bentonite used to treat NAP and PHE from aqueous solutions. The adsorption efficiency of modified clays towards NAP and PHE depends upon the amount of dodecylpyridinium and an enhancement in adsorption efficiency onto organoclays was observed as the more dodecylpyridinium was added. The high efficiency of these dodecylpyridinium-modified bentonite is due to the high affinity between PAHs and organoclay via pyridinium ring in dodecylpyridinium through π-π interactions, (Changchaivong & Khaodhiar, 2009).

Many authors have recently reported the use of metal organic framework (MOF)-based adsorbents to eliminate PAHs from the atmosphere. Zango et al. (2019) created three MOF-based adsorbents MIL-88(Fe), NH_2-MIL-88(Fe), and mixed-MIL-88(Fe) for ANT removal from water. The qe values of these MOFs were reported 23.0, 21.5, and 21.9 mg g^{-1}, with removal efficiencies of 98.3, 92.4, and 95.8% (Zango et al., 2019). Okoro et al. (2019) created Zn and Cu-based MOFs for the treatment of NAP and ANT. The as synthesized Zn-BDC and Cu-BDC MOFs had excellent potential performance and reported 97% and 50% removal of NAP and ANT, respectively (Okoro et al., 2019).

A literature review revealed that the nanoparticles and nanocomposite materials are extra capable in removing organic and inorganic contaminants through the adsorption treatment process. In this context, Dai et al. (2020) synthesized organic montmorillonite sodium alginate nanocomposites adsorbent to eliminate ACE, FL, and PHE from water. The nanocomposite was synthesized by mixing montmorillonite with cetyltrimethylammonium bromide and then coagulated with sodium alginate using $CaCl_2$ as a cross-linking agent. The findings of the study showed that the nanocomposite efficiently eliminate the PAHs and adsorption capacity observed as PHE (2.50 mg/g) > ACE (1.20 mg/g) > FL (0.90 mg/g) (Dai et al., 2020). Balati et al. (2015) created organic–inorganic nanohybrid NH2-SBA-15 adsorbent for NAP, ACE, and PHE removal from synthetic wastewater (real petroleum refinery industry effluent). To obtain maximum adsorption efficiency they were optimized the various parameters which affect the adsorption performance. The percent removal of NAP, ACE, and PHE and adsorption capacity (mg g^{-1}) of NH2-SBA-15 was reported as 93% (1.67 mg g^{-1}), 97% (1.06 mg g^{-1}), and 85% (0.24 mg g^{-1}), for NAP, ACE, and PHE respectively (Balati et al., 2015). Bayazit et al. (2017) created graphene nanoplatelet supported nanocomposite material (denoted as MIL-101) by hydrothermal process for purification of NAP contaminated water. They optimized various reaction parameters and reported approximately 93% removal of NAP with 0.075 g dosages of adsorbent at 298 K temperature (Bayazit et al., 2017). Mirzaee and Sartaj (2022) prepared a magnetic powder activated carbon (MPAC) using Fe_3O_4 nanoparticles for the treatment of selected PAHs. The authors reported that the MPAC achieved the adsorption equilibrium of PAHs in 6 h, with removal efficiencies varying between 95.6 and 100.0%. MPAC particles have adsorption capacities of 8.74–11.37 μg/mg for LMW-PAHs and 8.43–20.21 μg/mg for HMW-PAHs (Mirzaee & Sartaj, 2022). Table 13.4 outlines studies on removing LMW-PAHs from water.

13.6 Concluding Remarks and Future Perspectives

PAHs are present in water due to human activities, accidents, or natural events and they harm both the environment and human health. Therefore, methods to monitor and eliminate these pollutants have gained significant attention. Research papers

Table 13.4 Summary of adsorptive removal of LMW-PAHs from aqueous solution by various adsorbents

Adsorbents	PAHs	Removal efficiency (%) or adsorption capacity (mg/g)	Sources	References
Modified bentonite (organoclay)	NAP	–	Stormwater runoff	(Shang & Sun, 2019)
β-cyclodextrin functionalized magnetic hydrochar	ANT	60.27 mg/g	Aqueous solution	(Qu et al., 2021)
Metal organic framework (MOF) based adsorbents	ANT	95.8%	Aqueous solution	(Zango et al., 2019)
Activated carbon	PHE	~ 90%	Aqueous solution	(Liu et al., 2014)
Granular activated carbon	NAP	16.2 mg/g	Aqueous solution	(Eeshwarasinghe et al., 2018)
	ACE	20.0 mg/g		
	ACY	31.0 mg/g		
Vehicular tyre based activated carbon	ANT	> 99%	Aqueous solution	(Gupta, 2018)
Soybean stalk-based activated carbon	NAP	100%	Aqueous solution	(Kong et al., 2011)
	ACE	95.64%		
	PHE	99.89%		
Coal-based activated carbon	NAP	180.87 mg/g	Aqueous solution	(Ge et al., 2015)
Pinewood sawdust biochar	NAP	~ 90%	Synthetic wastewater	(Rashad et al., 2023)
	ANT			
	PHE			
Orange rind based activated carbon	PHE	70.92 mg/g	Aqueous solution	(Gupta, 2015)
Waste banana peel based activated carbon	NAP	333.33	Aqueous solution	(Gupta & Gupta, 2016)
	FL	285.71		
	PHE	217.39		
Klebsiella oxytoca biomass	NAP	13.2%	Wastewater	(Zhang et al., 2018)
	ACE	31.1%		
	FL	34.6%		
	PHE	59.7%		
	ANT	64.6%		
KOH-activated porous biochar	NAP	85.56% (450.43 mg/g)	Aqueous solution	(Qu et al., 2020)
Mesoporous MCM-41 molecular sieves	PHE	94%	Aqueous solution	(Hu et al., 2014)
Organobentonite	ACE	49.6%	Aqueous surfactant solution	(Zhou et al., 2013)
	FL	54.5%		

(continued)

Table 13.4 (continued)

Adsorbents	PAHs	Removal efficiency (%) or adsorption capacity (mg/g)	Sources	References
	PHE	69.4%		
Modified clinoptilolites	ANT	35%	Aqueous solution	(Hedayati & Li, 2020)
	FL	83.5%		
	PHE	93.8%		
Dodecylpyridinium-modified bentonite	NAP, PHE	–		(Changchaivong & Khaodhiar, 2009)
Zinc and copper based MOF	NAP	97%	Aqueous solution	(Okoro et al., 2019)
	ANT	50%		
Organic montmorillonite sodium alginate nanocomposites	ACE	1.20 mg/g	Aqueous solution	(Dai et al., 2020)
	FL	0.90 mg/g		
	PHE	2.50 mg/g		
Silica-based organic–inorganic nanohybrid adsorbent	NAP	93%	Synthetic wastewater	(Balati et al., 2015)
	ACE	97%		
	PHE	85%		
Graphene nanoplatelet supported MIL-101 composite	NAP	93%	Aqueous solution	(Bayazit et al., 2017)
Magnetic powder activated carbon nano composite	ACE	94.2 mg/g	Aqueous solution	(Mirzaee & Sartaj, 2022)
	ANT	98.8 mg/g		
	PHE	86.4 mg/g		

on PAHs presence in water sources and removal methods were reviewed in this chapter. There has been significant progress in developing various techniques and strategies for cleaning up these contaminations. However, many challenges remain unresolved, such as in-situ PAHs remediation is difficult. The scientific literature reviewed here suggests that membrane filtration and adsorptive removal are a viable solution for LMW-PAHs remediation. Modern techniques like combined treatment methods and adsorption based on nanocomposites have greatly increased the effectiveness of cleaning up environmental contamination. This should be explored further in future research.

References

Abdel-shafy, H. I., & Mansour, M. S. M. (2015). A review on polycyclic aromatic hydrocarbons: Source, environmental impact, effect on human health and remediation. *Egyptian Journal of Petroleum, 25*, 107–123.

Adeola, A. O., & Forbes, P. B. C. (2020). Advances in water treatment technologies for removal of polycyclic aromatic hydrocarbons: Existing concepts, emerging trends, and future prospects. *Water Environment Research, 93*, 343–359.

Akinpelu, A. A., Ali, E., Johan, M. R., Saidur, R., Qurban, M. A., & Saleh, T. A. (2018). Polycyclic aromatic hydrocarbons extraction and removal from wastewater by carbon nanotubes: A review of the current technologies, challenges and prospects. *Process Safety and Environment Protection, 122*, 68–82.

Alegbeleye, O. O., Opeolu, B. O., & Jackson, V. A. (2017). Polycyclic aromatic hydrocarbons: A critical review of environmental occurrence and bioremediation. *Environmental Management, 60*, 758–783.

Ali, R. M., Hamad, H. A., Hussein, M. M., & Malash, G. F. (2016). Potential of using green adsorbent of heavy metal removal from aqueous solutions: Adsorption kinetics, isotherm, thermodynamic, mechanism and economic analysis. *Ecological Engineering, 91*, 317–332.

Balati, A., Shahbazi, A., Amini, M. M., & Hashemi, S. H. (2015). Adsorption of polycyclic aromatic hydrocarbons from wastewater by using silica-based organic–inorganic nanohybrid material. *Journal of Water Reuse and Desalination, 05*, 50–63.

Barman, S. R., Banerjee, P., Mukhopadhayay, A., & Das, P. (2022). Biopolymer linked activated carbon-nano-bentonite composite membrane for efficient elimination of PAH mixture from aqueous solutions. *Biomass Conversion and Biorefinery, 14*, 359–373.

Bayazit, ŞS., Yildiz, M., Aşçi, Y. S., Şahin, M., Bener, M., Eğlence, S., & Salam, M. A. (2017). Rapid adsorptive removal of naphthalene from water using graphene nanoplatelet/MIL-101 (Cr) nanocomposite. *Journal of Alloys and Compounds, 701*, 740–749.

Behera, B. K., Das, A., Sarkar, D. J., Weerathunge, P., Parida, P. K., Das, B. K., Thavamani, P., Ramanathan, R., & Bansal, V. (2018). Polycyclic Aromatic Hydrocarbons (PAHs) in inland aquatic ecosystems: Perils and remedies through biosensors and bioremediation. *Environmental Pollution, 241*, 212–233.

Busetti, F., Heitz, A., Cuomo, M., Badoer, S., & Traverso, P. (2006). Determination of sixteen polycyclic aromatic hydrocarbons in aqueous and solid samples from an Italian wastewater treatment. *Plant Journal of Chromatography A, 1102*, 104–115.

Celebioglu, A., Topuz, F., Yildiz, Z. I., & Uyar, T. (2019). Efficient removal of polycyclic aromatic hydrocarbons and heavy metals from water by electrospun nanofibrous polycyclodextrin membranes. *ACS Omega, 4*, 7850–7860.

Changchaivong, S., & Khaodhiar, S. (2009). Adsorption of naphthalene and phenanthrene on dodecylpyridinium-modified bentonite. *Applied Clay Science, 43*, 317–321.

Chebbi, S., Allouache, A., Schwarz, M., Belkacemi, H., & Merabet, D. (2019). Treating produced water using induced air flotation: The effect of ethanol on conditioning and flotation of PAHs in the presence of Tween 80. *Polish Journal of Environmental Studies, 28*, 1–9.

Cheng, H., Bian, Y., Wang, F., Jiang, X., Ji, R., Gu, C., Yang, X., & Song, Y. (2019). Green conversion of crop residues into porous carbons and their application to efficiently remove polycyclic aromatic hydrocarbons from water: Sorption kinetics, isotherms and mechanism. *Bioresource Technology, 284*, 1–8.

Dąbrowski, A. (2001). Adsorption-from theory to practice. *Advances in Colloid and Interface Science, 93*, 135–224.

Dai, W., Wu, P., Liu, D., Hu, J., Cao, Y., Liu, T., Okoli, C. P., Wang, B., & Li, L. (2020). Adsorption of polycyclic aromatic hydrocarbons from aqueous solution by organic montmorillonite sodium alginate nanocomposites. *Chemosphere, 251*, 126074.

Eeshwarasinghe, D., Loganathan, P., Kalaruban, M., Sounthararajah, D. P., Kandasamy, J., & Vigneswaran, S. (2018). Removing polycyclic aromatic hydrocarbons from water using granular activated carbon: Kinetic and equilibrium adsorption studies. *Environmental Science and Pollution Research, 25*, 13511–13524.

Eeshwarasinghe, D., Loganathan, P., & Vigneswaran, S. (2019). Simultaneous removal of polycyclic aromatic hydrocarbons and heavy metals from water using granular activated carbon. *Chemosphere, 223*, 616–627.

Fatone, F., Fabio, S. D., Bolzonella, D., & Cecchi, F. (2011). Fate of aromatic hydrocarbons in Italian municipal wastewater systems: An overview of wastewater treatment using conventional activated-sludge processes (CASP) and membrane bioreactors (MBRs). *Water Research, 45*, 93–104.

Ge, X., Tian, F., Wu, Z., Yan, Y., Cravotto, G., & Wu, Z. (2015). Adsorption of naphthalene from aqueous solution on coal-based activated carbon modified by microwave induction: Microwave power effects. *Chemical Engineering and Processing: Process Intensification, 91*, 67–77.

Gong, C., Huang, H., Qian, Y., Zhang, Z., & Wu, H. (2017). Integrated electrocoagulation and membrane filtration for PAH removal from realistic industrial wastewater: Effectiveness and mechanisms. *RSC Advances, 7*, 52366–52374.

González-Pérez, D. M., Garralón, G., Plaza, F., Pérez, J. I., Moreno, B., & Gómez, M. A. (2012). Removal of low concentrations of phenanthrene, fluoranthene and pyrene from urban wastewater by membrane bioreactors technology. *Journal of Environmental Science and Health, Part A, 47*, 2190–2197.

Gupta, B., & Gupta, H. (2015). Iron oxide mediated degradation of mutagen pyrene and determination of degradation products. *International Journal of Environmental Science and Development, 6*(12), 908.

Gupta, H. (2015). Removal of phenanthrene from water using activated carbon developed from orange rind. *International Journal of Scientific Research in Environmental Sciences, 3*, 0248–0255.

Gupta, H. (2016). Photocatalytic degradation of phenanthrene in the presence of akaganeite nanorods and the identification of degradation products. *RSC Advances, 6*(114), 112721–112727.

Gupta, H. (2018). Anthracene removal from water onto activated carbon derived from vehicular tyre. *Separation Science and Technology, 53*(4), 613–625.

Gupta, H., & Gupta, B. (2016). Adsorption of polycyclic aromatic hydrocarbons on banana peel activated carbon. *Desalination and Water Treatment, 57*(20), 9498–9509.

Gupta, H., & Kumar, R. (2020). Distribution of selected polycyclic aromatic hydrocarbons in urban soils of Delhi, India. *Environmental Technology & Innovation, 17*, 100500.

Hedayati, M. S., & Li, L. Y. (2020). Removal of polycyclic aromatic hydrocarbons from aqueous media using modified clinoptilolite. *Journal of Environmental Management, 273*, 111113.

Honda, M., & Suzuki, N. (2020). Toxicities of polycyclic aromatic hydrocarbons for aquatic animals.

Hosseini, P. K., Liu, L., Hosseini, M. K., Bhattacharyya, A., Miao, J., & Wang, F. (2023). Treatment of a synthetic decanted oily seawater in a pilot-scale hollow fiber membrane filtration process: Experimental investigation. *Journal of Hazardous Materials, 441*, 129928.

Hu, Y., He, Y., Wang, X., & Wei, C. (2014). Efficient adsorption of phenanthrene by simply synthesized hydrophobic MCM-41 molecular sieves. *Applied Surface Science, 311*, 825–830.

Huang, L., Xi, Z., Wang, C., Zhang, Y., Yang, Z., Zhang, S., Chen, Y., & Zuo, Z. (2016). Phenanthrene exposure induces cardiac hypertrophy via reducing miR-133a expression by DNA methylation. *Science and Reports, 6*, 1–15.

Juhasz, A. L., & Naidu, R. (2000). Bioremediation of high molecular weight polycyclic aromatic hydrocarbons: a review of the microbial degradation of benzo[a]pyrene. *International Biodeterioration & Biodegradation, 45*.

Katibi, K. K., Yunos, K. F., Man, H. C., Aris, A. Z., Nor, M. Z. M., Azis, R. S., & Umar, A. M. (2021). Contemporary techniques for remediating endocrine-disrupting compounds in various water sources: Advances in treatment methods and their limitations. *Polymers (Basel), 13*, 1–46.

Kim, K., Jahan, S. A., Kabir, E., & Brown, R. J. C. (2013). A review of airborne polycyclic aromatic hydrocarbons (PAHs) and their human health effects. *Environment International, 60*, 71–80.

Kong, H., He, J., Gao, Y., Han, J., & Zhu, X. (2011). Removal of polycyclic aromatic hydrocarbons from aqueous solution on soybean stalk-based carbon. *Journal of Environmental Quality, 40*, 1737–1744.

Kumar, A., & Gupta, H. (2020). Activated carbon from sawdust for naphthalene removal from contaminated water. *Environmental Technology and Innovation, 20*, 101080.

Kumar, N., Kumar, B., Gupta, H., & Kumar, A. (2023). Development and evaluation of cellulose/graphene-oxide based composite for removing phenol from aqueous solutions. *Polymers, 15*(3), 572.

Kyzas, G. Z., & Matis, K. A. (2018). Flotation in water and wastewater treatment. *Processes, 6*, 1–16.

Lamichhane, S., Krishna, K. C. B., & Sarukkalige, R. (2016). Polycyclic aromatic hydrocarbons (PAHs) removal by sorption: A review. *Chemosphere, 148*, 336–353.

Lawal, A. T. (2017). Polycyclic aromatic hydrocarbons. A review. *Cogent Environmental Science, 3*, 1–89.

Li, S., Luo, J., Hang, X., Zhao, S., & Wan, Y. (2019). Removal of polycyclic aromatic hydrocarbons by nanofiltration membranes: Rejection and fouling mechanisms. *Journal of Membrane Science, 582*, 264–273.

Liu, J., Chen, J., Jiang, L., & Yin, X. (2014). Adsorption of mixed polycyclic aromatic hydrocarbons in surfactant solutions by activated carbon. *Journal of Industrial and Engineering Chemistry, 20*, 616–623.

Liu, J., Zhang, L., Winterroth, L. C., Garcia, M., Weiman, S., Wong, J. W., Sunwoo, J. B., & Nadeau, K. C. (2013). Epigenetically mediated pathogenic effects of phenanthrene on regulatory T cells. *Journal of Toxicology, 2013*, 1–13.

Ma, H., Wang, H., Zhang, H., Guo, H., Zhang, W., Hu, F., Yao, Y., Wang, D., Li, C., & Wang, J. (2019). Effects of phenanthrene on oxidative stress and inflammation in lung and liver of female rats. *Environmental Toxicology*, 1–10.

Madhav, S., Ahamad, A., Singh, A. K., Kushawaha, J., Chauhan, J. S., Sharma, S., & Singh, P. (2020). Water pollutants: sources and impact on the environment and human health. *Sensors in water pollutants monitoring: Role of material* (pp. 43–62.)

Madhav, S., Mishra, R., Kumari, A., Srivastav, A. L., Ahamad, A., Singh, P., Ahmed, S., Mishra, P. K., & Sillanpää, M. (2024). A review on sources identification of heavy metals in soil and remediation measures by phytoremediation-induced methods. *International Journal of Environmental Science and Technology, 21*(1), 1099–1120.

Mallick, S. (2019). Biodegradation of acenaphthene by *Sphingobacterium* sp. strain RTSB involving trans-3-carboxy-2-hydroxybenzylidenepyruvic acid as a metabolite. *Chemosphere, 219*, 748–755.

Manoli, E., & Samara, C. (2008). The removal of Polycyclic Aromatic Hydrocarbons in the wastewater treatment process: Experimental calculations and model predictions. *Environmental Pollution, 151*.

Masih, J., Masih, A., Kulshrestha, A., Singhvi, R., & Taneja, A. (2010). Characteristics of polycyclic aromatic hydrocarbons in indoor and outdoor atmosphere in the North central part of India. *Journal of Hazardous Materials, 177*, 190–198.

Meng, Y., Liu, X., Lu, S., Zhang, T., Jin, B., Wang, Q., Tang, Z., Liu, Y., Guo, X., Zhou, J., & Xi, B. (2019). A review on occurrence and risk of polycyclic aromatic hydrocarbons (PAHs) in lakes of China. *Science of the Total Environment, 651*, 2497–2506.

Mirzaee, E., & Sartaj, M. (2022). Activated carbon-based magnetic composite as an adsorbent for removal of polycyclic aromatic hydrocarbons from aqueous phase: Characterization, adsorption kinetics and isotherm studies. *Journal of Hazardous Materials Advances, 6*, 100083.

Mohammad, A. W., Teow, Y. H., Ang, W. L., Chung, Y. T., Oatley-radcliffe, D. L., & Hilal, N. (2014). Nanofiltration membranes review: Recent advances and future prospects. *Desalination, 356*, 226–254.

Mojiri, A., Zhou, J. L., Ohashi, A., Ozaki, N., & Kindaichi, T. (2019). Comprehensive review of polycyclic aromatic hydrocarbons in water sources, their effects and treatments. *Science of the Total Environment, 696*, 133971.

Mueller, J. G., Chapman, P. J., & Pritehard, P. H. (1989). Creosote-contaminated sites. *Environmental Science and Technology, 23*, 1197–1201.

Nwaichi, E. O., & Ntorgbo, S. A. (2016). Assessment of PAHs levels in some fish and seafood from different coastal waters in the Niger Delta. *Toxicological Reports, 3*, 167–172.

Nzila, A., Sankara, S., Al-momani, M., & Musa, M. M. (2018). Isolation and characterisation of bacteria degrading polycyclic aromatic hydrocarbons: Phenanthrene and anthracene. *Archives of Environmental Protection, 44*, 43–54.

Okoro, H. K., Tella, A. C., Ajibola, O. A., Zvinowanda, C., & Ngila, J. C. (2019). Adsorptive removal of Naphthalene and Anthracene from aqueous solution with Zinc and copper-terephthalate metal-organic frameworks. *Bulletin of the Chemical Society of Ethiopia, 33*, 229–241.

Ozerova, D.A., Zolkin, A.L., Kalyakina, V.M., Koval, Y.N., & Shamina, S. V. (2021). Analysis of methods for treatment of industrial wastewaters containing polycyclic aromatic hydrocarbons. *IOP Conference Series: Earth and Environmental Science*, 839.

Pal, P., Corpuz, A. G., Hasan, S. W., Sillanpää, M., & Banat, F. (2022). Treatment of polycyclic aromatic hydrocarbons (PAHs) from aqueous solutions by flotation using colloidal gas aphrons. *Separation and Purification Technology, 285*, 120367.

Patel, A. B., Shaikh, S., Jain, K. R., Desai, C., & Madamwar, D. (2020). Polycyclic aromatic hydrocarbons: Sources, toxicity, and remediation approaches. *Frontiers in Microbiology, 11*, 1–23.

Piccioli, M., Aanesen, S. V., Zhao, H., Dudek, M., & Øye, G. (2020). Gas flotation of petroleum produced water: a review on status, fundamental aspects, and perspectives. *Energy & Fuels, 34*, 15579–15592.

Prince, U., Aljerf, L., Obike, A., Emmanuel, O., & Joel, N. (2021). Groundwater for Sustainable Development Evaluation of physicochemical characteristics and health risk of polycyclic aromatic hydrocarbons in borehole waters around automobile workshops in Southeastern Nigeria. *Groundwater for Sustainable Development, 14*, 100615.

Qu, J., Wang, S., Wang, Y., Tian, X., Jiang, Z., Tao, Y., Wang, L., Deng, F., & Zhang, Y. (2021). Removal of Cd (II) and anthracene from water by β-cyclodextrin functionalized magnetic hydrochar: Performance, mechanism and recovery. *Bioresource Technology, 337*, 125428.

Qu, J., Wang, Y., Tian, X., Jiang, Z., Deng, F., Tao, Y., Jiang, Q., Wang, L., & Zhang, Y. (2020). KOH-activated porous biochar with high specific surface area for adsorptive removal of chromium (VI) and naphthalene from water: Affecting factors, mechanisms and reusability exploration. *Journal of Hazardous Materials*, 123292.

Ram, S., Awasthi, S., Mahdi, A. A., Patel, D. K., Singh, V. K., & Rita, M. (2009). Assessment of association of exposure to polycyclic aromatic hydrocarbons with bronchial asthma and oxidative stress in children: A case control study. *Indian Journal of Occupational and Environmental Medicine, 13*, 33–38.

Rani, C. N., & Karthikeyan, S. (2019). Investigation of naphthalene removal from aqueous solutions in an integrated slurry photocatalytic membrane reactor: Effect of operating parameters, identification of intermediates, and response surface approach. *Polycyclic Aromatic Compounds*, 1–20.

Rashad, E., Saleh, H. N., Eltaweil, A. S., Saleh, M. E., Sillanpaa, M., & Mostafa, A. R. (2023). Pinewood sawdust biochar as an effective biosorbent for PAHs removal from wastewater. *Biomass Conversion and Biorefinery, 13*, 13443–13459.

Ravindra, K., Sokhi, R., & Grieken, R. V. (2008). Atmospheric polycyclic aromatic hydrocarbons: Source attribution, emission factors and regulation. *Atmospheric Environment, 42*, 2895–2921.

Rengarajan, T., Rajendran, P., Nandakumar, N., Lokeshkumar, B., Rajendran, P., & Nishigaki, I. (2015). Exposure to polycyclic aromatic hydrocarbons with special focus on cancer. *Asian Pacific Journal of Tropical Biomedicine, 5*, 182–189.

Rubio-Clemente, A., Torres-Palma, R. A., & Peñuela, G. A. (2014). Removal of polycyclic aromatic hydrocarbons in aqueous environment by chemical treatments: A review. *Science of the Total Environment, 478*, 201–225.

Rubio, J., Souza, M. L., & Smith, R. W. (2002). Overview of flotation as a wastewater treatment technique. *Minerals Engineering, 15*, 139–155.

Sanches, S., Penetra, A., Granado, C., Cardoso, V. V., Ferreira, E., Benoliel, M. J., Crespo, M. T. B., Pereira, V. J., & Crespo, J. G. (2011). Removal of pesticides and polycyclic aromatic

hydrocarbons from different drinking water sources by nanofiltration. *Desalination and Water Treatment, 27*, 141–149.
Shang, H., & Sun, Z. (2019). PAHs (naphthalene) removal from stormwater runoff by organoclay amended pervious concrete. *Construction and Building Materials, 200*, 170–180.
Sharma, A., & Lee, B. (2015). Adsorptive/photo-catalytic process for naphthalene removal from aqueous media using in-situ nickel doped titanium nanocomposite. *Journal of Environmental Management, 155*, 114–122.
Sher, S., Waseem, M., & Leta, M. K. (2023). Review of techniques for the removal of polycyclic aromatic hydrocarbons from produced water. *Environments, 10*, 1–22.
Shimada, T., Takenaka, S., Murayama, N., Yamazaki, H., Kim, J., Kim, D., Yoshimoto, F. K., Guengerich, F. P., & Komori, M. (2015). Oxidation of acenaphthene and acenaphthylene by human cytochrome P450 enzymes. *Chemical Research in Toxicology, 28*, 268–278.
Siddens, L. K., Larkin, A., Krueger, S. K., Brad, C. A., Waters, K. M., Tilton, S. C., Pereira, C. B., Löhr, C. V., Arlt, V. M., Phillips, D. H., Williams, D. E., & Baird, W. M. (2012). Polycyclic aromatic hydrocarbons as skin carcinogens: Comparison of benzo[a]pyrene, dibenzo [def,p] chrysene and three environmental mixtures in the FVB/N mouse. *Toxicology and Applied Pharmacology, 264*, 377–386.
Silva, S. S. da, Chiavone-filho, O., de Barros Neto, E. L., & Foletto, E. L. (2014). Oil removal from produced water by conjugation of flotation and photo-Fenton processes. *Journal of Environmental Management*, 1–7.
Singh, S., Gupta, H., Dhiman, S., & Sahu, N. K. (2022). Decontamination of cationic dye brilliant green from the aqueous media. *Applied Water Science, 12*(4), 61.
Singh, S., Kumar, A., & Gupta, H. (2020). Activated banana peel carbon: A potential adsorbent for Rhodamine B decontamination from aqueous system. *Applied Water Science, 10*, 185.
Singh, S., Parveen, N., & Gupta, H. (2018). Adsorptive decontamination of rhodamine-B from water using banana peel powder: A biosorbent. *Environmental Technology and Innovation, 12*, 189–195.
Smol, M., & Włodarczyk-makuła, M. (2016). The EFFECTIVENESS IN THE REMOVAL OF PAHs from aqueous solutions in physical and chemical processes: a review. *Polycyclic Aromatic Compounds, 37*, 292–313.
Smol, M., & Włodarczyk-Makuła, M. (2012). Effectiveness in the removal of polycyclic aromatic hydrocarbons from industrial wastewater by ultrafiltration technique. *Archives of Environmental Protection, 38*, 49–58.
Smol, M., Włodarczyk-Makuła, M., Mielczarek, K., & Bohdziewicz, J. (2014). Comparison of the retention of selected PAHs from municipal landfill leachate by RO and UF processes. *Desalination and Water Treatment, 52*, 3889–3897.
Smol, M., Włodarczyk-Makuła, M., Mielczarek, K., Bohdziewicz, J., & Włóka, D. (2015). The use of reverse osmosis in the removal of PAHs from municipal landfill leachate. *Polycyclic Aromatic Compounds*, 37–41.
Srogi, K. (2007). Monitoring of environmental exposure to polycyclic aromatic hydrocarbons: A review. *Environmental Chemistry Letters, 5*, 169–195.
Tran, V. S., Ngo, H. H., Guo, W., Zhang, J., Liang, S., Ton-that, C., & Zhang, X. (2015). Typical low cost biosorbents for adsorptive removal of specific organic pollutants from water. *Bioresource Technology, 182*, 353–363.
Ukiwe, L. N., Egereonu, U. U., Njoku, P. C., Nwoko, C. I. A., & Allinor, J. I. (2013). Polycyclic aromatic hydrocarbons degradation techniques: a review. *International Journal of Chemistry, 5*, 43–55.
Wang, J., Chen, Z., & Chen, B. (2014). Adsorption of polycyclic aromatic hydrocarbons by graphene and graphene oxide nanosheets. *Environmental Science and Technology, 48*, 4817–4825.
Wang, J., Liu, X., Liu, G., Zhang, Z., Wu, H., Cui, B., Bai, J., & Zhang, W. (2019). Size effect of polystyrene microplastics on sorption of phenanthrene and nitrobenzene. *Ecotoxicology and Environmental Safety, 173*, 331–338.

Wang, L. K., Shammas, N. K., Selke, W. A., & Aulenbach, D. B. (2012). *Flotation technology*. Humana Press.

Xu, Q., Yuan, H., Wang, H., Xu, Y., & Yang, D. (2022). A review on modification methods of adsorbents for naphthalene in environment. *Catalysts, 12*, 1–17.

Yahya, M. S., & Lau, E. V. (2021). Graphene oxide (GO)-coated microbubble flotation for polycyclic aromatic hydrocarbon (PAH) removal from aqueous solutions. *Journal of Environmental Chemical Engineering, 9*, 106508.

Yuan, M., Tong, S., Zhao, S., & Jia, C. Q. (2010). Adsorption of polycyclic aromatic hydrocarbons from water using petroleum coke-derived porous carbon. *Journal of Hazardous Materials, 181*, 1115–1120.

Zango, Z. U., Jumbri, K., Sambudi, N. S., Bakar, N. H. H. A., Abdullah, N. A. F., Basheer, C., & Saad, B. (2019). Removal of anthracene in water by MIL-88(Fe), NH2-MIL-88(Fe), and mixed-MIL-88(Fe) metal–organic frameworks. *RSC Advances, 9*, 41490–41501.

Zhang, D., Lu, L., Zhao, H., Jin, M., Lü, T., & Lin, J. (2018). Application of *Klebsiella oxytoca* biomass in the biosorptive treatment of PAH-bearing wastewater: Effect of PAH hydrophobicity and implications for prediction. *Water, 10*, 1–14.

Zhang, Y., Dong, S., Wang, H., Tao, S., & Kiyama, R. (2016). Biological impact of environmental polycyclic aromatic hydrocarbons (ePAHs) as endocrine disruptors. *Environmental Pollution, 213*, 809–824.

Zhang, Y., Zhang, L., Huang, Z., Li, Y., Li, J., Wu, N., He, J., Zhang, Z., Liu, Y., & Niu, Z. (2019). Ecotoxicology and Environmental Safety Pollution of polycyclic aromatic hydrocarbons (PAHs) in drinking water of China: Composition, distribution and influencing factors. *Ecotoxicology and Environmental Safety, 177*, 108–116.

Zhao, Z., Zhang, L., Cai, Y., & Chen, Y. (2014). Distribution of polycyclic aromatic hydrocarbon (PAH) residues in several tissues of edible fishes from the largest freshwater lake in China, Poyang Lake, and associated human health risk assessment. *Ecotoxicology and Environmental Safety, 104*, 323–331.

Zhou, W., Wang, X., Chen, C., & Zhu, L. (2013). Removal of polycyclic aromatic hydrocarbons from surfactant solutions by selective sorption with organo-bentonite. *Chemical Engineering Journal, 233*, 251–257.

Chapter 14
Bioremediation Strategies for Pharmaceutical and Personal Care Products Contaminants: Microbial and Plant-Based Approaches

Manish Pant, Jabrinder Singh, Naveen Singhal, and Divya Bisen

Abstract In recent decades, the simultaneous impact of growing urbanisation, poor sanitation, and industrialization has caused a negative influence on global natural water reserves. The uncontrolled release of pollutants poses a significant hazard to both public health and environmental integrity. A scant 3% of the world's 326 million cubic miles of water qualifies as fresh water, with a large part encased in glaciers. The introduction of pharmaceutical/personal care products (PPCPs), which span a wide range of organic categories, has raised concerns due to their widespread presence in numerous environmental compartments. This hazard affects surface water, sewage, sludge, aquatic bodies, waste treatment plants, soil, and sediments, as well as human and wildlife populations. The growing awareness of PPCPs' ubiquity, biomagnification, destiny, and ecological risk highlights the need for thorough investigation. Water purification/treatment and the replenishment of natural aquifers are well known truths. Diverse strategies including physical, biological, and chemical approaches have been thoroughly investigated for PPCP eradication. This chapter focuses on the present state of bioremediation techniques including microbial and phytoremediation strategies adapted for PPCPs. Plant–microbe interactions show promise for synergistic bioremediation processes, perhaps leading to more successful cleanup of contaminated locations as it gains popular acceptance. These treatments aim to promote a shift to a cleaner, healthier environment that benefits all forms of life. Combating difficulties and exploring opportunities for improving bioremediation processes must be tackled by various SDG regulating agencies.

Keywords Bioremediation · Micro pollutants · Pharmaceuticals · Phytoremediation · Wastewater

M. Pant (✉) · J. Singh · N. Singhal
Department of Chemistry—School of Physical Sciences, DIT University, Dehradun, Uttarakhand 248009, India
e-mail: manishpant66@gmail.com

D. Bisen
School of Chemical Engineering and Physical Sciences, Lovely Professional University-Phagwara, Punjab 144411, India

H. Gupta et al. (eds.), *Emerging Contaminants in Water and Wastewater*, Environmental Science and Engineering, https://doi.org/10.1007/978-3-031-82579-8_14

14.1 Introduction

14.1.1 PPCPs

Water is a crucial component of ecosystems, sustaining human existence. However, as the world's population expands and industrial and agricultural operations expand, so does the demand for safe drinking water (Jurado et al., 2012). The dramatic increase in water demand places immense strain on aquatic ecosystems, which is exacerbated by the discharge of massive amounts of contaminated effluent, both organic and inorganic. These toxic substances frequently exceed the environment's natural ability to cleanse itself which resulting in a significant drop in water quality (Gupta, 2018; Khan et al., 2023; Kumar & Gupta, 2020; Kumar et al., 2017; Singh et al., 2022). As a result, water bodies become reservoirs teaming with a toxic blend of chemicals and contaminants necessitating rapid action to restore biological balance and ensure a reliable water supply for human consumption. Furthermore, alarming statistics portray a bleak picture stating that nearly half of indigenous freshwater fish species and according to Vie et al. (2009) around one-third of the global amphibian population is on the verge of extinction.

In 2004, it was observed that there were around seven million organic and inorganic substances globally. Only a scant three percent underwent proper inventory or regulation with the US—Environmental Protection Agency identifying just 126 compounds as priority pollutants (Daughton et al., 2004; Piwowarska et al., 2024; Hoenicke et al., 2007). The American Chemical Society's Registry has grown significantly to include more than 180 million compounds demonstrating ongoing developments in research and development that continue to enhance the possibility of discovering new compounds (ACS, 2021). In recent years, there has been a notable increase in studies revealing the presence of newly discovered contaminants across a variety of environmental domains. These pollutants have been detected at concentrations as minimal as nanograms to micrograms per liter. These contaminants or pollutants, known as emerging micro-pollutants, currently lack regulatory oversight (Kümmerer, 2009). According to 2007, numerous common PPCPs were added to the list including ibuprofen, diclofenac, metoprolol, cetirizine, carbamazepine, amoxicillin, musks, triclosan, parabens and phthalates. Some phthalates have been linked to cancer while others are known for their toxicology or ability to disrupt endocrine function (Duenas-Moreno et al., 2024; Ebele et al., 2017).

Despite being a small quantity environmental entry PPCPs have the potential to negatively impacts the water quality, health of humans and ecology as indicated by the studies (Yuan et al., 2009; Sires and Brillas, 2012). The enduring health implications linked to extended exposure to the collective low concentrations of diverse PPCPs, in addition to their conjugates & metabolites found in fresh water, still lack comprehensive understanding (Rivera-Utrilla et al., 2013; Stackelberg et al., 2004). As a result, there is an emergent call within the realms of environment science and engineering sectors to enhance as well as modify, or integrate advanced techniques into centralized wastewater treatment plants (CWWTPs) to effectively eliminate

PPCPs (along with other regulated pollutants) from wastewater prior to its release into natural ecosystems.

Wastewater treatment plants -WWTPs serve as a primary route for addressing the PPCPs presence in environment (Celle-Jeanton et al., 2014). However, the conventional treatment approach are often inefficient in removing many PPCPs due to their focus on degrading biodegradable, non-polar, and large molecules, which contrasts with the small and polar nature of PPCP substances. Furthermore, chemical reactions during treatment can result in the formation of conjugates-metabolites and secondary products in addition to the original compounds (Daughton, 2001; Farre et al., 2008; Ratola et al., 2012). As a result, researching advanced treatment options becomes critical to reducing their environmental impact (Gupta & Gupta, 2016; Kumar et al., 2023a, 2023b; Singh et al., 2020). Traditionally, PPCP treatment methods are divided into preliminary, secondary & tertiary stages categories. Preliminary treatment encompasses chemical or physicochemical techniques (like ozonation, electrocoagulation), while secondary treatment primarily employs biological processes like active bio sludge. Tertiary treatment encompasses advanced oxidation processes like catalytic oxidation (Pant et al., 2025a, b). The promising approach for the treatment, bioremediation as emerged as viable solution (Rana et al., 2017). Unlike some chemical methods, which may produce toxic byproducts, bioremediation employs the activities of fungal, bacterial, algae (micro-remediation), or plant species (phytoremediation), along with aerobic and anaerobic processes and membrane bioreactors, in order to preserve a balanced chemical ecosystem in water (Pant et al., 2025a, b; Misal et al., 2011).

In addition to the profusion of organic-degrading organisms such as bacteria and plants surviving in a variety of conditions, there is a rising interest in increasing biological activity to improve degradation efficiency. Numerous research has been conducted to investigate approaches to improve biological processes for remediating persistent organic pollutants (Gupta et al., 2024). This chapter goes into major discoveries from microbial and phytoremediation techniques designed particularly to treat PPCPs. The objectives opted in the work includes: (a) identify the sources of PPCPs and their categorization, (b) highlight the significance of mixed microbial culture in PPCP pollutants removal, (c) suitability of various plant species in bioremediation of these pollutants in wetlands, (d) to explore the synergistic impacts of plant–microbe association in pollutant remediation, (e) spot light the challenges and prospects in bioremediation strategies.

Researchers have investigated the many ecological niches where these species thrive, discovering novel techniques to maximize their remediation potential.

The new processes for the degradation of PPCPs are developed by utilizing the microbial communities and the different characteristics of specific plant species (Singh et al., 2024). Certain strategies and ecosystem engineering are applied to improve natural phytoremediation and biodegradation processes. This chapter offers an overview of current developments in PPCP remediation findings and techniques which is important for change the remediation sector. Sustainable options for PPCP remediation continue to be highly valued in environmental research and innovation (Hossein et al., 2023; Singh et al., 2024).

14.2 PPCPs in Wastewater: Source, Categories and Types

PPCPs have a major effect on soil, surface water and groundwater which are three crucial environmental domains (Abdallat et al., 2022). There entry into the aquatic ecology is essential (Balkrishna et al., 2017). These micropollutants and their byproducts permeate the water cycle through diverse pathways. Human metabolism stands out as a major contributor: following drug ingestion, the body undergoes partial metabolism of active pharmaceutical ingredients (APIs), releasing metabolites or unchanged parent compounds into the water cycle, predominantly through urine and to a lesser extent through fecal matter (Al Aukidy et al., 2012). Figure 14.1 visually outlines the potential origin and routes of PPCPs in nature, as elucidated by Kurade et al., (2021). WWTPs are recognized as significant environmental contributors of PPCPs. The inadequacy of customary WWTPs in eliminating PPCPs is a leading factor in their contamination of surface water and unintended adverse effects on plant and animal life in diverse environmental contexts (Petrovic et al., 2003). PPCPs, encompassing pharmaceuticals and active ingredients from personal care products, have received scant attention over time despite their notable role in the manifestation of persistent and commonly detected ECs-emerging contaminants in surface & ground aquifer systems globally (Lopez–Serna et al., 2013; Balkrishna et al., 2017). These products are consistently introduced into aquatic systems via household sewage and industrial effluents. Consequently, PPCPs are extensively found in sewage/effluent treatment units, waste water treatment plants, ground-water, storm water runoff, and sporadically in potable water (Kasprzyk-Hordern et al., 2009; Lopez–Serna et al., 2013). Particulate matter that exists in suspension and deposits of sediment along riverbeds also include substances with long-term residence (Wilkinson et al., 2017). Because of their persistent nature, these compounds can re-enter the water cycle, causing health concerns through drinking water. They are thought to contribute to reproductive abnormalities, a rise in cancer cases, and the establishment of antibiotic resistance. The drawn PPCPs are shown in Table 14.1.

14.3 Application of Bio-Remediation Strategies for PPCPs

14.3.1 Microbial-Remediation

Bioremediation is the practical, cost-effective and ecologically acceptable process of using microorganisms to remediate pollutants (Gao et al., 2020; Singh et al., 2021). In several contaminated sites in industrialised and developing nations. The bioremediation technique has been employed with varying degrees of effectiveness. Among the particular conditions for bioremediation are as follows (1) the polluted region must be in a state that supports the growth and metabolism of the microbial population and (2) the amount of pollutants present must be sufficient to support the microbial population's expansion. The environment, which includes soil kinds and

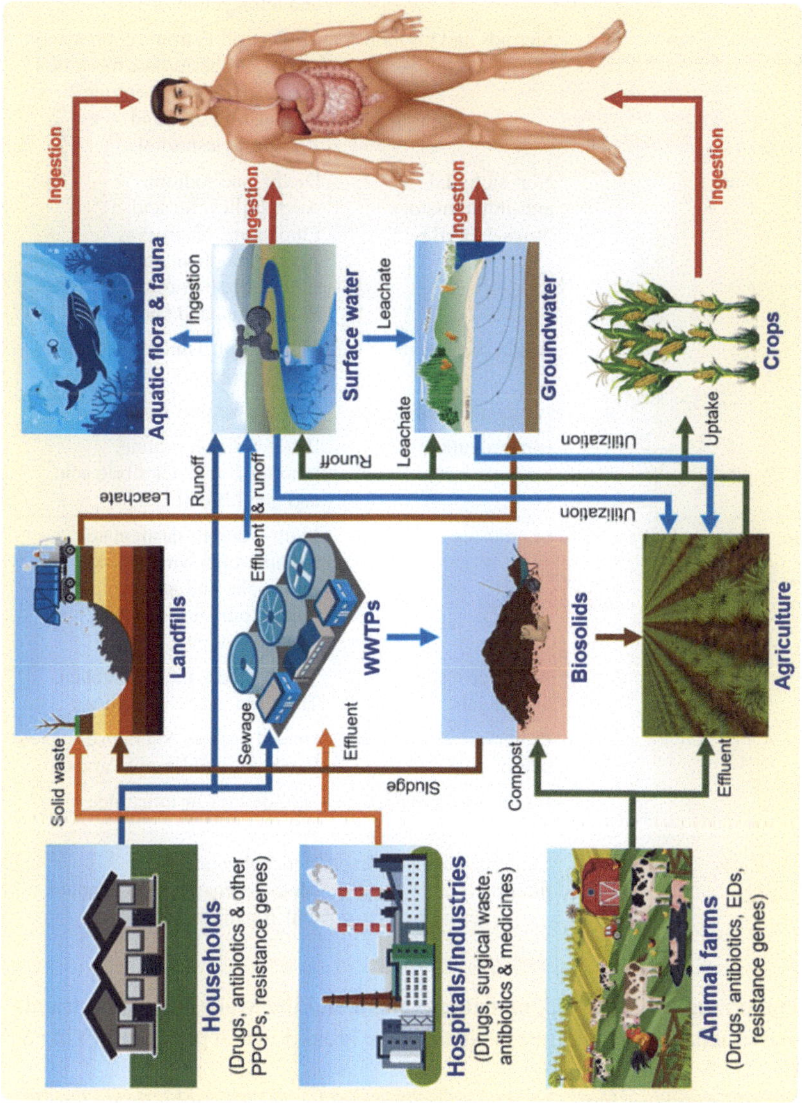

Fig. 14.1 An illustration showing the origin and paths of PPCPs in the environment (Reprinted from original article by Kurade et al. (2021) with permission from Elsevier)

Table 14.1 Categorization of PPCPs and illustrative examples

Pharmaceuticals	Antibiotics	Amoxicillin, Lincomycin, Sulfamethoxazole, Choloramphenicol Erythromycin and Trimethoprim and tetracycline
	Steroids and hormones	Cholestrol, Estrone, coprostanol, estriol, Progesterone, mestranol prednisolone, testosterone, diethylstilbestrol and 17α-ethinylestradiol
	Non-steroidal anti-inflammatory drugs–NSAIDs	Diclofenac sodium, Acetylsalicyclic acid, Fluoxetine, Naprosyn, Acetaminophen, Salicyclic acid, Celecoxib, Ibuprofen, Fenoprofen and ketoprofen
	Antihypertensives	Metoprolol, Timolol, Atenolol and Propranolol
	Anticonvulsants	Carbamazepine, Caffeine
	Lipid regulators	Bezafibrate, etofibrate, fenofibric acid, Clofibric acid and gemfibrozil
Personal care products (PCPs)	Fragrances	Methyl dihydrojasmonate, Isoquinoline, Nitro musks, Polycyclic and macrocyclic musks, pthlates, tonalide, lyral and galaxolide
	Antimicrobials/ antifungals	Triclocarban, methylparaben, Triclosan, and chlorophene
	Preservatives and food additives	Benzyl acetate, Sucralose, Parabens, and Propyl Paraben
	Insect-repellents	N,N-diethyl-m-toluamide (DEET)
	Sunscreen ultra-violet filters	Benzophenone, Methyl-benzylidene Camphor and Tonalide (AHTN)

textures, temperature, pH and EC, moisture content and the availability of nutrients and oxygen is crucial for the microbes that aid in the breakdown of pollutants (Ying & Wei, 2019).

A number of organisms have shown competence in degrading sophisticated pollutants into simpler forms and utilising the generated byproducts into their metabolism. Algae, bacteria, fungus and other microorganisms are among these creatures (Bhatt et al., 2022). *Ex-situ* and *in-situ* are two primary forms of bioremediation. Ex-situ bio-remediation involves removing the contaminated material from the site and treating it somewhere else using techniques including composting, agricultural land, bio-piles, windrows and bioreactors. Conversely, *in-situ* bioremediation entails starting the bioremediation process within the contaminated area using techniques like bioventing and bioslurp (Sharma et al., 2018; Azubuike et al., 2016).

Among the most often employed medicinal species are bacteria, fungus and microalgae. The cleaning processes that bacteria help with might be either anaerobic or aerobic when they are present. Aditionally, *Alcaligenes sp., Achromobacter sp., Comamonas sp., Burkholderia sp., Dehalococcoides sp., Pseudomonas sp., Rhodococcus sp., Ralstonia sp.* and *Sphingomonas sp.*, etc., are among the bacteria species that are frequently used. Similiarly fungi like *Pleurotus ostreatus* variants, *Fusarium sp., Phanerochaete chrysosporium, Aspergillus sp., Pythium ultimum, Fomitopsis palustris, Trametes versicolor, Candida aquatextoris and Stereum hirsutum,* etc. among others have demonstrated involvement in the degradation of numerous PPCPs (Bhatt et al., 2022; Cajthaml, 2015; Mohammadi et al., 2021; Zhang et al., 2016).

Pseudomonas sp. (16 SrRNA) which is obtained from activated sludge has been used in research endeavours aimed at eliminating carbamazepine (Li et al., 2013). Furthermore, it has been shown that adding mixed microbial cultures to activated sludge improves the removal of PPCPs (Zhou et al., 2014). Even with several PPCPs present, this blended culture showed remarkable removal efficacy which may have been caused by some PPCPs acting as these mixed cultures sources of energy and carbon (Vasiliadou et al., 2013). These findings emphasize the potential of mixed microbial cultures in augmenting PPCP removal. While extensively researched for PPCP removal in the activated-sludge during the WWTPs (Khunjar et al., 2011). It is important to acknowledge that mixed cultures have displayed ineffectiveness in certain investigations.

Macromycetes like polypores or mushrooms, have been shown to be useful in the remediation of a variety of pharmaceuticals like hormones, anti-inflammatory agents, beta-blockers, antibiotics and psychotropic drugs. Specifically, *Trametes versicolor* has exhibited promising capabilities in remediating NSAIDs (Non—Steroidal Anti-inflammatory Drugs) while efforts to remediate NSAIDs using bacterial monoculture have met with limited success. Moreover, Olicon—Hermandez et al., (2020) explored the breakdown of a mixture comprising ketoprofen, acetaminophen, naprosyn, diclofenac and ibuprofen with *Penicillium oxalicum* in bench fluidized bio-reactors. *P. oxalicum* showed the enhanced digestion percentages in the bioreactor compared to flask experiments. Furthermore, *Trametes-versicolor* has proven efficient in degrading the antibiotic ofloxacin, achieving the approximately 80% of removal efficiency. In a bioreactor employing fluidized air-pulse technology, the treatment of hospital waste containing high levels of ofloxacin achieved a remarkable 99% remediation rate. Moreover, approximately 50 fungal insulates from hatcheries of salmon, have shown degradation ability for oxytetracycline. Filamented fungi like *Penicillium communae, Epicoccum nigrum, Beauveria bassiana, Aspergillus terreus,* and *T. harzianum* have been identified as having preferable remediation rates, reaching around 78% (Ahumada-Rudolph et al., 2021).

Microalgae have emerged as pivotal agents in the absorption and degradation of various emerging micropollutants. The utilization of chemically—amended biomass from *Scaenedesmus obliquus* (green-algae), microalgae-based bioremediation has proven successful in eliminating pharmaceutical contaminants from synthetic wastewater (Ali et al., 2018). By harnessing microbial syndicates, acclimatized microbes, and employing the co-metabolic bio-techniques, the feasibility of micro-algae based solutions can be significantly enhanced. These systems have demonstrated the capability to remove 30–80% of all pharmaceutical active compounds (PhACs) (Hom–Diaz et al., 2017). Particularly noteworthy is their enhanced remediation potential, achieving up to 98% removal achieved for anti—inflammatory medications like codeine, acetaminophen, salicylic acid and ibuprofen, along with other contaminants like the diuretics hydrochlorothiazide (~ 84%) & furosemide (~100%).

These findings are effective for algal systems and photo bioreactors in biologically treating PhACs. The physicochemical properties of micropollutants as well as culture and operating parameters like as temperature and hydraulic retention time (HRT) shows significant impact on their elimination. Microbial decomposition has been primary method for removing PPCPs in CCWs with carbamazepine being utilized as a reference molecule for investigation. One researcher studied the microbial community composition within the intermix filler. High sequencing demonstrated that the prevalent bacteria were highly flexible to PPCPs (Shi et al., 2023).

14.3.2 Phytoremediation

The plant based remediation or phytoremediation is a potential bioremediation technique which uses green plants to absorb and metabolize along with detoxify or change environmental toxins. The approach has serious disadvantages despite praise for its environmental friendliness and sustainability. Increased concentrations of newly discovered contaminants or their dangerous byproducts may hinder plant growth reducing total physiological output and biosystem effectiveness (Kurade et al., 2021). This method is known to be time consuming because to its slow degradation and limited ability to absorb volatile pollutants from the surrounding air. Despite these challenges, phytoremediation is a viable and appealing way to restore damaged habitats at a reasonable cost. It shows great potential especially for pollutants that are concentrated around plant roots (Mojiri et al., 2013). Phytodegradation, phytoextraction, rhizosphere degradation, rhizofiltration, phytovolatilization, phytostabilization and endophytic degeneration are some of the methods by which phytoremediation functions. During the phytoremediation process, these processes frequently overlap at different phases (Arslan et al., 2017). The phytoremediation process and associated procedures are shown in Fig. 14.2.

In summary, phytoremediation is a tried-and-true, effective technique for removing PPCPs from the environment and gained popularity as an eco-technology for eliminating PPCPs from wastewater especially when implemented through

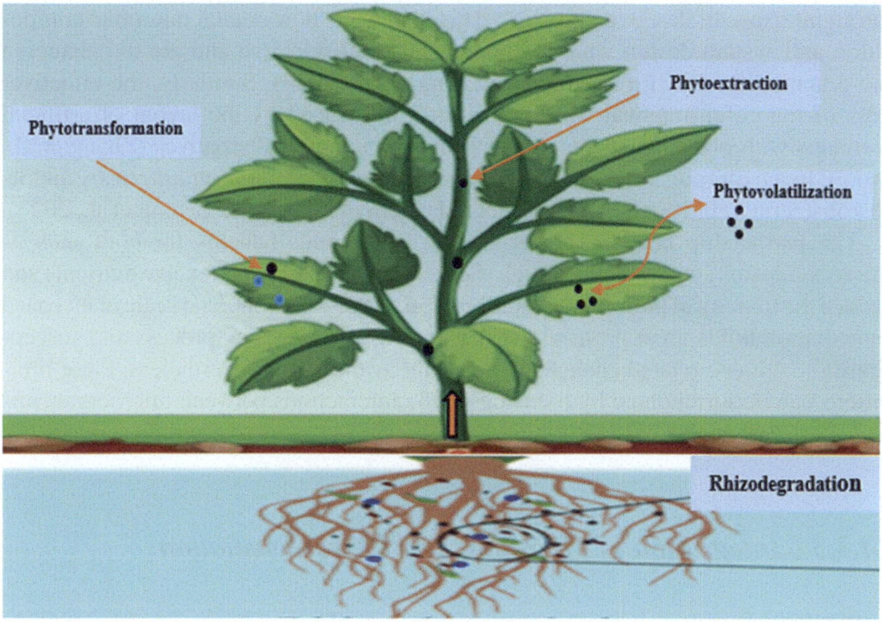

Fig. 14.2 Phytoremediation process and their mechanisms (Nguyen et al., 2019)

constructed wetlands (CWs) (Ferreira et al., 2017). There are several restrictions attached to the phytoremediation method of removing PPCPs, though. Firstly, in nations where land is limited, the use of CWs for the treatment of pollutants like PPCPs frequently necessitates a vast land area. Hence, the phytoremediation approach may not be practical for the elimination of PPCPs (Verhoeven & Meuleman, 1999). Secondly, wetland ageing or clogging may make long-term phytoremediation treatment difficult; as a result, PPCP removal efficiency may gradually decline (Wu et al., 2015). Thirdly, using a phytoremediation approach to address PPCPs may pose dangers to the environment (Verlicchi and Zambello, 2014).

Zhang et al. (2017) research study illustrates the effectiveness of wetland mesocosms planted with a variety of plant species (*Phragmites australis, Iris pseudacorus, Berula erecta, Typha latiflolia* and *Juncus effusus*) in removing IBU (Ibuprofen) and iohexol-IOH was investigated. Mesocosms containing *Juncus effusus* and *Boerrhavia erecta* which exhibited the best removal efficiency and eradicating 90% of IBU and IOH from wetlands with baseline concentrations of 100 μg L^{-1} (Zhang et al., 2017).

One study conducted in Denmark on activated sludge wastewater treatment facilities discovered that carbamazepine was the most persistent chemical of all the PPCPs handled. Comparing diclofenac and carbamazepine to ketoprofen, fenoprofen, triclosan and naproxen, the study found that they were more persistent. The remediation efficiencies ranged from 77 to 94%, 55% to 90%, 61% to 91%, and 46% to 90% respectively (Chen et al., 2015). It is noteworthy that the biological composition of the sludge significantly influenced the efficacy of PPCP removal exceeding the significance of design parameters. This highlights how crucial microbial composition and system design which including material selection and are to enhancing the potential for PPCP remediation (Zhang et al., 2018). Similarly, the effectiveness of ibuprofen removal was significantly influenced by the layout of artificial wetlands with planted systems outperforming unplanted alternatives (Zhang et al., 2017). In a hydroponic environment, researcher examined *Alternanthera sp.* and its rhizobacteria's ability to absorb and degrade methylparaben-MP (Mutar et al., 2023).

The partnership between bacteria and plants yields benefits for both such as microorganisms get nutrients through root exudates while plants receive nutrients and reduce the toxicity of pollution (Chaudhry et al., 2005). Commercial scale plant based remediation holds great promise but it is hindered by its slow process and susceptibility to climate related changes in pollutant availability. Nevertheless, these limitations can be surmounted by focusing on the interactions between microorganisms and plants (Chaudhry et al., 2005).

14.3.3 Integration of Microbial and Phytoremediation

In PPCP bioremediation, the interactions between plants and bacteria (Khalaf et al., 2023) and fungus (Asemoloye et al., 2017) are critical. These kinds of plants function as buffer zones and can reduce the spread of pollutants from many environmental sources by utilising processes like metabolism and bioaccumulation (Morillo & Villaverde, 2017). Considerable work has been done in the last 20 years to improve the efficacy of phytoremediation in contaminated areas by utilising the symbiotic link between bacteria and plants (Gomez-Sagasti et al., 2021). Rhizofiltration, phytovolatilization, phytoextraction, and rhizoremediation are just a few of the phytoremediation processes in which the plants in these synergistic partnerships actively engage (Gutierrez et al., 2021). Furthermore, plants provide their associated microbial communities with nutrition, habitat, and other metabolites (Pant & Singh, 2024a; Backer et al., 2018). In exchange, bacteria encourage plant development and adaptation in contaminated settings, reduce phyto-toxicity, and lower the evapotranspiration of dangerous pollutants through their metabolic processes (Thijs et al., 2018). This method encourages the restoration of natural habitats and the on-site removal of environmental contaminants in areas damaged by waste (Hassen et al., 2018).

Microbial and phytoremediation approaches show promise in addressing emerging pollutants (EPs) but their long processing times remain a key barrier (Supreeth, 2022). Literature reports highlight the plants efficiency in removing PPCP pollutants from polluted areas (Chen et al., 2021; Sauvetre et al., 2018; Saiyood et al., 2010). Plants form symbiotic interactions with rhizospheric and endophytic microorganisms, providing them with shelter and nourishment while these microbes help with pollutant breakdown and detoxification, which promotes plant growth (Shehzadi et al., 2014). Plants and rhizobacteria have a synergistic interaction, whereas plant-endophyte associations involve bacterial communities that live in plant tissues. Plants maintain endophytes, which protect them from pollutant toxicity through actions that promote plant growth. According to Lugtenberg and Kamilova (2009), root exudates attract microorganisms, allowing for successful colonisation and proliferation.

Endophytes are increasingly recognised as important contributors to host plant adaptability in polluted environments (Doty, 2008). Endophytic bacteria, which possess internal enzymes, play an important role in pollution degradation (Zhu et al., 2016). Endophytes get carbohydrates from plants through symbiotic partnerships while boosting plants' resilience against numerous environmental challenges, such as pollutants, temperature fluctuations, water and nutrient deficits, salinity, and infections (Santoyo et al., 2016). Consequently, leveraging endophytes with the ability to degrade and detoxify noxious organic conjugates in aggregation with carefully selected plant species tailored for the remediation site holds as an promise avenue for efficient and sustainable remediation in the contemporary times (Sun et al., 2014). Numerous endophytic microbes engage in symbiotic relationships with their vegetative counterparts, facilitating the phytoremediation process. For instance, *Brassica juncea, Sedum alfredii,* and *Pteris vittate* are recognized hyper-accumulators, whereas *Brassica napus, Arabidopsis thaliana* and *Glycine max.*, which are comsidered non-hyper-accumulators. In their research, Babu and his colleagues, 2013 extracted a BT-*Bacillus thuringiensis* variant from the roots of *Pinus sylvestris.* This particular strain was found to produce phyto-hormones that not only stimulated plant development but also facilitated the expedited elimination of harmful toxic compounds from mining tailings.

A study observed that the introduction of ciprofloxacin into constructed wetlands initially harmed bacterial communities. While these communities recovered within 2–5 weeks of acclimation, resulting in successful colonisation and subsequent pollution clearance, hence restoring plant health (Weber et al., 2011). However, it should be noted that the existing data are mostly from laboratory-scale trials. Further enhancements and modifications in remediation system design may result in more definitive conclusions in pilot-scale experiments. Although growing interest in the symbiotic interaction between plants and root-associated microorganisms, this topic is mostly unexplored. Existing research focuses primarily on bacterial species enriched in systems tailored to certain plant species, with little investigation into the inherent capabilities of bacterial communities naturally existing in plants and wetland components that use PPCPs (a carbon source). Several further occurrences of remediation driven by plant–microbe relationships are well documented in the literature, as shown in Table 14.2.

The relationship that exists between plants and bacteria is intrinsically tied to the environmental conditions to which they are suited. Alterations in precipitation and temperature regimes can have enormous repercussions for these interactions, both helpful and harmful (Rasmussen et al., 2020). This tendency grows importance in light of continuous climate change, which creates ecological uncertainties and complexities (Begum et al.., 2019). Figure 14.3 depicts plant–microbe-based remediation strategies, emphasising their symbiotic relationship in naturally purifying

Table 14.2 Utilizing plant—microbe synergy for PPCPs remediation

Plant used	Microbe used	PPCPs pollutant	References
Tobacco (*Nicotiana tabacum)*	*Coriolus versicolor*	Bisphenol-A and PCP	Divya and Kumar (2011)
Wheat—*Triticum aestivum*	*Sphingobium chlorophenolicum*	Pentachlorophenol (PCP)	Dams et al., (2007)
Wheat—*Triticum aestivum*	*Herbaspirillum sp.*	PCBs, Trichlorophenol (TCP)	Mannisto et al. (2001)
Vicia sativa	*Bacillus spp.*	Phenol	Ibáñez et al. (2014)
C. robusta	*B. megaterium*	Ampicillin	Vasilachi et al., (2021)
Festuca arundinecea Schreb	*Mycobacterium sp.*	Bisphenol A	Ncibi et al. (2017)
Alopecurus aequalis	*Massilia sp.*-Pn2	Phenanthrene	Liu et al., (2014)
Typha latifolia, Lemnaminor, Anthocerotae	Beta- and gamma-*Proteo-bacteria, Bacteroidetes*	Triclosan (PCP)	Liu et al., (2016)
Hydrilla verticillata, Typha angustifolia and *Salvinia natans*	Beta-, delta-, and gamma-*Proteobacteria, Cyanobacteria* and *Sphingobacteria,*	Triclosan (PCP)	Zhao et al., (2015)
Populus trichocarpa	*Pseudomonas putida*-VM1441	Naphthalene	Germaine et al., (2009)
Phragmites australis	*Diaphorobacter nitroreducens* and *Rhizobium radiobacter*	Carbamazepine	Sauvêtre et al., (2018)
Coffea Arabica	*Kluyvera cryocrescens*	Ranitidine	Li (2019)
Typha angustifolia	*Methylococcaceaea* (family)	Aceclofenac	Sivaranjanee et al. (2021)
Phalaris arundinacea	Rhizobacteria	Sulfamethoxazole Triclosan	Button et al., (2019)
Phragmites australis	*Chloroflexus sp.*	Diclofenac Sulfamethoxazole	Sochacki et al., (2018)
Typha angustifolia	Rhizobacteria	Ibuprofen	Li et al., (2016)
Streptomyces griseorubiginosus strains DS2 & DS4	Roots of *Miscanthus giganteus*	Diclofenac sod. and Sulfamethoxazole	Sauvetre et al. (2020)
S. validus	Plant	Sulfonamides	Zhou et al., (2024)

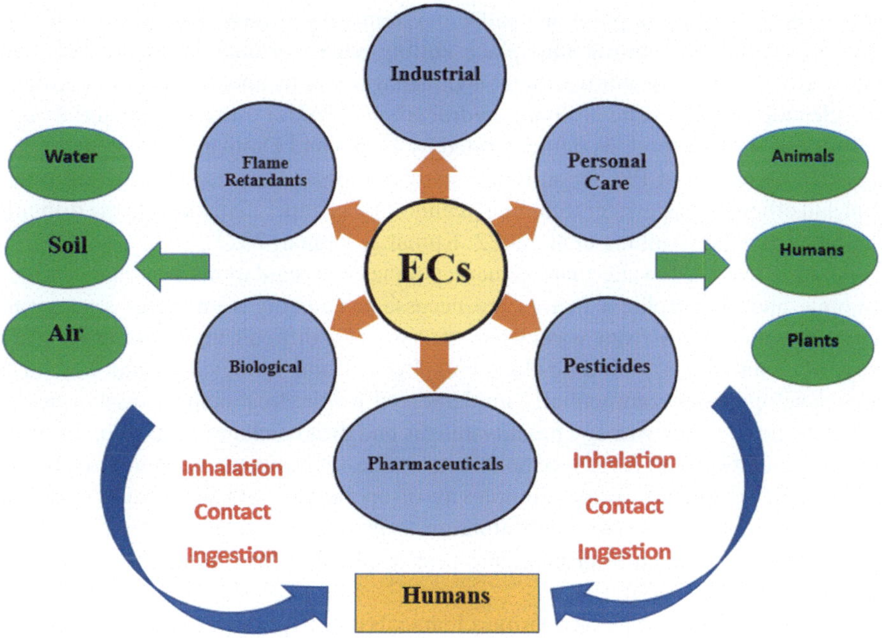

Fig. 14.3 Utilizing plant–microbe partnerships, remediation techniques for environmental cleanup of ECs are depicted (Kumar et al., 2022)

toxins from soil, water, and air, resulting in a sustainable approach to environmental cleanup.

14.4 Challenges and Financial Considerations Related to Remediation

One major environmental concern in the modern era is the discharge of PPCPs into waterways. There is a growing need for freshwater and clean drinking water sources due to the growing world population. This spike has resulted in a considerable increase in wastewater output from both home and industrial sources globally, along with rapid population growth and urbanisation. Global water scarcity has been made worse by concurrent forces including industrial growth, agricultural land development, and climate change, which have put enormous pressure on freshwater reserves. Wastewater handling practices that are poorly managed and lack adequate regulation provide serious dangers to the environment and public health, especially in metropolitan areas that are rapidly developing. The unregulated release of wastewater, encompassing both household and industrial effluents, presents considerable risks to neighbouring populations and ecosystems. As a result, increasing community knowledge and putting into place cutting-edge, sustainable, affordable, and highly effective wastewater treatment and management technologies are imperative. The cleanup of polycyclic aromatic hydrocarbons (PAHs) using nanomaterials is a well-established method, as demonstrated by Gupta and Dhiman (2023). Given the significant impacts of PPCPs on biodiversity and ecosystem health, creating practical and efficient wastewater treatment technology is imperative, despite the difficult challenge at hand (Ambika et al., 2022; Kumar & Kumar, 2022).

For effective wastewater management, policies and regulations as well as socioeconomic and geographical factors are necessary. Creating a universal strategy to remove contaminants from wastewater streams is a difficult undertaking. A wide range of technologies, including chemical, physical, and biological techniques, have been developed and commercially implemented for the treatment of wastewater in recent decades. Every strategy has advantages and disadvantages of its own, in addition to difficulties with regard to the effects on the environment, economy, viability, efficacy, production of sludge, complexity of operations, and practicality. Finding the best approach in a particular situation is therefore crucial. Combining many approaches can frequently produce the best results for water quality at the lowest cost (Kumar et al., 2020).

The prevailing literature on pharma. chemicals tend to prioritize site-explicit treatment procedures or the targeted removal of specific contaminants from effluents, rather than offering holistic solutions for real-time scenarios. Various factors, encompassing processing demands, electricity and labor expenditures, construction outlays, local considerations, recycling capacities, and longevity, all contribute significantly to the overall costs incurred. Moreover, the cost-effectiveness of a particular adsorbent is contingent upon factors such as availability, lifespan, recyclability, and processing requisites, elements frequently disregarded in current scholarly discourse (Baskar et al., 2022), aspects often neglected in existing literature (Kumar et al., 2023a, 2023b).

14.5 Environmental Sustainability and Impact Assessment

Pharmaceutical compounds have proven important in improving living standards, curing a great deal of disease, and changing patterns of production and consumption in contemporary human society. Nevertheless, after being consumed, eliminated pharmaceutical compounds may find their way into wastewater treatment plants (WWTPs), where they may enter unaltered or undergo biotransformation. Unfortunately, WWTPs are not very successful in eliminating pharmaceutical pollutants, and they are therefore regarded as point sources of pharmaceutical pollution along with aquaculture, agriculture, hospitals, and pharmaceutical manufacturing facilities. Because of their potential harm to human health and the environment, their presence in water bodies is a growing global concern. Pharmaceutical contaminants can have ecotoxicological effects on non-target animals belonging to different trophic levels, even at low quantities. The development of decontamination systems for the efficient removal of pharmaceutical contaminants is critically needed, given their hazardous effects and the inadequacy of WWTPs in doing so. Sensitizing authorities and communities about responsible resource use is necessary to preserve the health of life-supporting ecosystems. Aquifers that feed rivers must not spill and watersheds must be protected and land management must be addressed in order to meet SDG (Pant & Singh, 2024b).

Additionally, the potential applications of biosorption/phytoemediation and biochar in water treatment, biomedical, bioenergy, and agricultural fields can be expanded through plant–microbe and plant-NP interactions. Furthermore, the remediation of PPCP offers a promising field for microbial methods based on plants. These innovative treatments make use of the beneficial interaction between bacteria and plants to eliminate PPCP contamination. The particular metabolic capabilities of microorganisms associated with plant roots like as rhizobacteria and mycorrhizal fungi can be utilised to improve the breakdown and elimination of PPCP from the environment. This strategy offers the ability to both provide economically priced, ecologically suitable alternatives that peacefully coexist with natural ecosystems and reduce PPCP contamination. By integrating plant-based microbial remediation into our more thorough approaches to environmental sustainability and impact assessment, we are demonstrating our dedication to building a resilient, ecologically balanced, and healthy future (Gavrilescu et al., 2015; Kumar et al., 2022).

14.6 Conclusion and Future Perceptions

In conclusion, the widespread occurrence of PPCPs in aquatic ecosystems highlights the importance of addressing their pseudo-persistent nature, which is exacerbated by continual discharge. PPCPs include a wide range of compounds that are used on a daily basis, including pharmaceuticals and cosmetic products. Notably PPCPs spread in natural ecosystems as a result of factors such as increased industrialization and urbanisation, the necessity for quick environmental cleanup and pollutant stabilisation is growing increasingly critical to ensuring the sustainability of these ecosystems. Several studies have found that introducing mixed microbiological cultures (fungul insulates, bacterial, and microalgal sp.) into activated sludge enhances PPCP removal. However, plant-based remediation has been around as a popular remediation method, the pursuit of quicker pollution removal is a formidable challenge, necessitating ongoing research efforts in this field. Wetland-mesocosms planted with a variety of plant species show promise for bioremediation.

The mutually beneficial collaboration of indigenous bacteria and living plants in polluted areas boosts phytoremediation performance which indicating a promising possibility for addressing both organic and inorganic toxins (Madhav et al., 2024). Plants foster symbiotic relationships with rhizospheric and endophytic microbes providing them with shelter and nutrients while these microbes aid in contaminant degradation and detoxification and supporting the plant growth.

The combination of microorganisms and living plants in contaminated areas improves phytoremediation efficacy which demonstrating a viable solution for dealing with both organic and inorganic toxins. Along with supplying food and shelter, plants work with rhizospheric and endophytic bacteria to break down and detoxify pollutants, which promotes the growth of other plants. Even though there are a number of established techniques for removing PPCP, such as chemical and physical processes, their widespread acceptability is limited by their high implementation costs and lack of environmental friendliness. Biological approaches, on the other hand, are notable for their intrinsic benefits, which include a rapid rate of degradation, low energy requirements, and little secondary pollutant production. With respect to various biological treatment techniques, removal efficiency, operating parameters, and environmental implications, the study provides a true and quantitative assessment of the state of the field.

Prospects:

Additionally, there is a lot of potential for enhancing environmental remediation efforts with the application of nano-phytoremediation technology. However, the confusing path that nanoparticles take in the environment and their possible negative effects on species that are not their intended targets continue to be important subjects that need more investigation and study. In order to minimise the effects of PPCP pollution and foster environmental resilience, it is imperative that we continue to innovate and investigate biological treatment options and integrated technologies as we move towards a more sustainable future.

References

Abdallat, G. A., Salameh, E., Shteiwi, M., & Bardaweel, S. (2022). Pharmaceuticals as emerging pollutants in the reclaimed wastewater used in irrigation and their effects on plants, soils, and groundwater. *Water, 14*(10), 1560. https://doi.org/10.3390/w14101560

Ahumada-Rudolph, R., Novoa, V., Becerra, J., Cespedes, C., & Cabrera-Pardo, J. R. (2021). Mycoremediation of oxytetracycline by marine fungi mycelium isolated from salmon farming areas in the south of Chile. *Food and Chemical Toxicology, 152*, 112198. https://doi.org/10.1016/j.fct.2021.112198

Al Aukidy, M., Verlicchi, P., Jelic, A., Petrovic, M., & Barcelò, D. (2012). Monitoring release of pharmaceutical compounds: Occurrence and environmental risk assessment of two WWTP effluents and their receiving bodies in the Po Valley, Italy. *Science of the Total Environment, 438*, 15–25. https://doi.org/10.1016/j.scitotenv.2012.08.061

Ali, M. E., Abd El-Aty, A. M., Badawy, M. I., & Ali, R. K. (2018). Removal of pharmaceutical pollutants from synthetic wastewater using chemically modified biomass of green alga Scenedesmus obliquus. *Ecotoxicology and Environmental Safety, 151*, 144–152. https://doi.org/10.1016/j.ecoenv.2018.01.012

Ambika, S., Kumar, M., Pisharody, L., Malhotra, M., Kumar, G., Sreedharan, V., Singh, L., Nidheesh, P. V., Bhatnagar, A. (2022). Modified biochar as a green adsorbent for removal of hexavalent chromium from various environmental matrices: mechanisms, methods, and prospects. *Chemical Engineering Journal, 439*, 135716. https://doi.org/10.1016/j.cej.2022.135716

Arslan, M., Imran, A., Khan, Q. M., & Afzal, M. (2017). Plant–bacteria partnerships for the remediation of persistent organic pollutants. *Environmental Science and Pollution Research, 24*, 4322–4336. https://doi.org/10.1007/s11356-015-4935-3

Asemoloye, M. D., Ahmad, R., & Jonathan, S. G. (2017). Synergistic rhizosphere degradation of γ-hexachlorocyclohexane (lindane) through the combinatorial plant-fungal action. *PLoS One, 12*(8), e0183373.

Azubuike, C. C., Chikere, C. B., & Okpokwasili, G. C. (2016). Bioremediation techniques–classification based on site of application: Principles, advantages, limitations and prospects. *World Journal of Microbiology and Biotechnology, 32*, 1–18. https://doi.org/10.1007/s11274-016-2137-x

Babu, A. G., Kim, J. D., & Oh, B. T. (2013). Enhancement of heavy metal phytoremediation by Alnus firma with endophytic Bacillus thuringiensis GDB-1. *Journal of Hazardous Materials, 250*, 477–483. https://doi.org/10.1016/j.jhazmat.2013.02.014

Backer, R., Rokem, J. S., Ilangumaran, G., Lamont, J., Praslickova, D., Ricci, E., Subramanian, S., & Smith, D. L. (2018). Plant growth-promoting rhizobacteria: Context, mechanisms of action, and roadmap to commercialization of biostimulants for sustainable agriculture. *Frontiers in Plant Science, 9*, 402666. https://doi.org/10.3389/fpls.2018.01473

Balakrishna, K., Rath, A., Praveenkumarreddy, Y., Guruge, K. S., & Subedi, B. (2017). A review of the occurrence of pharmaceuticals and personal care products in Indian water bodies. *Ecotoxicology and environmental safety, 137*, 113–120. https://doi.org/10.1016/j.ecoenv.2016.11.014

Baskar, A. V., Bolan, N., Hoang, S. A., Sooriyakumar, P., Kumar, M., Singh, L., Jasemizad, T., Padhye, L. P., Singh, G., Vinu, A., & Sarkar, B. (2022). Recovery, regeneration and sustainable management of spent adsorbents from wastewater treatment streams: A review. *Science of the Total Environment, 822*, 153555. https://doi.org/10.1016/j.scitotenv.2022.153555

Begum, N., Qin, C., Ahanger, M. A., Raza, S., Khan, M. I., Ashraf, M., Ahmed, N., & Zhang, L. (2019). Role of arbuscular mycorrhizal fungi in plant growth regulation: Implications in abiotic stress tolerance. *Frontiers in Plant Science, 10*, 1068. https://doi.org/10.3389/fpls.2019.01068

Bhatt, P., Bhandari, G., & Bilal, M. (2022). Occurrence, toxicity impacts and mitigation of emerging micropollutants in the aquatic environments: Recent tendencies and perspectives. *Journal of Environmental Chemical Engineering, 10*(3), 107598. https://doi.org/10.1016/j.jece.2022.107598

Button, M., Cosway, K., Sui, J., & Weber, K. (2019). Impacts and fate of triclosan and sulfamethoxazole in intensified re-circulating vertical flow constructed wetlands. *Science of the Total Environment, 649*, 1017–1028. https://doi.org/10.1016/j.scitotenv.2018.08.395

Cajthaml, T. (2015). Biodegradation of endocrine-disrupting compounds by ligninolytic fungi: Mechanisms involved in the degradation. *Environmental Microbiology, 17*, 4822–4834. https://doi.org/10.1111/1462-2920.12460

Celle-Jeanton, H., Schemberg, D., Mohammed, N., Huneau, F., Bertrand, G., Lavastre, V., & Le Coustumer, P. (2014). Evaluation of pharmaceuticals in surface water: Reliability of PECs compared to MECs. *Environment International, 73*, 10–21. https://doi.org/10.1016/j.envint.2014.06.015

Chaudhry, Q., Blom-Zandstra, M., Gupta, S. K., & Joner, E. (2005). Utilising the synergy between plants and rhizosphere microorganisms to enhance breakdown of organic pollutants in the environment (15 pp). *Environmental Science and Pollution Research, 12*, 34–48. https://doi.org/10.1065/espr2004.08.213

Chen, X., Vollertsen, J., Nielsen, J. L., Gieraltowska Dall, A., & Bester, K. (2015). Degradation of PPCPs in activated sludge from different WWTPs in Denmark. *Ecotoxicology, 24*, 2073–2080. https://doi.org/10.1007/s10646-015-1548-z

Chen, J., Liu, S. S., He, L. X., Cheng, Y. X., Ye, P., Li, J., Ying, G. G., Wang, Y. J., & Yang, F. (2021). The fate of sulfonamides in the process of phytoremediation in hydroponics. *Water Research, 198*, 117145. https://doi.org/10.1016/j.watres.2021.117145

Cuellar-Bermudez, S. P., Aleman-Nava, G. S., Chandra, R., Garcia-Perez, J. S., Contreras-Angulo, J. R., Markou, G., Muylaert, K., Rittmann, B. E., & Parra-Saldivar, R. (2017). Nutrients utilization and contaminants removal. A review of two approaches of algae and cyanobacteria in wastewater. *Algal Research, 24*, 438–449. https://doi.org/10.1016/j.algal.2016.08.018

Dams, R. I., Paton, G. I., & Killham, K. (2007). Rhizoremediation of pentachlorophenol by Sphingobium chlorophenolicum ATCC 39723. *Chemosphere, 68*(5), 864–870. https://doi.org/10.1016/j.chemosphere.2007.02.014

Daughton, C. G. (2001). Emerging pollutants, and communicating the science of environmental chemistry and mass spectrometry: Pharmaceuticals in the environment. *Journal of the American Society for Mass Spectrometry, 12*(10), 1067–1076. https://doi.org/10.1016/S1044-0305(01)00287-2

Daughton, C. G. (2004). Non-regulated water contaminants: Emerging research. *Environmental Impact Assessment Review, 24*(7–8), 711–732. https://doi.org/10.1016/j.eiar.2004.06.003

Divya, B., & Deepak Kumar, M. (2011). Plant–microbe interaction with enhanced bioremediation. *Research Journal of Biotechnology, 6*, 4.

Doty, S. L. (2008). Enhancing phytoremediation through the use of transgenics and endophytes. *New Phytologist, 179*(2), 318–333. https://doi.org/10.1111/j.1469-8137.2008.02446.x

Dueñas-Moreno, J., Vázquez-Tapia, I., Mora, A., Cervantes-Avilés, P., Mahlknecht, J., Capparelli, M. V., Kumar, M., & Wang, C. (2024). Occurrence, ecological and health risk assessment of phthalates in a polluted urban river used for agricultural land irrigation in central Mexico. *Environmental Research, 240*, 117454. https://doi.org/10.1016/j.envres.2023.117454

Ebele, A. J., Abdallah, M. A. E., & Harrad, S. (2017). Pharmaceuticals and personal care products (PPCPs) in the freshwater aquatic environment. *Emerging contaminants, 3*(1), 1–16. https://doi.org/10.1016/j.emcon.2016.12.004

Ferreira, A. R., Ribeiro, A., & Couto, N. (2017). Remediation of pharmaceutical and personal care products (PPCPs) in constructed wetlands: applicability and new perspectives. *Phytoremediation: Management of Environmental Contaminants, 5*, 277–292. https://doi.org/10.1007/978-3-319-52381-1_9

Gao, X., Kang, S., Xiong, R., & Chen, M. (2020). Environment-friendly removal methods for endocrine disrupting chemicals. *Sustainability, 12*(18), 7615. https://doi.org/10.3390/su12187615

Gavrilescu, M., Demnerová, K., Aamand, J., Agathos, S., & Fava, F. (2015). Emerging pollutants in the environment: Present and future challenges in biomonitoring, ecological risks and bioremediation. *New Biotechnology, 32*(1), 147–156. https://doi.org/10.1016/j.nbt.2014.01.001

Germaine, K. J., Keogh, E., Ryan, D., & Dowling, D. N. (2009). Bacterial endophyte-mediated naphthalene phytoprotection and phytoremediation. *FEMS Microbiology Letters, 296*(2), 226–234. https://doi.org/10.1111/j.1574-6968.2009.01637.x

Gómez-Sagasti, M. T., Garbisu, C., Urra, J., Míguez, F., Artetxe, U., Hernández, A., Vilela, J., Alkorta, I., & Becerril, J. M. (2021). Mycorrhizal-assisted phytoremediation and intercropping strategies improved the health of contaminated soil in a peri-urban area. *Frontiers in Plant Science, 12*, 693044. https://doi.org/10.3389/fpls.2021.693044

Gupta, H. (2018). Anthracene removal from water onto activated carbon derived from vehicular tyre. *Separation Science and Technology, 53*(4), 613–625. https://doi.org/10.1080/01496395.2017.1405038

Gupta, H., & Gupta, B. (2016). Adsorption of polycyclic aromatic hydrocarbons on banana peel activated carbon. *Desalination and Water Treatment, 57*(20), 9498–9509. https://doi.org/10.1080/19443994.2015.1029007

Gupta, H., & Dhiman, S. (2023). Synthetic nanoparticle-based remediation of soils contaminated with polycyclic aromatic hydrocarbons. In *Handbook of green and sustainable nanotechnology*. https://doi.org/10.1007/978-3-030-69023-6_102-1

Gupta, H., Dhiman, S., Kumar, A., Paliwal, S. K., Khan, G., Singh, H., Mishra, A., & Mishra, A. K. (2024). Photo-and bio-degradation of selected persistent organic pollutants. *Separation and Purification Reviews*, 1–14. https://doi.org/10.1080/15422119.2024.2332953

Gutierrez-Gines, M. J., Alizadeh, H., Alderton, E., Ambrose, V., Meister, A., Robinson, B. H., Halford, S., Prosser, J. A., & Horswell, J. (2021). Phytoremediation of microbial contamination in soil by New Zealand native plants. *Applied Soil Ecology, 167*, 104040. https://doi.org/10.1016/j.apsoil.2021.104040

Hassen, W., Neifar, M., Cherif, H., Najjari, A., Chouchane, H., Driouich, R. C., Naili, F., Mosbah, A., Souissi, Y., Raddadi, N., & Cherif, A. (2018). Pseudomonas rhizophila S211, a new plant growth-promoting rhizobacterium with potential in pesticide-bioremediation. *Frontiers in Microbiology, 9*, 327718. https://doi.org/10.3389/fmicb.2018.00034

Hoenicke, R., Oros, D. R., Oram, J. J., & Taberski, K. M. (2007). Adapting an ambient monitoring program to the challenge of managing emerging pollutants in the San Francisco Estuary. *Environmental Research, 105*(1), 132–144. https://doi.org/10.1016/j.envres.2007.01.005

Hom-Diaz, A., Jaén-Gil, A., Bello-Laserna, I., Rodríguez-Mozaz, S., Vicent, T., Barceló, D., & Blánquez, P. (2017). Performance of a microalgal photobioreactor treating toilet wastewater: Pharmaceutically active compound removal and biomass harvesting. *Science of the Total Environment, 592*, 1–11. https://doi.org/10.1016/j.scitotenv.2017.02.224

Hossein, M., Asha, R., Bakari, R., Islam, N. F., Jiang, G., & Sarma, H. (2023). Exploring eco-friendly approaches for mitigating pharmaceutical and personal care products in aquatic ecosystems: A sustainabilityassessment. *Chemosphere, 316*, 137715. https://doi.org/10.1016/j.chemosphere.2022.137715

Ibañez, S. G., Merini, L. J., Barros, G. G., Medina, M. I., & Agostini, E. (2014). Vicia sativa–rhizospheric bacteria interactions to improve phenol remediation. *International Journal of Environmental Science and Technology, 11*, 1679–1690. https://doi.org/10.1007/s13762-013-0357-2

Jurado, A., Vàzquez-Suñé, E., Carrera, J., de Alda, M. L., Pujades, E., & Barceló, D. (2012). Emerging organic contaminants in groundwater in Spain: A review of sources, recent occurrence and fate in a European context. *Science of the Total Environment, 440*, 82–94. https://doi.org/10.1016/j.scitotenv.2012.08.029

Kasprzyk-Hordern, B., Dinsdale, R. M., & Guwy, A. J. (2009). The removal of pharmaceuticals, personal care products, endocrine disruptors and illicit drugs during wastewater treatment and its impact on the quality of receiving waters. *Water research, 43*(2), 363–380. https://doi.org/10.1016/j.watres.2008.10.047

Khalaf, D. M., Cruzeiro, C., Siani, R., Kublik, S., & Schröder, P. (2023). Resilience of barley (Hordeum vulgare) plants upon exposure to tramadol: Implication for the root-associated bacterial community and the antioxidative plant defence system. *Science of the Total Environment, 892*, 164260. https://doi.org/10.1016/j.scitotenv.2023.164260

Khan, N., Ali, S., Shahid, M. A., Mustafa, A., Sayyed, R. Z., & Curá, J. A. (2021). Insights into the interactions among roots, rhizosphere, and rhizobacteria for improving plant growth and tolerance to abiotic stresses: A review. *Cells, 10*(6), 1551. https://doi.org/10.3390/cells10061551

Khan, W. U., Ahmed, S., Dhoble, Y., & Madhav, S. (2023). A critical review of hazardous waste generation from textile industries and associated ecological impacts. *Journal of the Indian Chemical Society, 100*(1), 100829. https://doi.org/10.1016/j.jics.2022.100829

Khunjar, W. O., Mackintosh, S. A., Skotnicka-Pitak, J., Baik, S., Aga, D. S., & Love, N. G. (2011). Elucidating the relative roles of ammonia oxidizing and heterotrophic bacteria during the biotransformation of 17α-ethinylestradiol and trimethoprim. *Environmental Science and Technology, 45*(8), 3605–3612. https://doi.org/10.1021/es1037035

Kumar, A., & Gupta, H. (2020). Activated carbon from sawdust for naphthalene removal from contaminated water. *Environmental Technology and Innovation, 20*, 101080. https://doi.org/10.1016/j.eti.2020.101080

Kumar, V., & Kumar, M. (2022). *Integrated environmental technologies for wastewater treatment and sustainable development*. Elsevier.

Kumar, R., Rani, M., Gupta, H., Gupta, B., Park, D., & Jeon, B. H. (2017). Distribution of trace elements in flowing surface waters: Effect of seasons and anthropogenic practices in India.

International Journal of Environmental Analytical Chemistry, 97(7), 637–656. https://doi.org/10.1080/03067319.2017.1339035

Kumar, M., Xiong, X., Sun, Y., Yu, I. K., Tsang, D. C., Hou, D., Gupta, J., Bhaskar, T., & Pandey, A. (2020). Critical review on biochar-supported catalysts for pollutant degradation and sustainable biorefinery. *Advanced Sustainable Systems, 4*(10), 1900149. https://doi.org/10.1002/adsu.201900149

Kumar, V., Agrawal, S., Bhat, S. A., Américo-Pinheiro, J. H. P., Shahi, S. K., & Kumar, S. (2022). Environmental impact, health hazards, and plant-microbes synergism in remediation of emerging contaminants. *Cleaner Chemical Engineering, 2*, 100030. https://doi.org/10.1016/j.clce.2022.100030

Kumar, N., Kumar, B., Gupta, H., & Kumar, A. (2023a). Development and evaluation of cellulose/graphene-oxide based composite for removing phenol from aqueous solutions. *Polymers, 15*(3), 572. https://doi.org/10.3390/polym15030572

Kumar, M., Sridharan, S., Sawarkar, A. D., Shakeel, A., Anerao, P., Mannina, G., Sharma, P., & Pandey, A. (2023b). Current research trends on emerging contaminants pharmaceutical and personal care products (PPCPs): A comprehensive review. *Science of the Total Environment, 859*, 160031. https://doi.org/10.1016/j.scitotenv.2022.160031

Kümmerer, K. (2009). Antibiotics in the aquatic environment–a review–part I. *Chemosphere, 75*(4), 417–434. https://doi.org/10.1016/j.chemosphere.2008.11.086

Kurade, M. B., Ha, Y. H., Xiong, J. Q., Govindwar, S. P., Jang, M., & Jeon, B. H. (2021). Phytoremediation as a green biotechnology tool for emerging environmental pollution: A step forward towards sustainable rehabilitation of the environment. *Chemical Engineering Journal, 415*, 129040. https://doi.org/10.1016/j.cej.2021.129040

La Farre, M., Pérez, S., Kantiani, L., & Barceló, D. (2008). Fate and toxicity of emerging pollutants, their metabolites and transformation products in the aquatic environment. *TrAC Trends in Analytical Chemistry, 27*(11), 991–1007. https://doi.org/10.1016/j.trac.2008.09.010

Li, M. H. (2019). Comparative toxicities of 10 widely used biocides in three freshwater invertebrate species. *Chemistry and Ecology, 35*(5), 472–482. https://doi.org/10.1080/02757540.2019.1579311

Li, X., Zheng, W., & Kelly, W. R. (2013). Occurrence and removal of pharmaceutical and hormone contaminants in rural wastewater treatment lagoons. *Science of the Total Environment, 445*, 22–28. https://doi.org/10.1016/j.scitotenv.2012.12.035

Li, Y., Zhang, J., Zhu, G., Liu, Y., Wu, B., Ng, W. J., Appan, A., & Tan, S. K. (2016). Phytoextraction, phytotransformation and rhizodegradation of ibuprofen associated with Typha angustifolia in a horizontal subsurface flow constructed wetland. *Water Research, 102*, 294–304. https://doi.org/10.1016/j.watres.2016.06.049

Liu, J., Liu, S., Sun, K., Sheng, Y., Gu, Y., & Gao, Y. (2014). Colonization on root surface by a phenanthrene-degrading endophytic bacterium and its application for reducing plant phenanthrene contamination. *PLoS ONE, 9*(9), e108249. https://doi.org/10.1371/journal.pone.0108249

Liu, J., Wang, J., Zhao, C., Hay, A. G., Xie, H., & Zhan, J. (2016). Triclosan removal in wetlands constructed with different aquatic plants. *Applied Microbiology and Biotechnology, 100*, 1459–1467. https://doi.org/10.1007/s00253-015-7063-6

Lo´pez-Serna, R., et al. (2013). Occurrence of 95 pharmaceuticals and transformation products in urban groundwaters underlying the metropolis of Barcelona, Spain. Environ. Pollut. https://doi.org/10.1016/j.envpol.2012.11.022

Lugtenberg, B., & Kamilova, F. (2009). Plant-growth-promoting rhizobacteria. *Annual Review of Microbiology, 63*(1), 541–556.

Madhav, S., Mishra, R., Kumari, A., Srivastav, A. L., Ahamad, A., Singh, P., Ahmed, S., Mishra, P. K., & Sillanpää, M. (2024). A review on sources identification of heavy metals in soil and remediation measures by phytoremediation-induced methods. *International Journal of Environmental Science and Technology, 21*(1), 1099–1120. https://doi.org/10.1007/s13762-023-04950-5

Männistö, M. K., Tiirola, M. A., & Puhakka, J. A. (2001). Degradation of 2, 3, 4, 6-tetrachlorophenol at low temperature and low dioxygen concentrations by phylogenetically different groundwater and bioreactor bacteria. *Biodegradation, 12*, 291–301. https://doi.org/10.1023/A:1014362508447

Misal, S. A., Lingojwar, D. P., Shinde, R. M., & Gawai, K. R. (2011). Purification and characterization of azoreductase from alkaliphilic strain Bacillus badius. *Process Biochemistry, 46*(6), 1264–1269. https://doi.org/10.1016/j.procbio.2011.02.013

Mohammadi, M., Gheibi, M., Fathollahi-Fard, A. M., Eftekhari, M., Kian, Z., & Tian, G. (2021). A hybrid computational intelligence approach for bioremediation of amoxicillin based on fungus activities from soil resources and aflatoxin B1 controls. *Journal of Environmental Management, 299*, 113594. https://doi.org/10.1016/j.jenvman.2021.113594.

Mojiri, A., Aziz, H. A., Zahed, M. A., Aziz, S. Q., & Selamat, M. R. B. (2013). Phytoremediation of heavy metals from urban waste leachate by southern cattail (Typha domingensis). *International Journal of Scientific Research in Environmental Sciences, 1*(4), 63–70. https://doi.org/10.12983/ijsres-2013-p063-070

Morillo, E., & Villaverde, J. (2017). Advanced technologies for the remediation of pesticide-contaminated soils. *Science of the Total Environment, 586*, 576–597. https://doi.org/10.1016/j.scitotenv.2017.02.020

Mutar, Z. H., Al-Baldawi, I. A., Mohammed, A. A., Hameed, H. K., Abdullah, S. R. S., & Ismail, N. I. (2023, July). Efficacy of phytoremediation for methylparaben removal from contaminated water: A study by Alternanthera spp and Associated Rhizobacteria. In *IOP Conference Series: Earth and Environmental Science* (Vol. 1215, No. 1, pp. 012002). IOP Publishing. https://doi.org/10.1088/1755-1315/1215/1/012002

Ncibi, M. C., Mahjoub, B., Mahjoub, O., Sillanpää, M. (2017). Remediation of emerging pollutants in contaminated wastewater and aquatic environments: Biomass-based technologies. *CLEAN–Soil, Air, Water, 45*(5), 1700101. https://doi.org/10.1002/clen.201700101

Nguyen, P. M., Afzal, M., Ullah, I., Shahid, N., Baqar, M., & Arslan, M. (2019). Removal of pharmaceuticals and personal care products using constructed wetlands: Effective plant-bacteria synergism may enhance degradation efficiency. *Environmental Science and Pollution Research, 26*, 21109–21126. https://doi.org/10.1007/s11356-019-05320-w

Olicón-Hernández, D. R., Ortúzar, M., Pozo, C., González-López, J., & Aranda, E. (2020). Metabolic capability of Penicillium oxalicum to transform high concentrations of anti-inflammatory and analgesic drugs. *Applied Sciences, 10*(7), 2479. https://doi.org/10.3390/app10072479

Pant, M., & Singh, J. (2024a). Distribution of heavy metals contamination in the environment and their toxicological impacts. In *Heavy Metal Contamination in the Environment* (pp. 95–113). CRC Press.

Pant, M., & Singh, J. (2024b). Seasonal variation of surface water quality and streamflow in Rispana: A tributary of Ganges river, India. *Environmental Quality Management*, 1–14.https://doi.org/10.1002/tqem.22199

Pant, M., Singh, J., Bisen, D., Manori, S. & Shukla, R. K. (2025a). Nanocomposite-based photocatalysis: Tackling Pharmaceutical and Personal Care Products (PPCPs) pollutants in environmental remediation. In *Sustainable Development Goals Towards Environmental Toxicity and Green Chemistry* (pp. 83–105). Springer.

Pant, M., Naik, S. C. S., Singhal, N., & Singh, J. (2025b). Bioremediation approaches for mitigation of emerging water contaminants in wastewater. In *Biotechnologies for Wastewater Treatment and Resource Recovery* (pp. 93–108). Elsevier.

Petrović, M., Gonzalez, S., & Barceló, D. (2003). Analysis and removal of emerging contaminants in wastewater and drinking water. *TrAC Trends in Analytical Chemistry, 22*(10), 685–696. https://doi.org/10.1016/S0165-9936(03)01105-1

Piwowarska, D., Kiedrzyńska, E., & Jaszczyszyn, K. (2024). A global perspective on the nature and fate of heavy metals polluting water ecosystems, and their impact and remediation. *Critical*

Reviews in Environmental Science and Technology, 1–23. https://doi.org/10.1080/10643389.2024.2317112

Rana, R. S., Singh, P., Kandari, V., Singh, R., Dobhal, R., & Gupta, S. (2017). A review on characterization and bioremediation of pharmaceutical industries' wastewater: An Indian perspective. *Applied Water Science, 7*, 1–12. https://doi.org/10.1007/s13201-014-0225-3

Rasmussen, P. U., Bennett, A. E., & Tack, A. J. (2020). The impact of elevated temperature and drought on the ecology and evolution of plant–soil microbe interactions. *Journal of Ecology, 108*(1), 337–352. https://doi.org/10.1111/1365-2745.13292

Ratola, N., Cincinelli, A., Alves, A., & Katsoyiannis, A. (2012). Occurrence of organic microcontaminants in the wastewater treatment process. A mini review. *Journal of Hazardous Materials, 239*, 1–18. https://doi.org/10.1016/j.jhazmat.2012.05.040

Rivera-Utrilla, J., Sánchez-Polo, M., Ferro-García, M. Á., Prados-Joya, G., & Ocampo-Pérez, R. (2013). Pharmaceuticals as emerging contaminants and their removal from water. A review. *Chemosphere, 93*(7), 1268–1287.https://doi.org/10.1016/j.chemosphere.2013.07.059

Saiyood, S., Vangnai, A. S., Thiravetyan, P., & Inthorn, D. (2010). Bisphenol A removal by the Dracaena plant and the role of plant-associating bacteria. *Journal of Hazardous Materials, 178*(1–3), 777–785. https://doi.org/10.1016/j.jhazmat.2010.02.008

Santoyo, G., Moreno-Hagelsieb, G., del Carmen Orozco-Mosqueda, M., & Glick, B. R. (2016). Plant growth-promoting bacterial endophytes. *Microbiological Research, 183*, 92–99. https://doi.org/10.1016/j.micres.2015.11.008

Sauvêtre, A., May, R., Harpaintner, R., Poschenrieder, C., & Schröder, P. (2018). Metabolism of carbamazepine in plant roots and endophytic rhizobacteria isolated from Phragmites australis. *Journal of Hazardous Materials, 342*, 85–95. https://doi.org/10.1016/j.jhazmat.2017.08.006

Sauvêtre, A., Węgrzyn, A., Yang, L., Vestergaard, G., Miksch, K., Schröder, P., & Radl, V. (2020). Enrichment of endophytic Actinobacteria in roots and rhizomes of Miscanthus × giganteus plants exposed to diclofenac and sulfamethoxazole. *Environmental Science and Pollution Research, 27*, 11892–11904. https://doi.org/10.1007/s11356-020-07609-7

Sharma, B., Dangi, A. K., & Shukla, P. (2018). Contemporary enzyme based technologies for bioremediation: A review. *Journal of Environmental Management, 210*, 10–22. https://doi.org/10.1016/j.jenvman.2017.12.075

Shehzadi, M., Afzal, M., Khan, M. U., Islam, E., Mobin, A., Anwar, S., & Khan, Q. M. (2014). Enhanced degradation of textile effluent in constructed wetland system using Typha domingensis and textile effluent-degrading endophytic bacteria. *Water Research, 58*, 152–159. https://doi.org/10.1016/j.watres.2014.03.064

Shi, Q., Yuan, Y., Zhou, Y., Yuan, Y., Liu, L., Liu, X., Li, F., Leng, C., & Wang, H. (2023). Pharmaceutical and personal care products (PPCPs) degradation and microbial characteristics of low-temperature operation combined with constructed wetlands. *Chemosphere, 341*, 140039. https://doi.org/10.1016/j.chemosphere.2023.140039

Singh, S., Kumar, A., & Gupta, H. (2020). Activated banana peel carbon: A potential adsorbent for Rhodamine B decontamination from aqueous system. *Applied Water Science, 10*(8), 1–8. https://doi.org/10.1007/s13201-020-01274-4

Singh, A. K., Bilal, M., Iqbal, H. M., Meyer, A. S., & Raj, A. (2021). Bioremediation of lignin derivatives and phenolics in wastewater with lignin modifying enzymes: Status, opportunities and challenges. *Science of the Total Environment, 777*, 145988. https://doi.org/10.1016/j.scitotenv.2021.145988

Singh, S., Gupta, H., Dhiman, S., & Sahu, N. K. (2022). Decontamination of cationic dye brilliant green from the aqueous media. *Applied Water Science, 12*(4), 61. https://doi.org/10.1007/s13201-022-01596-5

Singh, S., Pant, A., Dutta, K., Rani, R., Vithanage, M., & Daverey, A. (2024). Phytoremediation of pharmaceuticals and personal care products using the constructed wetland. *Environmental Chemistry and Ecotoxicology, 6*, 104–116. https://doi.org/10.1016/j.enceco.2024.04.001

Sirés, I., & Brillas, E. (2012). Remediation of water pollution caused by pharmaceutical residues based on electrochemical separation and degradation technologies: A review. *Environment International, 40*, 212–229. https://doi.org/10.1016/j.envint.2011.07.012

Sivaranjanee, R., & Kumar, P. S. (2021). A review on remedial measures for effective separation of emerging contaminants from wastewater. *Environmental Technology and Innovation, 23*, 101741. https://doi.org/10.1016/j.eti.2021.101741

Sochacki, A., Nowrotek, M., Felis, E., Kalka, J., Ziembińska-Buczyńska, A., Bajkacz, S., Ciesielski, S., & Miksch, K. (2018). The effect of loading frequency and plants on the degradation of sulfamethoxazole and diclofenac in vertical-flow constructed wetlands. *Ecological Engineering, 122*, 187–196. https://doi.org/10.1016/j.ecoleng.2018.08.003

Stackelberg, P. E., Furlong, E. T., Meyer, M. T., Zaugg, S. D., Henderson, A. K., & Reissman, D. B. (2004). Persistence of pharmaceutical compounds and other organic wastewater contaminants in a conventional drinking-water-treatment plant. *Science of the Total Environment, 329*(1–3), 99–113. https://doi.org/10.1016/j.scitotenv.2004.03.015

Sun, K., Liu, J., Jin, L., & Gao, Y. (2014). Utilizing pyrene-degrading endophytic bacteria to reduce the risk of plant pyrene contamination. *Plant and Soil, 374*, 251–262. https://doi.org/10.1007/s11104-013-1875-x

Supreeth, M. (2022). Enhanced remediation of pollutants by microorganisms–Plant combination. *International Journal of Environmental Science and Technology, 19*(5), 4587–4598. https://doi.org/10.1007/s13762-021-03354-7

The American Chemical Society. CAS Registry—The Gold Standard for Chemical Substance Information. Available online: https://www.cas.org/support/documentation/chemical-substances. (Accessed on 31 March 2021).

Thijs, S., Sillen, W., Truyens, S., Beckers, B., Van Hamme, J., Van Dillewijn, P., & Vangronsveld, J. (2018). The sycamore maple bacterial culture collection from a TNT polluted site shows novel plant-growth promoting and explosives degrading bacteria. *Frontiers in Plant Science, 9*, 396836. https://doi.org/10.3389/fpls.2018.01134

Validov, S. Z., Kamilova, F., & Lugtenberg, B. J. (2009). Pseudomonas putida strain PCL1760 controls tomato foot and root rot in stonewool under industrial conditions in a certified greenhouse. *Biological Control, 48*(1), 6–11. https://doi.org/10.1016/j.biocontrol.2008.09.010

Vasilachi, I. C., Asiminicesei, D. M., Fertu, D. I., & Gavrilescu, M. (2021). Occurrence and fate of emerging pollutants in water environment and options for their removal. *Water, 13*(2), 181. https://doi.org/10.3390/w13020181

Vasiliadou, I. A., Molina, R., Martínez, F., & Melero, J. A. (2013). Biological removal of pharmaceutical and personal care products by a mixed microbial culture: Sorption, desorption and biodegradation. *Biochemical Engineering Journal, 81*, 108–119. https://doi.org/10.1016/j.bej.2013.10.010

Verhoeven, J. T., & Meuleman, A. F. (1999). Wetlands for wastewater treatment: Opportunities and limitations. *Ecological Engineering, 12*(1–2), 5–12. https://doi.org/10.1016/S0925-8574(98)00050-0

Verlicchi, P., & Zambello, E. (2014). How efficient are constructed wetlands in removing pharmaceuticals from untreated and treated urban wastewaters? A review. *Science of the Total Environment, 470*, 1281–1306. https://doi.org/10.1016/j.scitotenv.2013.10.085

Vié, J. C., Hilton-Taylor, C., & Stuart, S. N. (eds) . (2009). *Wildlife in a changing world: an analysis of the 2008 IUCN Red List of threatened species*. IUCN.

Weber, K. P., Mitzel, M. R., Slawson, R. M., & Legge, R. L. (2011). Effect of ciprofloxacin on microbiological development in wetland mesocosms. *Water Research, 45*(10), 3185–3196. https://doi.org/10.1016/j.watres.2011.03.042

Wei, H., Tang, M., & Xu, X. (2023). Mechanism and influence factors of plant uptake, accumulation, transport, metabolism pathways of pharmaceuticals and personal care products and their phytotoxicity: A review. *Science of The Total Environment*, 164413. https://doi.org/10.1016/j.scitotenv.2023.164413

Wilkinson, J. L., Hooda, P. S., Swinden, J., Barker, J., & Barton, S. (2017). Spatial distribution of organic contaminants in three rivers of Southern England bound to suspended particulate material and dissolved in water. *Science of the Total Environment, 593*, 487–497. https://doi.org/10.1016/j.scitotenv.2017.03.167

Wu, H., Zhang, J., Ngo, H. H., Guo, W., Hu, Z., Liang, S., Fan, J., & Liu, H. (2015). A review on the sustainability of constructed wetlands for wastewater treatment: Design and operation. *Bioresource Technology, 175*, 594–601. https://doi.org/10.1016/j.biortech.2014.10.068

Ying, T., & Wei, C. (2019). Soil microbiomes—A promising strategy for contaminated soil remediation: A review. *Pedosphere, 29*(3), 283–297. https://doi.org/10.1016/S1002-0160(18)60061-X

Yuan, F., Hu, C., Hu, X., Qu, J., & Yang, M. (2009). Degradation of selected pharmaceuticals in aqueous solution with UV and UV/H_2O_2. *Water Research, 43*(6), 1766–1774. https://doi.org/10.1016/j.watres.2009.01.008

Zhang, C., Li, Y., Wang, C., et al. (2016). Occurrence of endocrine-disrupting compounds in aqueous environment and their bacterial degradation: A review. *Critical Reviews in Environment Science and Technology, 46*, 1–59. https://doi.org/10.1080/10643389.2015.1061881

Zhang, L., Lv, T., Zhang, Y., Stein, O. R., Arias, C. A., Brix, H., & Carvalho, P. N. (2017). Effects of constructed wetland design on ibuprofen removal–A mesocosm scale study. *Science of the Total Environment, 609*, 38–45. https://doi.org/10.1016/j.scitotenv.2017.07.130

Zhang, L., Lyu, T., Zhang, Y., Button, M., Arias, C. A., Weber, K. P., Brix, H., & Carvalho, P. N. (2018). Impacts of design configuration and plants on the functionality of the microbial community of mesocosm-scale constructed wetlands treating ibuprofen. *Water Research, 131*, 228–238. https://doi.org/10.1016/j.watres.2017.12.050

Zhao, C., Xie, H., Xu, J., Xu, X., Zhang, J., Hu, Z., Liu, C., Liang, S., Wang, Q., & Wang, J. (2015). Bacterial community variation and microbial mechanism of triclosan (TCS) removal by constructed wetlands with different types of plants. *Science of the Total Environment, 505*, 633–639. https://doi.org/10.1016/j.scitotenv.2014.10.053

Zhou, N. A., Lutovsky, A. C., Andaker, G. L., Ferguson, J. F., & Gough, H. L. (2014). Kinetics modeling predicts bioaugmentation with Sphingomonad cultures as a viable technology for enhanced pharmaceutical and personal care products removal during wastewater treatment. *Bioresource Technology, 166*, 158–167. https://doi.org/10.1016/j.biortech.2014.05.028

Zhou, T., Yu, Z., Zhang, L., Gong, C., & Yan, C. (2024). Removal of sulfonamides from water by wetland plants: Performance, microbial response and mechanism. *Science of the Total Environment, 916*, 170181. https://doi.org/10.1016/j.scitotenv.2024.170181

Zhu, X., Ni, X., Waigi, M. G., Liu, J., Sun, K., & Gao, Y. (2016). Biodegradation of mixed PAHs by PAH-degrading endophytic bacteria. *International Journal of Environmental Research and Public Health, 13*(8), 805. https://doi.org/10.3390/ijerph13080805

The manufacturer's authorised representative in the EU is Springer Nature Customer Service Centre GmbH, Europaplatz 3, 69115 Heidelberg, Germany. If you have any concerns regarding our products, please contact ProductSafety@springernature.com

Printed and bound by CPI Group (UK) Ltd, Croydon, CR0 4YY
01/10/2026
02225960-0014